Astrophysics Through Computation
With Mathematica® Support

This new text surveys a series of fundamental problems in astrophysics, both analytically and computationally, for advanced students in physics and astrophysics. The contents are supported by more than 110 class-tested *Mathematica*® notebooks, allowing rigorous solutions to be explored in a visually engaging way. Topics covered include many classical and historically interesting problems, enabling students to appreciate the mathematical and scientific challenges that have been overcome in the subject's development. The text also shows the advantages and disadvantages of using analytical and computational methods. It will serve students, professionals, and capable amateurs to master the quantitative details of modern astrophysics and the computational aspects of their research projects.

Downloadable *Mathematica*® resources available at www.cambridge.org/koberlein.

Brian Koberlein is Senior Lecturer of physics and astronomy at the Rochester Institute of Technology.

David Meisel is Distinguished Professor of physics and astronomy at the State University of New York, Geneseo.

DDM thanks Carolyn for 56 years of shared entropy that made his contribution possible.
BDK thanks Julia and Douglas for bringing love and joy to his universe.

Astrophysics Through Computation

With Mathematica® Support

BRIAN KOBERLEIN

Rochester Institute of Technology, New York

DAVID MEISEL

State University of New York, Geneseo

CAMBRIDGE
UNIVERSITY PRESS

CAMBRIDGE
UNIVERSITY PRESS

32 Avenue of the Americas, New York NY 10013-2473, USA

Cambridge University Press is part of the University of Cambridge.

It furthers the University's mission by disseminating knowledge in the pursuit of
education, learning and research at the highest international levels of excellence.

www.cambridge.org
Information on this title: www.cambridge.org/9781107010741

© Brian Koberlein and David Meisel 2013

First published 2013

A catalogue record for this publication is available from the British Library

Library of Congress Cataloguing in Publication data

Koberlein, Brian.
Astrophysics through computation : with Mathematica® support / Brian Koberlein, David Meisel.
p. cm.
Includes bibliographical references and index.
ISBN 978-1-107-01074-1 (hardback)
1. Astrophysics – Data processing – Textbooks. 2. Mathematica (Computer program language)
I. Meisel, David D. II. Title.
QB462.2.K63 2013
523.010285'53–dc23   2012051612

ISBN 978-1-107-01074-1 Hardback

Additional resources for this publication at www.cambridge.org/koberlein

Contents

Preface

Why another book on astrophysics? Undergraduates and first-year graduate students are deluged with complex information that they are expected to "know" at least qualitatively, and there are plenty of texts that present just such a broad comprehensive survey of astronomy and astrophysics (either observationally or theoretically) and do that just fine. In fact, there are many things being taught in introductory astronomy classes today that 50 years ago appeared only in doctoral theses. But in those same times there were those (Chandrasekhar, Einstein, Hubble, and Spitzer, for example) who, having no access to the powerful computers of today, developed the elegant analytical and observational theories upon which our modern ideas are based. Not only did these early people work with incomplete data sets and poorly understood physical concepts, but they also had to invent their own mathematical and computational methodologies to make their concepts quantitative. The scientific progress of those times was largely the product of sheer intellect from beginning to end.

Astronomy and astrophysics now make such immense strides almost continuously that undergraduate and beginning graduate students are rarely aware of the extraordinary quantitative foundations given to these disciplines during the nineteenth and twentieth centuries, and this knowledge gap widens with each passing day. There are two factors at work here. First, the analytical mathematics used by these early masters was quite above that usually considered suitable for undergraduate instruction, and second the actual computational programming required to produce realistic modern models these days is considered too sophisticated to be meaningfully approached by all but the most advanced undergraduates and beginning graduate students. Yet NASA and other space agencies have often honored the intellectual giants of this era by naming spacecraft after them without explaining to modern students those lines of mathematical/quantitative reasoning that made possible the revolutions in thought of those scientists.

The purpose of this text is to use a modern computer mathematics system to give undergraduate and first-year graduate students a quantitative bridge between the old and the new. We do not intend for this volume to be a comprehensive survey of astronomy and astrophysics, either observationally or theoretically. Instead we cover a series of topics where it is evident (at least to us) that the mathematical (analytical or numerical) development in the hands of a skilled "practitioner" was critical to the understanding of available observations and/or proposed models. Our selected quantitative tasks had to fit the ready availability of the mathematical tools in the chosen computer math system, and a number of topics were rejected as being unable to be solved conveniently because of an unreasonable amount of processing time, because the data were too big for a desktop computer, or because the complexity of the problem obscured the concept we were trying to illustrate. As a result, the level of presentation varies greatly with the difficulty of the problem. We do not avoid a discussion just because the concept is considered too difficult if the math can be used to produce intelligible results. In general, we do not expect students to write their own software from scratch, but we do hope that they can use the text or notebooks as templates for their own applications.

If we do not include a favorite "moment in history" for every reader of the text, please consider those chosen as steppingstones to an extraordinary wealth of material from which you can formulate your own examples. In every case we consider, we extend (and encourage the student to do likewise) the exploration of a cited early work by means of modern computer technology. This graphically illustrates the quantitative directions that the work might have taken had such technological marvels been available in those early times. These computational activities also reveal various shortcomings in the early work that might have been avoided had advanced computation been available. Students will learn to recognize these for themselves as they work through the text and notebooks. We do this so that students will develop a sense of connection between those days when most ideas started in the human brain and today's modern world of supercomputers where visualization is so complicated that the brain needs computer "filtering" to understand what the numerical results are saying. But just because we have tended to develop the ideas of astrophysics of the past (and some of the present) does not mean we neglect to show some of the connections with the present and future.

To achieve our goals we have taken the broadest definition of computation possible to include data analysis as well as theoretical modeling, as this is the way modern astrophysics has evolved. Scientific papers in all fields have become a synthesis of both theory and data analysis, and today's students need to be able to navigate equally well in either capacity. Throughout the book we leave "nuggets" for future computational exploration with only brief comments about them, almost in passing. Only in the suggested computational projects do we make any reference to some of these. Instructors and students alike are expected to think about these on their own.

It may be wondered why we picked Wolfram *Mathematica*® for our programming system when others are readily available. One of us (DM) has used *Mathematica* as a programming system for the last decade and has seen it evolve from a somewhat hard-to-learn and abstract mathematical programming language into a comprehensive mathematical analysis system that suits both theoretical development and extensive data analysis. Its coherent structure allows use for all research tasks without having to change software for different subtasks: first, it is practically the only system available that is transparently and completely cross platform; second, whereas other packages keep subdividing as new features are added, *Mathematica* gets more and more unified as befits modern research that is cross-disciplinary and interdisciplinary. The amount of scientific and computational capability in its current version is nothing short of spectacular, as illustrated in the more than 115 notebooks we have used as the foundation of this book. *Mathematica* is now available internationally with versions that can be purchased directly by students and interested laypersons even if their home institutions have no licensed versions. There is also a free reader that allows anyone to at least read each notebook and in some instances some of the multimedia included. If one has professional colleagues who do not own a copy, they can apply for a special *Mathematica* version that allows editing and reworking of selected documents. If you want a taste of *Mathematica* at no cost, then there is always *WolframAlpha* online as a massive public demonstration of *Mathematica*'s capabilities.

Although only snippets of the needed *Mathematica* expressions at critical junctures are given in the textbook, the full notebooks with comments are available online. Because each of these is a self-contained "program" or collection of related algorithms, there are various levels of complexity involved in examining the contents there.

1. **Beginner's level**: At this level, the best situation is to have a working copy of the regular, student or home *Mathematica* because in Version 9 there are several features that make learning the system much easier than in the past, including online direct connections to the Wolfram site and its new video tutorials for beginners or the availability of natural language commands. If full *Mathematica* capability is not available to the reader, we have provided .pdf copies of all the notebooks that can be

studied independently. An alternative to this is to get a free copy of the *Wolfram CDF Reader* from their website as it will the read .nb files directly and allow native plot display including rotating the 3D plots and printing them plus rapid scrolling, none of which is available in the .pdf copies.

2. **Intermediate level**: Running the notebooks or portions of notebooks with different data in "what-if" mode using some original *Mathematica* utilities provided in their own short programs or projects. If original routines or Wolfram-owned routines are used in publicly available notebooks, be sure to copy over the appropriate copyright notice into your own notebook. *Mathematica* or *Reader Pro* is required for this.

3. **Mastery level**: Use these notebooks as templates for generating new versions or extensions of the full notebooks for projects. Flow charts are recommended before starting drastic changes, and always keep double copies of all modifications separate from the original notebooks as downloaded. A full copy of *Mathematica* is required here.

Finally, we have resisted doing "fancy" programming in our notebooks, preferring to stay within products that instructors are likely to see within their own classroom. These notebooks are not necessarily efficient or compact, but they do work well enough to get the point across. Our experience is that program authors are never satisfied with any of the versions they produce, but in this case we are. We are sure that talented student programmers can review these notebooks and produce better versions, and they are welcome to try their hands at it. For them, these notebooks should provide templates upon which more refined versions can be developed to their own personal tastes while they sharpen their programming abilities. *Mathematica* can be self-documenting within limits, but because of its built-in LaTeX word processing properties, it has allowed us to provide much more documentation than a usual program listing contains. We have even provided some documentation to get students started in *Mathematica* programming and usage. But this book is not a programming manual, as for that there are plenty of texts available. We have concentrated instead on computational matters of direct concern in astronomy and astrophysics.

1   Introduction

Astrophysics draws upon a wide range of topics in astronomy and physics. Topics as widely ranging as observational techniques, thermodynamics, and general relativity are all central to the material covered in this text. Many readers, particularly undergraduate students, will have only passing experience with some of these foundational concepts. It is impossible for us to provide a comprehensive review of these within the scope of this text, but we begin with a basic overview of the most central topics necessary to approach the astrophysical subjects to be covered. In later chapters we build on these various terms in greater detail. Although we do assume more than a general background in physics and astronomy, we do not provide a comprehensive discussion of the basics, as these can be obtained elsewhere. Instead we present only those more advanced background concepts as needed to the task at hand.

1.1   Fundamental stellar properties

One of the central goals of astronomy is the specification of the properties of the sun, stars, and other self-luminous bodies in the universe. Learning about the ranges of these properties and how the quantities characterizing them are determined are a major part of any astronomy course, so we only briefly discuss them here.

1.1.1   Cosmic distance scales

The most fundamental property of a celestial object is its distance from another body. Usually the reference is the sun, the star nearest to Earth. Within the solar system, distances are specified using *astronomical units* (AU), equal to the mean distance between Earth and sun. Outside the solar system, distances are so large that one immediately switches to another unit, the *parsec*. The parsec was devised when distances to stars were first measured by the method of trigonometric parallax, and distance was determined from the annual shift of an object's angular position in the sky.[1]

At the time of Kepler and Copernicus, the difficulty in getting stellar distances was that the parallax angle was very small. Most astronomers of the time reasoned that the brightest stars would be the closest, and so early measurements focused on the brightest stars. But that assumption works only if all stars are about the same brightness, which is not the case. The brightest appearing stars most often also have the greatest luminous output, and therefore they can be seen over even greater distances than stars such as the sun. Hence the parallax searches using the brightest stars largely failed. When it was possible to measure parallaxes, the angles were always smaller than 1 second of arc (arcsecond) or 4.85×10^{-6} radians.

---

[1] The symbol π was often used for parallax angle, which is easy to confuse with the number π.

The distance unit for parsec is defined as 206, 265 AU, such that it is the distance at which parallax is 1 arcsecond. The practical parallax limit in modern times is on the order of a milliarcsecond, so the distance limit is about 1 kiloparsec. Objects have been measured at megaparsecs, but this uses other methods, as we shall see. At so-called cosmological distances, one relies on relative brightness to estimate distances.

1.1.2 Cosmic brightness scale

Because the earliest observations of stars were made with the unaided eye and date back beyond the time of ancient Greeks, the brightness is rarely specified using energy units. Instead a unit called *magnitude* is used. When referring to what the human eye would perceive, the quantity called *apparent magnitude*, *m*, is given.

In modern times magnitudes are often given outside the range of human vision, so a color or wavelength range is specified, such as m_{blue} for a magnitude estimated under a blue filter, or m_{red} for one estimated through a red filter. Magnitudes have some peculiar properties relative to energy measurements that have become standard for historical reasons. The Greeks had only integer magnitudes without 0. Their ranking was such that the unity symbol had the highest value (think first prize) and the largest number had the lowest ranking. Because stars seemed to fall visually on this scale from 1 to 6, the magnitude scale difference in the visual range was 5. Later in history, it was determined that the total magnitude difference of visible stars corresponded to a brightness (luminous energy) ratio of about 100. Thus magnitudes are logarithmic measures of brightness, just as decibels measure audio loudness.

Decibels are base 10 logarithms, but magnitudes are not. Because a difference of $\Delta m = 5$ is a brightness ratio of 100, we have

$$100 = \left(\log_x\right)^5, \tag{1.1}$$

which yields a logarithmic base of 2.512.[2] This is typically expressed as a base 10 logarithm with a multiplying constant so that the brightness ratio (intensity ratio) I/I_0 becomes

$$\frac{I}{I_0} = 10^{0.4\Delta m}, \tag{1.2}$$

or

$$\Delta m = 2.5 \log_{10}\left(\frac{I}{I_0}\right). \tag{1.3}$$

The apparent magnitude itself does not give any indication of the actual brightness of an object because the intrinsic brightness is attenuated by the inverse square law of distance,

$$I_1 = I_0 \frac{r_1^2}{r_0^2}. \tag{1.4}$$

In astronomy, the standard distance for brightness measurement is not 1 parsec as you might expect, but rather 10 parsecs. The reasons for this are again found in early observations. There are no major stellar bodies within 1 parsec of the sun, but there are many within 10 parsecs, so this was chosen. The *absolute magnitude*, *M*, is thus defined as the apparent magnitude a star would have at a distance of 10 parsecs. The absolute magnitude can be calculated from the apparent magnitude by

$$M = m - 5\left(\log_{10} D - 1\right), \tag{1.5}$$

where *D* is the object's distance in parsecs.

---

[2] See **1-1.Stellar** for details.

1.1.3 Color index and temperature

The color of an object can be specified approximately by the *color index* (C.I.) such that

$$C.I. = m_{\text{blue}} - m_{\text{red}}. \tag{1.6}$$

The color index is related to the color temperature of the object. If it can be assumed that the object is a blackbody, then the Planck function is used to estimate this quantitatively.

The energy density u for a photon gas is defined by the Planck function with the energy in joules,

$$u(\omega, T) = \frac{\hbar \omega^3 / \pi^2 c^3}{e^{\hbar \omega / kT} - 1}, \tag{1.7}$$

where ω is the angular frequency observed and T is the blackbody temperature. The theoretical color index is then

$$C.I. = 2.5 \log_{10} \left(\frac{u(\omega_r, T)}{u(\omega_b, T)} \right). \tag{1.8}$$

In optical astronomy the wavelength form of Planck's radiation law is often used. Thus

$$u(\lambda, T) = \frac{2hc^2 / \lambda^5}{e^{hc/\lambda kT} - 1}. \tag{1.9}$$

The "blue" minus "red" is a bit of an exaggeration. Usually color indices span smaller wavelength ranges. Johnson and Morgan (1953) introduced a standardized system known as the UBV system. This system used color filters for observing magnitudes in the ultraviolet, blue, and "visible" ranges, from which one could generate $U - B$ and $B - V$ color indices. Modern astronomers use UBVRI standard with mean wavelengths of $U = 3600$ Å, $B = 4400$ Å, $V = 5500$ Å, $R = 7000$ Å, and $I = 9000$ Å.

By convention, 10 000 K is the temperature for which the color index is supposed to be 0 regardless of the wavelength difference. In practical terms, this means there is a correction constant that must be added to the preceding expressions to get the correct color index.[3]

1.1.4 Radius, temperature, and luminosity

Most stellar objects are excellent approximations to spheres, and as such their brightness properties are straightforward. Each luminous object radiates through its surface via the Planck law. To obtain the total energy over all wavelengths, one must integrate the Planck law in explicit form,

$$L = \int u(\lambda, T) \, dA d\Omega d\lambda = 4\pi R^2 \sigma T^4, \tag{1.10}$$

known as the Stefan–Boltzmann equation.

The absolute magnitude obtained with the entire Stefan–Boltzmann equation is called the *bolometric* absolute magnitude,

$$M_{bol} = -2.5 \log_{10} L. \tag{1.11}$$

For convenience this is often normalized to the values of the sun. Thus the radius of the sun ($R = 1$) and its temperature (5800 K) are assigned along with its absolute magnitude (4.8) to calculate the bolometric absolute magnitude function

$$M_{bol} - M_{\odot} = -2.5 \log_{10} \left(\frac{L}{L_{\odot}} \right). \tag{1.12}$$

[3] See **1-1Stellar** for examples.

Because observations are made with color band filters, it is the visual absolute magnitude that is most often determined. Because luminosity is defined over the whole spectrum, we need to correct M_v to obtain M_{bol}. The magnitude correction that is added is called the bolometric correction, and it is the magnitude equivalent of the ratio of the full spectrum to the partial spectrum.

1.2 Determination of stellar mass

Mass is the source of gravitational fields throughout the cosmos, and everything that has mass contributes to this field. In general relativity mass curves space–time, and this is where gravitational fields originate. Even on a less sophisticated scale humans intuitively, but perhaps qualitatively, know the effects of gravity and how moving masses are influenced by it. Because gravity is such a long-range force, its effects are felt across immense distances; therefore its importance within astronomy and astrophysics is paramount.

The determination of mass from orbital mechanics is a very classical pursuit and works surprisingly well as long as the orbiting object is not too close to its main body. It started when Newton re-derived Kepler's laws of motion based on the rules of Newtonian gravity, with the result that the new Kepler's laws had gravitational theory in terms of the mass of one or both binary bodies. If observational factors were favorable, the individual masses could be estimated. When objects are not orbiting, the main way to tell the mass of an object is to observe its perturbations by other masses or to observe perturbations of other masses on it.

For a binary star, the orbit in space is, at least without the presence of disturbing objects, just the usual two-body ellipse. Kepler's laws for the system can be expressed as

$$(m_1 + m_2)\, P^2 = R^3, \tag{1.13}$$

where P is the orbital period in years and $R = a_1 + a_2$ is the sum of the semi-major axes of the two stars in AU. From the center of mass,

$$m_1 a_1 = m_2 a_2. \tag{1.14}$$

These two equations can then be used to determine the masses m_1 and m_2 in solar mass units quite simply. Obtaining values for a_1 and a_2 is a much bigger challenge.

1.2.1 Visual binaries

Just as Galileo discovered moons orbiting Jupiter, later astronomers discovered that some of the stars that appeared as multiple through telescopes were in fact orbiting each other. Early measurements were often just "sketched" relative to points plotted on graph paper. The properties were then measured from the graph paper directly. Later visual measurements were made with a filar micrometer in a polar coordinate system centered on the primary star, or they were provided by photographic means. Even done visually with a micrometer, measurements were liable to much error because the distances between the stars were so small. In modern times, binary positions are measured using speckle interferometers with considerably more precision.

As observed from Earth, the orbit of a binary star around the primary star is an ellipse, but its semi-major axis length and foci positions are distorted. To find the orbital parameters we follow the derivations and conventions given by Smart (1960) based on an earlier method attributed to Kowalsky, as they represent the mid-20th century state of binary orbit determination. At that time, observations were heavily dependent on filar micrometers that produced polar measurements (ρ, θ) of the position of the secondary star relative to the primary star. Here ρ is the radial distance in seconds of arc and θ the position angle in degrees as measured eastward from north.

As was usual for that day, before analysis could be performed, the equation of the apparent ellipse was transformed to a pair (x, y) of coordinates where x was north at $0°$ and y east at $90°$. Thus

$$x = \rho \cos\theta, \qquad y = \rho \sin\theta. \tag{1.15}$$

In Cartesian coordinates, the general equation for an ellipse can be written as

$$Ax^2 + 2Hxy + By^2 + 2Gx + 2Fy + 1 = 0. \tag{1.16}$$

From the observed data one can then perform a least-squares analysis to find the coefficients. Smart suggests the way to derive a solution graphically, but a modern computational approach is more effective.[4] Once the values of these coefficients are obtained, the orbital elements can be derived in a nontrivial way:

1. The nodal angle Ω is obtained from

$$\left(F^2 - G^2 + A - B\right) \sin 2\Omega + 2\left(FG - H\right) \cos 2\Omega = 0. \tag{1.17}$$

2. The inclination angle i and the semi-latus rectum $p = a\left(1 - e^2\right)$ are found by solving two equations,

$$FG - H = -\frac{\sin 2\Omega \tan^2 i}{2p^2}, \qquad F^2 + G^2 - (A + B) - \frac{\tan^2 i}{p^2} = \frac{2}{p^2}. \tag{1.18}$$

3. The argument of periapsis ω is found from

$$\tan\omega = \frac{(F \cos\Omega - G \sin\Omega) \cos i}{F \sin\Omega + G \cos\Omega}. \tag{1.19}$$

4. The orbital eccentricity e is found from

$$e = \frac{(G \sin\Omega - F \cos\Omega) \, p \cos i}{\sin\omega}. \tag{1.20}$$

5. The semi-major axis a is found from e and p,

$$a = \frac{p}{1 - e^2}. \tag{1.21}$$

6. The true anomaly ν is found for any point where θ is available using

$$\tan(\nu + \omega) = \tan(\theta - \Omega) \sec i. \tag{1.22}$$

7. The eccentric anomaly E is found from

$$\tan\frac{\nu}{2} = \sqrt{\frac{1 + e}{1 - e}} \tan\frac{E}{2}. \tag{1.23}$$

8. With the mean motion given by $n = 2\pi/T$, where T is the orbital period, the time difference from periastron $(t - \tau)$ for each observation is obtained from Kepler's equation,

$$n(t - \tau) = E - e \sin E. \tag{1.24}$$

---

[4] See **6-0VisBin** for examples of such computational solutions.

Obtaining the period T is a central problem in astronomy and astrophysics. Past observations were rarely equally spaced in time; therefore sophisticated Fourier methods that are so effective today could not be applied directly. Long periods are notoriously hard to determine, particularly with unequal interval methods.

A more modern approach can be found in Green (1985). Green derives the orbit parameters using the more sophisticated Thiele–Innes method. The Thiele–Innes method starts with the projection properties of the physical orbit upon the sky, and although its derivations are not given here, its computational structure can be somewhat more compact and is generally preferred in modern times.

It is clear that the process of obtaining individual masses of binary stars is an arduous one from visual data alone without spectroscopic, eclipsing, and interferometric observations to help with resolving the various ambiguities of the solutions. We address this topic in more detail in Chapter 6 when we consider the stellar motions in the N-body problem and in Chapter 8 where we consider the motions of stars around the galactic center.[5]

1.2.2 Spectroscopic binaries

If the radial velocity of one of the visual binary components can be determined from spectroscopic observations, then the angular orbital properties of i and Ω can be resolved directly. If radial velocity information is available for both stars, the mass ratio can also be obtained as an alternative to using astrometric data on proper motions. The radial velocities are not sufficient on their own to determine the masses, because the inclination is not derivable unless the system is also either a visual or eclipsing binary.

It is worth noting that spectroscopic methods can obtain the orbital period T, the eccentricity e, the daily motion n, and the argument of periapsis ω uniquely. They can also determine which nodes are which and the sign of the inclination when used in conjunction with visual or interferometric observations. The equation for orbital radial velocity (where z is along the line of sight, and $n = 2\pi / T_{days}$ is the daily motion) is

$$\frac{\mathrm{d}z}{\mathrm{d}t} = \frac{na\sin i}{\sqrt{1 - e^2}}\left(\cos(v + \omega) + e\cos\omega\right). \tag{1.25}$$

The observed radial velocities will have the motion of the center of mass in each, and this is determined so that the line $v = $ constant divides the radial velocity curve into two equal areas. This has to be done so that $\mathrm{d}z/\mathrm{d}t$ is isolated.[6]

1.2.3 Other methods

Other examples of mass determination can be seen in a two-line binary (when neither star can be resolved, but a spectroscopic line can be obtained for both stars) or an eclipsing binary. However, detailed considerations of the spectroscopic methods used when there are two lines present or when a light curve is available for the star is beyond the scope of this book. In the case of two lines, the methods follow the one-line analysis in principle. Eclipsing binaries require an even more extensive elaboration

---

[5] Computational examples of visual binary orbits are given in **6-0VisBin** and in five notebooks in Chapter 8.
[6] Details can be found in **1-2SpectBin**.

and are not discussed here. See Smart (1960) or Green (1985) for descriptions. Green analyzes a binary pulsar as an interesting modern example and we consider this in Chapters 4 and 5.[7]

1.3 Kinetic theory

Many astrophysical models rely heavily on the behavior of fluid gases and plasmas. Although simple models often assume these to be an ideal gas, more sophisticated models require an examination of behavior at the particle level. Therefore an understanding of kinetic theory is central to many of these models. In this text we assume readers have at least a general understanding of thermodynamics and kinetic theory. For readers who have not taken a formal course in statistical thermodynamics or classical thermodynamics, we recommend going through the Wolfram *Mathematica*® notebooks on the subject in the appendix. They present the concepts of kinetic theory used in the text as well as demonstrating some basics of *Mathematica* programming.

1.3.1 Maxwell–Boltzmann statistics

To keep things as simple as possible, we consider a force-free monatomic dilute ideal gas. For a large collection of these "ideal" particles at a temperature T, the average kinetic energy of a particle is

$$\frac{1}{2}mv^2 = \frac{3}{2}kT, \tag{1.26}$$

where k is Boltzmann's constant. Although this relation defines temperature in terms of particle kinetic energy, speed v in this equation is an average speed of the particles. The collection of particles is distributed over many speeds, spread out about the average speed. The probability distribution for the particles is given by the Maxwell–Boltzmann distribution,

$$f(v) = 4\pi v^2 \left(\frac{m}{2\pi kT}\right)^{3/2} e^{-mv^2/2kT}, \tag{1.27}$$

such that the probability of finding a particle with a speed between v and $v + dv$ is

$$dp(v) = f(v)\, dv. \tag{1.28}$$

The function is normalized so that

$$\int_0^\infty f(v)\, dv = 1. \tag{1.29}$$

Random walk studies have shown that the Maxwell–Boltzmann equation is the equilibrium velocity distribution for dilute classical gases. McLennan (1989, p. 39) obtains the equation as a solution of the Fokker–Planck equation, whereas Mohling (1982) derives it from binary collision theory.[8]

If the system is not in thermodynamic equilibrium, changes in the velocity distribution due to external forces are described by the Boltzmann Transport Equation (BTE). The BTE allows in principle a complete specification of the transport equations required in most astrophysical situations although

---

[7] See, for example, **4-8ModelNS** and **5-6binarypulsar**.

[8] Computational examples of the Maxwell–Boltzmann distribution can be found in **7-5Maxwell**. Examples of the Boltzmann Transport Equation can be found in **7-6Boltzmann** and **7-7Collisions**.

there are hydrodynamic equations that will serve as well. Suitable expressions are derived in Reif (1965), Mohling (1982), and McLennan (1989). The standard form of the BTE is

$$\left[\frac{\partial}{\partial t} + \bar{v} \cdot \nabla_r + F_{ext} \cdot \nabla_p\right] f(\bar{r}, \bar{p}, t) = \left(\frac{\partial f}{\partial t}\right)_{coll}, \tag{1.30}$$

where $f(r, p, t)$ is the distribution function in position-momentum phase space. If the right-hand side of the equation vanishes, the system is said to be collisionless.

1.3.2 The partition function and Saha equation

The Maxwell–Boltzmann and related equations derive from the assumption that our gas particles are classical with no internal structure. However, at the quantum level the particles of a system can have discrete rather than continuous energy states. If one takes j as the index representing the possible discrete quantum states of a system, and E_j as the energy of the system in that state, then one may define the partition function for the system,

$$Z = \sum_j e^{-E_j/kT}, \tag{1.31}$$

which gives a distribution of particles in the quantum states. If there are multiple states that share the same energy E_j, then the system is said to be degenerate, and the partition function becomes

$$Z = \sum_j g_j e^{-E_j/kT}, \tag{1.32}$$

where g_j is known as the degeneracy factor. Partition functions are central to the Boltzmann–Gibbs–Helmholtz approach to thermal physics and thermodynamics. Essentially, it is assumed that one knows the quantum mechanical energy level structure for the most elementary component (say an "atom") of the system. Construction of the partition function for these "atoms" then leads to the macroscopic properties of ensembles consisting of those atoms.[9]

For a gas at high temperatures, thermal collisions can ionize a certain fraction of the atoms within the gas. Ionization equilibrium was a concept prevalent in astrophysics in the early part of the 20th century. It was given a quantitative status by the astrophysicist Saha through a derivation that is closely related to the law of mass action.[10] In thermal equilibrium, the excitation within the bound states is described by the Boltzmann distribution. The ionization is described by the Saha equation,

$$\frac{N_{y+1}N_e}{N_y} = \frac{Z_{y+1}Z_e}{Z_y}e^{-\chi/kT}, \tag{1.33}$$

where N_y represents the number density in the yth ionization (with y electrons removed), N_e is the electron density, the Zs are the respective partition functions, and χ is the ionization potential. The ionization potential corresponds to the molecular dissociation energies in the regular law of mass action. Although Saha's equation is fairly simple, it actually masks how complicated the pooled ionization from multiple atoms is to treat in practice.[11]

---

[9] Several examples of partition functions can be found in **6Partitionfs**, in the thermonotebooks directory of the appendix.
[10] For a discussion of the law of mass action, see **11MassAction**.
[11] See **12Ionization** for a more detailed discussion.

1.3.3 Fermi–Dirac statistics

At room temperature and higher the ideal gas law is usually a good description of a gas. However, once a gas is cooled beyond a certain point the classical rules no longer hold, even approximately. At low temperatures, known as the quantum degeneracy region, it is the particle spin forces that are dominant.

All simple (elementary) particles possess a quality known as spin. It has the same basic properties as angular momentum, except that it is quantized into discrete values. Particles that have even or zero spin states are called bosons. Particles that have total spins in multiples of $1/2$ are called fermions. The most important manifestation of these spin states is its effect on the thermal average of particle occupancy obtained from the so-called Gibbs sum for non-dilute gases,

$$\langle N(\epsilon) \rangle = \frac{1}{e^{(\epsilon - \mu)/kT} \pm 1}, \tag{1.34}$$

where ϵ is the energy of the state and μ is the chemical potential. The $+1$ form describes fermions, while the -1 form describes bosons. For ideal gases it is assumed that the exponential is much larger than 1; thus

$$\langle N(\epsilon) \rangle = e^{-(\epsilon - \mu)/kT}, \tag{1.35}$$

which is the Maxwell–Boltzmann case.

Fermions by the Pauli exclusion principle can have occupancy states of 1 or 0, with the average occupancy bound by that range. At absolute zero, all energy states with energies less than the Fermi energy ϵ_F will be filled, and all states above the Fermi energy will be empty.

A simple calculation of the Fermi energy can be found by assuming electrons that are in a rigid cubic box (infinite potential well) of side a.[12] The energy states for such a box can be indexed by $\mathbf{n} = (n_x, n_y, n_z)$, and the energy states are then

$$E_{\mathbf{n}} = \frac{\hbar^2 \pi^2}{2ma^2} \mathbf{n}^2. \tag{1.36}$$

The number of states with $E_{\mathbf{n}} < E_F$ are those that lie within a spherical volume of n_F; thus

$$N = 2 \left(\frac{1}{8} \right) \frac{4\pi}{3} n_F^3. \tag{1.37}$$

The factor 2 is due to the two allowed spin states, while the $1/8$ factor accounts for our need for only positive energy levels. From this one can express n_F in terms of the total number of electrons. From these we find

$$E_f = \frac{\hbar^2 \pi^2}{2ma^2} \mathbf{n}_{\mathbf{F}}^2 = \frac{\hbar^2}{2m} \left(3\pi^2 n \right)^{2/3}, \tag{1.38}$$

where $n = N/a^3$ is the number density of electrons. The total energy of the system is then

$$E = \int E_F \mathrm{d}n = \frac{3}{5} N E_F, \tag{1.39}$$

and the average energy of the electrons is

$$\bar{E} = \frac{3}{5} E_F. \tag{1.40}$$

---

[12] Surprisingly, this is a reasonable approximation for electrons in a metal.

From the Fermi energy one can define a Fermi temperature,

$$T_F = \frac{E_F}{k}.$$ (1.41)

In most instances of a many-particle system, the calculated Fermi temperature in Kelvin is many orders of magnitude higher than the actual prevailing temperature. When that condition holds, a good approximation of the properties of fermions is simply to set $T = 0$. A better approximation in the vicinity of $T = 0$ is to use series approximations of the chemical potential as shown by Laurendeau (2010). More complex solutions can be obtained computationally.[13]

1.3.4 Bose–Einstein statistics

Bosons are not limited in their occupancy states. For bosons the energy distribution function (similar to the partition function) is

$$B(\epsilon) = \frac{1}{e^{(\epsilon - \mu)/kT} - 1}.$$ (1.42)

The density of states for bosons of zero spin is

$$n(\epsilon) = \frac{V}{4\pi^2} \left(\frac{2m}{\hbar^2} \right) \sqrt{\epsilon}.$$ (1.43)

From this one can calculate the energy of the boson system,

$$E = \int_0^\infty B(\epsilon) n(\epsilon) \epsilon \, d\epsilon,$$ (1.44)

which yields

$$E = \frac{3V}{4\sqrt{2}} \left(\frac{m}{\pi \hbar^2} \right)^{3/2} (kT)^{5/2} \mathrm{Li}_{5/2}(e^{\mu/kT}),$$ (1.45)

where

$$\mathrm{Li}_s(x) = \sum_{i=1}^\infty \frac{x^i}{i^s}$$ (1.46)

is known as a polylog function. Although this is a complicated function it can be handled computationally fairly easily. *Mathematica*, for example, includes the function as $\mathrm{Li}_s(x) = $ `PolyLog[s,x]`.[14]

As the temperature of a boson gas approaches absolute zero, the system becomes degenerate. A degenerate boson gas collapses into the single lowest energy state. In a very real sense, they behave as if they are a single boson in the ground state, known as a Bose–Einstein condensate.

1.3.5 Black-body radiation

Photons are massless bosons and therefore follow Bose–Einstein statistics. The energy distribution for photons is usually expressed in terms of frequency or wavelength rather than energy; hence

$$B_\nu(T) = \frac{2h\nu^3}{c^2} \frac{1}{e^{h\nu/kT} - 1},$$ (1.47)

$$B_\lambda(T) = \frac{2hc^2}{\lambda^5} \frac{1}{e^{hc/\lambda kT} - 1},$$ (1.48)

---

[13] See **13FermionsBosons**.
[14] For more detailed examples see **13FermionsBosons**.

where $E = h\nu = hc/\lambda$ is the energy of a photon and h is Planck's constant. This distribution is commonly known as Planck's law or the black-body radiation function, and it describes the energy distribution of light radiated from a perfect blackbody.

The power radiated by a black body per unit area is found by integrating Planck's law over all frequencies (wavelengths) and over the solid angle of the visible hemisphere,

$$P = \int B_\nu(T)\cos\theta \; d\Omega \; d\nu = \sigma T^4, \tag{1.49}$$

where

$$\sigma = \frac{2\pi^5 k^4}{15c^2 h^3} \tag{1.50}$$

is known as the Stefan–Boltzmann constant. This T^4 dependence on power is known as the Stefan–Boltzmann law. From this one can calculate the luminosity of a star of radius R,

$$L = 4\pi R^2 \sigma T^4. \tag{1.51}$$

The wavelength at which the black-body curve is a maximum is given by Wien's displacement law. It can be found by taking the derivative of the Planck function,

$$\frac{d}{d\lambda}\left(\frac{2hc^2}{\lambda^5}\frac{1}{e^{hc/\lambda kT}-1}\right) = 0. \tag{1.52}$$

Setting $u = hc/\lambda kT$, this equations reduces to

$$\frac{ue^u}{e^u-1} - 5 = 0. \tag{1.53}$$

This does not have a simple analytical solution[15] but it can be solved computationally as

$$u_\lambda = 4.965\cdots. \tag{1.54}$$

The frequency form of Wien's law can be derived in the same way; however, it should be noted that the result is a different value. Setting $u_\nu = h\nu/kT$, one obtains

$$3\left(e^u - 1\right) - ue^u = 0, \tag{1.55}$$

which yields

$$u_\nu = 2.821\cdots. \tag{1.56}$$

One cannot convert between maximum wavelength and maximum frequency simply by converting between the two.[16] Often Wien's law is expressed as a direct relation to temperature as

$$\lambda_{\max} = \frac{2.897 \times 10^6 \text{ mm K}}{T}, \qquad \nu_{\max} = 5.879 \times 10^{10} \text{ Hz/k}T. \tag{1.57}$$

---

[15] The solution involves the Lambert W function.
[16] See **14PhotonProperties**.

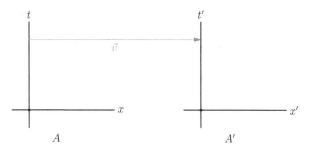

Fig. 1.1 Coordinate frames in relative motion.

1.3.6 Further reading

We have only scratched the surface of the topics in kinetic theory that are central to the topics covered in this text. Several of these are covered in more detail in the appendix notebooks. For example, discussion of other equilibrium equations that are sometimes of interest in astrophysics include Law of Mass action (**11MassAction**), Clausius–Clapyron (**18Phases**), diffusion (**17Transport**, **7Fluctuations**), and diatomic molecules properties (**9Diatomicmu**). The references to the Stat Thermo literature can be found in **CrossRefList.csv**.

1.4 Special relativity

Coordinate transformations have always played an important role in astronomy and astrophysics, and it is a mark of observational sophistication that standard textbooks now incorporate relativity as a normal part of coordinate specification and manipulation. Green (1985), in his update of Smart's spherical astronomy text (1960), develops relativistic expressions wherever necessary. In Goldstein's classic text updated by Poole and Safko (2002) there are chapters devoted to relativistic mechanics. But the crowning achievement for the introduction of relativity into everyday life has been in the area of global positioning systems (GPS). Here we explore some of the special relativistic aspects that modern students may encounter. We assume a passing familiarity with the elementary aspects of special relativity as taught in second-year undergraduate physics.

1.4.1 Coordinate transformations

The principle of relativity has been known since before Galileo. Stated in modern terms, the principle states that the laws of physics are the same in all inertial frames. Thus, if we consider two coordinate frames A and A', where A' moves with a constant velocity v relative to the x-axis of A as seen in figure (1.1), then the two coordinate systems are related by

$$t' = t, \qquad\qquad\qquad x' = x - vt, \qquad\qquad (1.58)$$
$$y' = y, \qquad\qquad\qquad z' = z.$$

In this way Newton's laws of motion are the same in both frames of reference. That is

$$\mathbf{F} = m\frac{\mathrm{d}^2x}{\mathrm{d}t^2} = m\frac{\mathrm{d}^2x'}{\mathrm{d}t'^2}. \tag{1.59}$$

A secondary consequence of Galilean relativity is that the velocity of an object depends upon your frame of reference. This includes the speed of light. If the speed of light is c in the A frame, then in the A' frame it is $c' = c - v$. This implies that there should be an optical experiment to distinguish between the A and A' frames. This is how sound works, so this prediction was not unexpected. However, Michelson and Morley (1887) found that the speed of light was the same in all inertial frames. This *was* as surprise, and it was not resolved until 1905 by Einstein. Light is governed by Maxwell's equations, which in a vacuum is

$$\nabla \cdot \mathbf{E} = 0, \qquad\qquad\qquad \nabla \cdot \mathbf{B} = 0,$$

$$\nabla \times \mathbf{E} = -\frac{\partial \mathbf{B}}{\partial t}, \qquad\qquad \nabla \times \mathbf{B} = \frac{1}{c^2}\frac{\partial \mathbf{E}}{\partial t}. \tag{1.60}$$

From these equations we find

$$\nabla \times (\nabla \times \mathbf{E}) = \nabla(\nabla \cdot \mathbf{E}) - \nabla^2\mathbf{E} = -\frac{\partial}{\partial t}(\nabla \times \mathbf{B}), \tag{1.61}$$

and thus

$$\nabla^2\mathbf{E} - \frac{1}{c^2}\frac{\partial^2\mathbf{E}}{\partial t^2} = 0. \tag{1.62}$$

This is the standard wave equation for a wave traveling at the speed of light c. Since the vacuum speed of light is the same in all frames of reference, it must be that

$$\nabla^2\phi - \frac{1}{c^2}\frac{\partial^2\phi}{\partial t^2} = \nabla'^2\phi - \frac{1}{c^2}\frac{\partial^2\phi}{\partial t'^2}. \tag{1.63}$$

The Galilean transformation fails this requirement because substituting equations (1.58) gives

$$\nabla^2\phi - \frac{1}{c^2}\frac{\partial^2\phi}{\partial t^2} = \nabla'^2\phi - \frac{1}{c^2}\frac{\partial^2\phi}{\partial t'^2} - \frac{v^2}{c^2}\frac{\partial^2\phi}{\partial x'^2} + 2v\frac{\partial^2\phi}{\partial x'\partial t'}, \tag{1.64}$$

so we consider a general linear transformation[17]:

$$x' = \alpha x + \beta t,$$
$$t' = \epsilon x + \delta t, \tag{1.65}$$

where α, β, ϵ, and δ are velocity-dependent coefficients. For simplicity we consider $\phi(x, t)$ so that the wave equation in the A coordinate frame reduces to

$$\frac{\partial^2\phi}{\partial x^2} - \frac{1}{c^2}\frac{\partial^2\phi}{\partial t^2} = 0. \tag{1.66}$$

---

[17] For this derivation, we consider only a transformation along the x direction. The general formulation yields the same result, but is significantly more cumbersome when derived in this way. As we see in Chapter 5, there are better ways to consider this transformation.

Using the chain rule for partial differentiation:

$$\frac{\partial \phi}{\partial x} = \frac{\partial \phi}{\partial x'}\frac{\partial x'}{\partial x} + \frac{\partial \phi}{\partial t'}\frac{\partial t'}{\partial x} = \alpha \frac{\partial \phi}{\partial x'} + \epsilon \frac{\partial \phi}{\partial t'}, \tag{1.67}$$

$$\frac{\partial \phi}{\partial t} = \frac{\partial \phi}{\partial x'}\frac{\partial x'}{\partial t} + \frac{\partial \phi}{\partial t'}\frac{\partial t'}{\partial t} = \beta \frac{\partial \phi}{\partial x'} + \delta \frac{\partial \phi}{\partial t'}. \tag{1.68}$$

In this way, equation (1.66) becomes

$$\left(\alpha^2 - \frac{\beta^2}{c^2}\right)\frac{\partial^2 \phi}{\partial x'^2} + 2\left(\alpha\epsilon - \frac{\delta\beta}{c^2}\right)\frac{\partial^2 \phi}{\partial x'\partial t'} + \left(\epsilon^2 - \frac{\delta^2}{c^2}\right)\frac{\partial^2 \phi}{\partial t'^2} = 0. \tag{1.69}$$

To satisfy equation (1.63) it must be that

$$\begin{aligned} \alpha^2 c^2 - \beta^2 &= c^2, \\ c^2 \alpha\epsilon - \delta\beta &= 0, \\ \delta^2 - \epsilon^2 c^2 &= 1. \end{aligned} \tag{1.70}$$

From equation (1.65) the origin of A' is given by

$$x' = 0 = \alpha x + \beta t. \tag{1.71}$$

The velocity of A' in the A frame is given by

$$v = \frac{\mathrm{d}x}{\mathrm{d}t} = \frac{\mathrm{d}}{\mathrm{d}t}\left(-\frac{\beta}{\alpha}t\right) = -\frac{\beta}{\alpha}; \tag{1.72}$$

thus $\beta = -v\alpha$. This together with equations (1.70) give the solutions

$$\begin{aligned} \alpha = \delta &= \frac{1}{\sqrt{1 - v^2/c^2}}, \\ \beta &= \frac{-v}{\sqrt{1 - v^2/c^2}}, \\ \epsilon &= \frac{-v/c^2}{\sqrt{1 - v^2/c^2}}. \end{aligned} \tag{1.73}$$

The transformations are therefore

$$x' = \frac{x - vt}{\sqrt{1 - v^2/c^2}} \qquad t' = \frac{t - vx/c^2}{\sqrt{1 - v^2/c^2}}. \tag{1.74}$$

This is known as the Lorentz transformation, which describes the connection between relatively moving coordinate frames under special relativity. Because this transformation mixes spatial and temporal coordinates it radically changes our view of space and time.[18]

Consider the simple measurement of the length of an object. In their respective coordinate frames, $L = x_2 - x_1$ and $L' = x'_2 - x'_1$. Under Galilean relativity these are identical, but under the Lorentz transformation

$$L' = x'_2 - x'_1 = \frac{x_2 - vt}{\sqrt{1 - v^2/c^2}} - \frac{x_1 - vt}{\sqrt{1 - v^2/c^2}} = \frac{x_2 - x_1}{\sqrt{1 - v^2/c^2}}; \tag{1.75}$$

---

[18] For the general equations of a 3D transformation and other computational aspects of special relativity to follow, see **1-3Srelativity**.

thus

$$L = L'\sqrt{1 - v^2/c^2}. \tag{1.76}$$

In other words, the length of an object moving relative to us appears shorter than its measured length when not moving, an effect known as length contraction. The same is true for the observation of a ticking clock. If the time between ticks in our two reference frames is $T = t_2 - t_1$ and $T' = t'_2 - t'_1$, then

$$T' = t'_2 - t'_1 = \frac{t_2 - vx/c^2}{\sqrt{1 - v^2/c^2}} - \frac{t_1 - vx/c^2}{\sqrt{1 - v^2/c^2}} = \frac{t_2 - t_1}{\sqrt{1 - v^2/c^2}}, \tag{1.77}$$

and thus

$$T' = \frac{T}{\sqrt{1 - v^2/c^2}}, \tag{1.78}$$

which means that the ticks of a clock measured as from a moving coordinate frame appear slower than when measured in the clock's reference frame, known as time dilation.[19]

It is tempting to presume that L and T are the real measurements for an object while L' and T' are merely illusory. However, central to the concept of relativity is the assertion that all inertial reference frames are equally valid. Thus we can only say L and T are the co-moving or "rest" frame measurements.

To place all inertial reference frames on the same footing, what is needed is a description that is universal and thus the same in all coordinate systems. This can be achieved following Minkowski (1909). Because the Lorentz transformation mixes space and time coordinates, it is useful to define a new coordinate for time, $x^0 = ct$, having units of length. This places time on roughly the same footing as the spatial coordinates.

The distance between two points $G(r_1)$ and $H(r_2)$ in space can be defined by the length element dr, where by the Pythagorean theorem

$$dr^2 = dx^2 + dy^2 + dz^2. \tag{1.79}$$

In the same way, we can define a "distance" between two points $G(r_1, t_1)$ and $H(r_2, t_2)$ separated in space and time by the length element ds where

$$ds^2 = c^2 dt^2 - dx^2 - dy^2 - dz^2. \tag{1.80}$$

The summation is not positive because the four dimensions of space and time do not form a Cartesian space. Under a Lorentz transformation

$$ds'^2 = ds^2. \tag{1.81}$$

Defined in this way the length element ds is the same in all frames of reference. That is, ds is Lorentz invariant.

This allows us to define a proper distance between points in space and time that is universal. In the same way, we can also define a proper time

$$d\tau = \frac{ds}{c}, \tag{1.82}$$

which is also Lorentz invariant.

---

[19] The effects of length contraction and time dilation are quite subtle. We present only a cursory overview here.

For two points separated only by time $dr = 0$; thus $d\tau = dt$. The proper time of an object is therefore the time in the object's rest frame. For two points separated in time and space the line element ds can be rewritten as

$$ds^2 = dt^2 \left[c^2 - \left(\frac{dx}{dt} \right)^2 - \left(\frac{dy}{dt} \right)^2 - \left(\frac{dz}{dt} \right)^2 \right] = c^2 dt^2 \left(1 - \frac{v^2}{c^2} \right), \tag{1.83}$$

and proper time then becomes

$$d\tau = dt\sqrt{1 - v^2/c^2}. \tag{1.84}$$

For general Lorentz transformations it is useful to express equations in their matrix form. That is, one denotes

$$(ct, x, y, z) \rightarrow \left(x^0, x^1, x^2, x^3 \right), \tag{1.85}$$

such that the line element can be expressed as the matrix product[20]

$$ds^2 = \sum_{\mu,\nu} \eta_{\mu\nu} dx^\mu dx^\nu, \tag{1.86}$$

where

$$\eta_{\mu\nu} = \begin{pmatrix} -1 & 0 & 0 & 0 \\ 0 & 1 & 0 & 0 \\ 0 & 0 & 1 & 0 \\ 0 & 0 & 0 & 0 \end{pmatrix} \tag{1.87}$$

is known as the metric.

The matrix version of the Lorentz transformation works because it assumes the column "vector" defining the coordinate frame has its time portion (ct) differing in sign from its space portion (x, y, z) and that its transformation obeys the rules of standard matrix multiplication. In mathematical terms that means the coordinates are a vector field whose events depend on the 4-vector in flat (Minkowski) space–time whose magnitude is

$$ds^2 = c^2 (\text{time interval})^2 - (\text{space interval})^2. \tag{1.88}$$

Unlike standard Cartesian geometry, Minkowski vectors can have magnitudes that are both real and imaginary. There are therefore three types of events:

1. Light-like: $ds^2 = 0$,
2. Time-like: $ds^2 > 0$,
3. Space-like: $ds^2 < 0$.

Only event positions can be any one of these. All other physical vector fields (velocity, momentum, current density) with the exception of force are time-like.

---

[20] Typically for such products the summation is simply assumed rather than written explicitly, following a convention known as summation notation or Einstein notation.

1.4.2 Energy and momentum

The conservation of energy and momentum in special relativity, just like in ordinary mechanics, can be used to derive a number of important relationships. But here we hit a conceptual difficulty. In elementary physics we make no distinction between vectors of position and others such as velocity or force vectors. In special relativity we can no longer be so cavalier about that simplification.

Although we have demonstrated the Lorentz transformation in a rectangular coordinate system, one can do the same in a general basis $\mathbf{e}$ (e_0, e_1, e_2, e_3), where the metric is given as $g_{\mu\nu} = \mathbf{e} \cdot \mathbf{e}$. In this way path $P(\tau)$ in the 4D space–time can be expressed as

$$P(\tau) = \sum_{\mu} x^{\mu}(\tau) e_{\mu}. \tag{1.89}$$

In practice, the repeating of μ as both sub- and superscripts is sufficient to indicate the summation over μ, so the summation sign is omitted.

The velocity of this path is given by

$$\mathbf{u}(\tau) = \frac{P(\tau)}{\mathrm{d}\tau} = \frac{\mathrm{d}x^{\mu}(\tau)}{\mathrm{d}\tau} e_{\mu}. \tag{1.90}$$

For the components,

$$u^0 = \frac{\mathrm{d}t}{\mathrm{d}\tau} = \gamma c, \qquad u^i = \frac{\mathrm{d}x^i}{\mathrm{d}\tau} = \gamma v^i, \qquad v^i = \frac{\mathrm{d}x^i}{\mathrm{d}t}, \tag{1.91}$$

where $\gamma = 1/\sqrt{1 - v^2/c^2}$. This represents the velocity 4-vector.

From the velocity 4-vector we can define the momentum 4-vector simply by multiplying by the object's mass m_o; thus

$$\mathbf{p} = m_o \mathbf{u} = (m_o \gamma c, m_o \gamma \mathbf{v}). \tag{1.92}$$

Since Newton's laws must be the same in each frame we require

$$F = ma \Rightarrow F' = m'a'. \tag{1.93}$$

This can be expressed by writing the equations in their Lorentz invariant form. Thus Newton's second law can be written as

$$\mathbf{F} = \frac{\mathrm{d}\mathbf{p}}{\mathrm{d}\tau}. \tag{1.94}$$

The force equation thus becomes

$$\mathbf{F} = \frac{1}{\sqrt{1 - v^2/c^2}} \frac{\mathrm{d}\mathbf{p}}{\mathrm{d}t} = \frac{\mathbf{F}_o}{\sqrt{1 - v^2/c^2}}, \tag{1.95}$$

where $\mathbf{F}_o$ is the force on an object in its rest frame. This means that for a moving object, the effective force $\mathbf{F}$ necessary to exert a force $\mathbf{F}_o$ on the object increases at larger speeds. Thus

$$\frac{\mathbf{F}}{\mathbf{a}} = \frac{m_o}{\sqrt{1 - v^2/c^2}}, \tag{1.96}$$

where m_o is the mass of the object when measured at rest, also known as its proper mass.[21]

---

[21] The $\mathbf{F}/\mathbf{a}$ quantity is sometimes referred to as the relativistic mass. It should be emphasized, however, that this terminology is an outdated misnomer. The term mass is correctly used only to mean proper mass.

The right-hand side of this equation can be written as a series expansion,

$$m_o c^2 + \frac{1}{2} m_o v^2 + \frac{3}{8} \frac{m_o}{c^2} v^4 + \cdots \tag{1.97}$$

The second term in this series is simply the classical kinetic energy of the particle. This means every term must have units of energy. The first term can then be seen as the potential energy of the mass itself, since if $v = 0$, we have

$$E = m_o c^2. \tag{1.98}$$

In other words, there is an equivalence between mass and energy, and one can be converted into the other.

The general equation can be written as $E = mc^2$. Using this notation the self product of the momentum 4-vector is given by

$$\mathbf{p} \cdot \mathbf{p} = \frac{E^2}{c^2} = m^2 c^2 + m^2 v^2 \tag{1.99}$$

and is invariant under a Lorentz transformation. It is commonly written as the relativistic energy equation

$$E^2 = m^2 c^4 + p^2 c^2, \tag{1.100}$$

where p here is the usual 3-momentum. For this reason the momentum 4-vector is often known as the energy-momentum 4-vector.

The equivalence between mass and energy has significant consequences in astrophysics. Suppose then that we had some type of reaction such as

$$A_{m_1} + B_{m_2} \rightarrow C_{m_3}. \tag{1.101}$$

Special relativity requires that any change of mass results in a change of energy, thus we have

$$A_{m_1} + B_{m_2} \rightarrow C_{m_3} + \Delta E \tag{1.102}$$

where

$$\Delta E = \Delta mc^2 = (m_1 + m_2 - m_3)\, c^2. \tag{1.103}$$

If $m_1 + m_2 > m_3$, then the reaction releases energy, whereas if $m_1 + m_2 < m_3$, the reaction loses energy. This mass–energy conversion is central to nuclear reactions. It is the energy source for both fusion and fission of nuclear materials, which drives the life cycle of the stars.

1.4.3 The aberration of light

The 4-velocity of a photon has an energy

$$E = h\nu_o = mc^2, \tag{1.104}$$

where h is Planck's constant, ν_o is its rest frequency, and m is its "mass".[22] Its momentum is

$$p = \frac{h\nu_o}{c} = mc, \tag{1.105}$$

---

[22] It should be emphasized that although we use the term "mass" here, this is not to say photons have mass in any meaningful sense. Rather, the notation is a convenient way of deriving observed results.

and the direction of its vector velocity is given by $\mathbf{n}$. One way to combine the energy and momentum expressions is through a form known as the k-vector,

$$\mathbf{k} = \frac{h\nu_o}{c}(c, n_x, n_y, n_z). \tag{1.106}$$

This notation allows us to derive the relativistic aberration of light. Following Zimmerman and Olness (2002), we consider light reflecting off a moving mirror. When the mirror is at rest, the incident and reflected angles are identical ($\theta_i = \theta_r = \theta_o$), but if the mirror is moving in the direction of the normal to the mirror, the angles are different. This difference of angle produces what is known as the aberration of light.

For simplicity, we will assume the 2D case; thus the incident and reflected k-vectors for the rest case are

$$\mathbf{k}_i = \frac{h\nu_o}{c}(c, \cos\theta_o, \sin\theta_o, 0), \tag{1.107}$$

$$\mathbf{k}_r = \frac{h\nu_o}{c}(c, -\cos\theta_o, \sin\theta_o, 0). \tag{1.108}$$

Taking a Lorentz transformation of these for motion along the normal (x-axis) we find

$$\mathbf{k}'_i = \frac{h\nu_o}{c}(c - \gamma v \cos\theta_o/c, -\gamma v + \gamma \cos\theta_o, \sin\theta_o, 0), \tag{1.109}$$

$$\mathbf{k}'_r = \frac{h\nu_o}{c}(c + \gamma v \cos\theta_o/c, -\gamma v - \gamma \cos\theta_o, \sin\theta_o, 0). \tag{1.110}$$

Comparing these two cases, the incident and reflected angles are related by

$$\cos\theta_r = \frac{2c^2 v + (c^2 + v^2)\cos\theta_i}{c^2 + v^2 + 2v\cos\theta_i}, \qquad \sin\theta_r = \frac{(c-v)(c+v)\sin\theta_i}{c^2 + v^2 + 2v\cos\theta_i}. \tag{1.111}$$

This is often written as

$$\tan\theta_r = \frac{(c-v)(c+v)\sin\theta_i}{c^2 + v^2 + 2v\cos\theta_i}, \tag{1.112}$$

from which it becomes clear that the aberration is smallest for small angles and largest for angles approaching $\theta = \pi/2$. At small angles ($\sin\theta \sim \theta$) and low velocities ($v/c << 1$) the relation can be expressed as

$$\theta_r = \frac{1 - v/c}{1 + v/c}\theta_i, \tag{1.113}$$

which yields the classical equation for the change of angles,

$$\Delta\theta = -\frac{v\theta_i}{c}. \tag{1.114}$$

In addition to the aberration of reflected angle, there is also a Doppler shift of the reflected frequency. From the first terms of our transformed k-vectors we find

$$\nu_r = \frac{c^2 + v^2 + 2v\cos\theta_i}{c^2 - v^2}\nu_i. \tag{1.115}$$

For the special case of $\theta_i = 0$ this reduces to

$$\nu_r = \frac{1 + v/c}{1 - v/c}\nu_i. \tag{1.116}$$

This is not the usual form given for the redshift or blueshift of light, as it is a comparison of incident and reflected frequencies. The light reaching the mirror is shifted due to the motion of the mirror relative to the source, and the reflected light is shifted again due to the motion of the mirror relative to the observer. If one considers the simpler case of relative motion between source and observer, the observed Doppler relationship is

$$\nu = \sqrt{\frac{1 + v/c}{1 - v/c}}\nu_o.$$ (1.117)

In astronomy and astrophysics, the observed Doppler shift is expressed by the quantity z, where

$$z = \frac{\nu_o - \nu}{\nu} = \frac{\lambda = \lambda_o}{\lambda_o}.$$ (1.118)

In this way, light is redshifted for $z > 0$ and blueshifted for $z < 0$.

1.4.4 The Compton effect

In the preceding derivations a mirror was a convenient device because its large assumed mass relative to a photon meant the recoil effect could be ignored. There are many astrophysical situations in which the recoils are not 0. Once such case is the Compton effect, where a "free" (but stationary) particle of small mass and a gamma ray photon collide. Its inverse is where an infrared photon strikes a relativistic particle (such as an electron) and the scattered photon becomes an X-ray. It is ironic that this extremely important effect is not given its own index entry in either Zimmerman and Olness (2002) or Goldstein, Poole, and Safko (2002), while the somewhat more esoteric and mathematical treatment by Naber (1992) mentions and discusses the Compton effect explicitly. Most undergraduate physics and astrophysics students have had some experience with the Compton effect either theoretically in a modern physics course or experimentally in an upper level physics lab. It is so pervasive because it offers a convincing test of special relativity.

We here consider only the simple case of a photon colliding with an "at rest" particle and scattering at an angle θ from its incident direction. A more general and computational approach can be found in **1-3Srelativity**.

In Compton scattering, both energy and momentum are conserved. With the particle at rest, the initial momentum[23] is that of the incident photon, $\mathbf{p}_{\gamma i}$. After the collision, the momentum of the scattered particle is then

$$\mathbf{p}_m = \mathbf{p}_{\gamma i} - \mathbf{p}_{\gamma s},$$ (1.119)

where $\mathbf{p}_{\gamma s}$ is the momentum of the scattered photon. Thus

$$p_m^2 = p_{\gamma i}^2 + p_{\gamma s}^2 - 2p_{\gamma i}p_{\gamma s}\cos\theta.$$ (1.120)

Because for the photon $cp = h\nu$, this can be written as

$$p_m^2 c^2 = (h\nu_i)^2 + (h\nu_s)^2 - (h\nu_i)(h\nu_s)\cos\theta,$$ (1.121)

where ν_i and ν_s are the frequencies of the incident and scattered photon respectively.

[23] We here use the usual 3-momentum.

For the initial energy we must consider both the energy of the incident photon and the rest energy of the mass,

$$E_o = h\nu_i + m_o c^2. \tag{1.122}$$

The final energy is

$$E_f = h\nu_s + \sqrt{p_m^2 c^2 + m_o^4 c^4}. \tag{1.123}$$

Equating these, one can solve for the final momentum of the mass

$$p_m^2 c^2 = \left(m_o c^2 + h\nu_i - h\nu_s\right) - m_o^2 c^4. \tag{1.124}$$

From equations (1.121) and (1.124) we find

$$\frac{\nu_i - \nu_s}{\nu_i \nu_s} = hm_o c^2 \left(1 - \cos\theta\right). \tag{1.125}$$

This is more commonly written in terms of the wavelength,

$$\lambda_s - \lambda_i = hm_o c \left(1 - \cos\theta\right). \tag{1.126}$$

Because it is a high-energy phenomenon, both the Compton effect and its inverse process play a large role in high-energy astrophysics. Shu (1991, p. 184) discusses primarily inverse Compton losses and compares Compton scattering with Thompson scattering that occurs in hot stars. Harwit (1988) compares both processes and includes the equation for the Compton cross-section (computed quantum mechanically), which is called the Klein–Nishina formula.

1.5 Image processing

When one moves from working with stellar objects (whose images are basically points) to galaxies including the Milky Way and its nebular contents, data processing tasks have to be expanded to cover objects that have appreciable 2D extent on the sky (and correspondingly 3D in space). Encountering exquisite images of objects such as those obtained with the ground-based radio VLA, the optical Hubble telescope, or infrared Spitzer space telescope is taken very much for granted without realizing that they are the product of an intensive computational effort. The properties of today's imaging devices, namely the linearity of the detecting element response to radiation and the linearity and constancy of the detecting element spacings relative to each other, facilitate the choice and complexity of mathematical techniques that can be used for the processing images such devices produce.

In today's world, the term image processing usually means taking digital images that are quite good and making them even better. But it was not always so. Image processing in the 1970s was called image restoration, and a great deal of effort was spent trying to remove instrumental and environmental image degradation. Perhaps no greater lesson in image restoration was experienced after the launch of the Hubble space telescope when it was learned that the main mirror was incorrectly figured. That flaw has been corrected in the telescope instrumentation itself by a Space Shuttle servicing mission, but that was after many years of flawed operation where the raw images were terribly distorted. Computational imaging processing was able to restore the flawed images so that they could be used as intended, even if many could not be considered "pretty" in the artistic sense.

1.5.1 Handling the data glut

Astronomers and astrophysicists have always shared data, and it is more true today than ever before. The electronic data revolution for astrophysics and astronomy took real hold in the pre-Hubble spacecraft days with the establishment of the NASA/ADC (Astronomical Data Center) when catalogs listing data about a number of stars and other objects were digitized, recorded on CDs, and distributed at national and international meetings for free. Once Internet speeds became fast enough, the number of successful spacecraft producing a plethora of information exceeded hand manageability, and cooperative international ventures became really numerous, these catalogs were kept online with numerous links coordinated through the space agencies and the International Astronomical Union (IAU). Eventually both scientific papers and catalogs were made available, and so it has remained.

Most undergraduate students have already had some exposure to Internet search engines such as Google, Bing, Yahoo!, etc. to sort through the many extraneous items stored on the millions of accessible websites. But those engines are dominated by commercial interests who pay to enhance their site ratings so that their ads are found in preference to the items you want to find. Of course, users can develop strategies to get around such biases, but it is more economical if such engines can be avoided altogether. We have found it easiest to locate and store URLs for several of the more reliable and stable sites that we have found useful, and recommend starting there. Below is a list (current 2013) of useful links where astronomical/astrophysical information is available for download, including tables and images.[24]

1. Solar-Terrestrial, Solar Corona, Solar Wind, Aurora
   1. SpaceWeather: http://www.swpc.noaa.gov
   2. Solar/Corona: http://sohowww.nascom.nasa.gov
   3. Solar Dynamic Observatory: http://sdo.gsfc.nasa.gov
   4. SolarWind Predictions: http://www.swpc.noaa.gov/wsa-enlil
   5. SpaceWeather Archive Data: http://www.swpc.noaa.gov/ftpmenu/index.html
   6. Ovation AURORA: http://helios.swpc.noaa.gov/ovation
2. Stellar, Galactic, Extragalactic
   1. ADS US Portal: http://adswww.harvard.edu
   2. IAU/CDS/ADS: http://cds.u-strasbg.fr
   3. Virtual Observatory US: http://www.usvao.org
   4. HighEnergy Astrophysics: http://heasarc.gsfc.nasa.gov
   5. ATNF Pulsars: http://www.atnf.csiro.au/research/pulsar/psrcat/
   6. NED/IPAC Extragalactic: http://ned.ipac.caltech.edu
   7. European Space Agency: http://www.eso.org/public
   8. Space Telescope Data Portal MAST: http://archive.stsci.edu
   9. Hubble Images: http://hubblesite.org/gallery
3. Fundamental Reference Data
   1. NIST: http://www.nist.gov/pml/index.cfm
   2. X-ray data: http://physics.nist.gov/PhysRefData/XrayTrans/Html/search.html
   3. X-ray energy Lists: http://xdb.lbl.gov/Section1/Sec_1-2.html

---

[24] For a current list of websites, see **1-4Info**.

4. SPIE: http://spie.org
5. Hubble Imaging: http://hubblesite.org/gallery/behind_the_pictures

1.5.2 The FITS data format

The Flexible Image Transport System (FITS) is the preferred data format for professional use. The FITS was invented by the astronomical community early in the electronic image business when most of the present image and graphics formats did not exist in any form, and the image processing business was in its infancy. It is still in use by the astronomical community because it is simple, efficient, and logical in its structure, and it is not compressed. Being uncompressed means it has its original resolution and requires no fancy software to decompress it. Astronomical images in the early days were often photographic and often took hours to days of exposure to create. After digitizing the accumulated images, most could fit into available computer storage without worry. Enthusiasm for file compression came years later when commercial digital cameras became widely available and raw images could be gigabytes in size in one short exposure.

File compression is of two varieties: lossy and lossless. JPEG is an example of lossy compression, while TIFF is considered to be lossless. Inexpensive digital cameras often use something like JPEG as a native format to save memory, and because JPEG is lossy, the images lose resolution each time they are copied. More expensive cameras often have proprietary raw formats, saving the image in a lossless format while allowing the option of a secondary output such as TIFF that will maintain resolution as copies are made. The catch is that TIFF copies of the same original image can be 100s if not 1000s of times larger than JPEG copies.

In combing the Internet for data and images for research projects, one will encounter FITS files that can be downloaded at will. In addition, charge-coupled device (CCD) cameras intended mainly for the astronomical market (both professional and amateur) have FITS file output. Thus encounters with the FITS format can be expected to be rather regular whether a professional astronomical career is pursued or not. CCDs designed for scientific use (unlike their commercial counterparts) are strictly linear devices without nonlinear gain circuits; hence the images they produce usually have poor dynamic ranges and require a considerable amount of image processing just to view the image, let alone measure and process it for scientific purposes.

Because Unix and Linux computers are still the de facto standards for astronomy and astrophysics, there are a number of stand-alone applications that deal with FITS formats for those operating systems. Several applications also offer Windows versions of their software offerings, including a FITS add-on for the commercial *Photoshop* program. On the other hand, commercial packages are expensive and are rarely intended for the scientific market, so that route is not recommended. Instead we recommend the open source scientific processor known as ImageJ (http://rsb.info.nih.gov/ij/), which is written in Java with Linux, Windows, and OS X versions available. This software has more than a million users, and it has a number of features that are customizable by the user.

Working with FITS files has its pitfalls even if the applications claim to be able to read the format. For example, it turns out that spectral data are not well handled in FITS files, so there are many variations of data arrangements within the standard FITS format. This has created problems for the usual FITS files (Version 2), particularly those to be used in the Virtual Observatory project. A FITS version 3 has been devised, and many of the old files may need to be converted and placed back in the archives. Another problem arises in that the FITS format can be used to store any 2D tabular data whether or not

Unprocessed image (upper right) versus changes in brightness, gamma, and contrast.

it is intended to represent an image, and how well present systems can interpret this is not known in advance. It takes a bit of trial and error as we will see.

1.5.3   Exploring FITS images through *Mathematica*

As an initial example of dealing with images of extended objects, we work with a small CCD image of Comet Hyakutake (1996) taken with an SBIG CCD with a 35-mm camera lens by Andy Gerrard, then at SUNY-Geneseo. This is imported into *Mathematica* by

image1 = Import["comet8.fits"]

On import it shows, as seen in figure (1.2), certainly that the file is intended to be an image, but the brightness and contrast need adjustments to see that it is actually a comet.

In *Mathematica* one often uses ImageAdjust, but its use here initially fails because the file contains header information *Mathematica* does not like. Deconstruction of the file's data list via list component notation shows the image array is composed of floating point numbers, but the initial problem seems to be that the data are enclosed two levels below the imported file structure delimiters. This is verified by checking the maximum and minimum values of image1[[1,1]], which are 1 and 0 respectively. We can see that the image in figure (1.2) is a comet by enhancing the range by a factor of 100.

image1A = Image[100 image1[[1]]]

In ImageAdjust the parameters are {constrast, brightness, gamma}. The parameters for changing the contrast and brightness are arithmetic, with 0 being no change, so that both positive and negative values are allowed. The gamma is the contrast slope (gain) so that no change is 1, meaning the original contrast slope of the CCD will be preserved. In the days of photographic film gamma was rarely unity.

We can adjust the image further such as reducing contrast by 50%, brightness by 50%, or gamma by 50%.

```
image1B = ImageAdjust[image1A, -0.5]
image1C = ImageAdjust[image1A, {0, -0.5}]
image1D = ImageAdjust[image1A, {0, 0, 0.5}]
```

The processing of this image is straightforward because the lens focal length was short, along with a small focal ratio so that turbulence broadening and guiding errors are not evident. It can be handled in a fashion similar to that of today's best amateur and professional images, where processing is often more to taste rather than specific scientific requirements.

As a more complex example, consider a full CCD image containing a small sub-image of C/Hartley 2 (2010) taken by Aaron Steinhauer and his undergraduate students at SUNY Geneseo. This was a much fainter object appearing on a stellar rate driven (but not well polar aligned) 10-second exposure through the prime focus of an 8-inch Celestron telescope. The effects of "seeing" and image trailing are much more evident in this exposure. Because of these defects, more processing will be required than for the previous image.

Compared with modern techniques, our attempt at de-blurring a comet image may seem a bit primative, but the methods are accessible in the open literature. Our algorithm is based on Andrews and Hunt (1977, p. 83) and Hunt (1978, p. 204) and in principle works with any image processing program that allows the 2D Fourier transform and its inverse to be taken of an image.

Most digital enhancement filters are matrix multiplicative, and some may use what is called a convolution. Convolution is a mathematical process where the output quantity is determined by a "smearing" process acting on the input quantity occurring within an integral. Convolution is what an instrument response does to an input signal whether the instrument involves a lens, mirror, or an electronic amplifier. Mathematically, Fourier transforms (FTs) are integrals and hence are capable of being involved in convolutions; thus one of the main uses of the Fourier transform and its relatives is performing convolutions (Bracewell, 1965). Computationally the mathematical process of convolution is carried out numerically through the Fast Fourier Transform (FFT) (Brigham, 1974).

So how is a convolution calculated? First take the FTs of the two functions you want to convolve. Multiply the transforms together and then take the inverse transform of the product, which yields the convolution. *Mathematica* has the capability of doing both numerical and symbolic Fourier transforms.

Signal restorations (image de-blurring) are the inverse of a convolution known as a deconvolution where one must divide the transform of the output image by the FT of the instrument function (the instrument's response to a single pulse input) or some suitable variant on this. Division is the problem with deconvolution because many instrument functions have transforms with roots in the frequency domain, and one often has to be very clever to avoid them.[25] In practice one still uses Fourier or other transform reconstruction, but the theory makes deconvolution into a matter of multiplication of two FTs rather than a division of one by the other.

We again start by importing the image. Enhancement by a factor of 14 shows the details of the comet, as seen in figure (1.3).

```
image2 = Import["Comet16.fits"]
image2A = Image[14 image2[[1]]]
ImageHistogram[image2A]
```

---

[25] See Hunt (1978, p. 204) for a basic discussion.

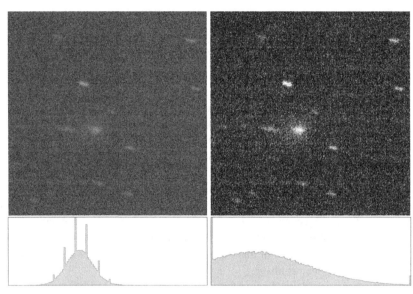

Fig. 1.3 Original image (left) and de-blurred image (right) with their histograms.

From the image histogram it is clear the image has a narrow range of brightness. Notice also that the image has a granular appearance, not unlike traditional photographic film. This occurs because the processing has driven some pixels beyond the brightest or darkest that can be represented by the digitization range, in this case 16 bits on a gray scale.

There are almost as many texts on image processing as there are on chaos theory, so we cannot provide a comprehensive discussion.[26] However, there are three methods that are both useful and basic:

1. The direct inverse filter – includes dividing by the FT of the response function, which may lead to problems
2. The Wiener filter – a modification of the direct filter to compensate for the random noise in the image as well as the particular systematic defect described by the inverse filter
3. The power spectrum filter – uses the magnitude of the Wiener filter as the FT of the instrument function

We here focus on the third method because it seems to provide the best enhancement of this particular image.[27]

We start with a Fourier transform of the image:

```
image2B = ImageAdjust[image2A, 1.3];
ftimage2B = Fourier[ImageData[image2B]];
```

According to Andrews and Hunt (p. 81), the uniform blur function is $\sin f / f$, and this transform agrees with that. The next step is to specify how long the blur is in pixels. From computer screen measurements of the transform, we find is is about 21.5 pixels. The frequency scale of the spectrum is found from the

---

[26] For more up to date references, consult the offerings of the SPIE (http://spie.org).
[27] For examples of all three methods, see **1-5ImageRestore**.

image dimensions (540×540) and the Nyquist frequency; thus we let

$$f_x = \frac{x_i - 280}{280}, \qquad f_y = \frac{y_i - 280}{280}. \tag{1.127}$$

The power spectrum approach includes "noise" compensation in the filter. We do not give details of this here, but simply construct a function based on the inverse filter. The power spectrum transform multiplies rather than divides. We assume the power spectrum of the noise is given by an exponential whose parameters are adjustable. The "noise" is the amplitude, while $\omega0$ is the frequency width. If $\omega0$ is large then the noise is relatively flat with frequency. If $\omega0$ is small then the noise is approximately low-frequency noise. If it is negative then the noise is high frequency. These parameters can be adjusted to "taste." One can also make the noise be Gaussian in frequency, and that simulates band-limited noise in the time domain.

```
ftBlurPSF[xii_, yii_, θθ_, aa_, noise_, ω0_] :=
   Module[{ω}, ω = Cos[θθ] (xii - 280) / 280 + Sin[θθ] (xii - 280) / 280;
```

$$1 \Big/ \sqrt{\left((aa\ Sinc[aa\ \omega])^2 + \frac{noise\ e^{-Abs[\omega/\omega0]}}{(aa\ Sinc[aa\ \omega])^2} \right)}\ ];$$

```
ftimagedeblur3 = Table[{0, 0}, {ii, 1, 560}, {jj, 1, 560}];
Do[ftimagedeblur3[[ii, jj]] = ftimage2B[[ii, jj]]
     ftBlurPSF[ii, jj, 90°, 21.5, 0.15, 1000], {ii, 1, 560}, {jj, 1, 560}];
final3 = ImageAdjust[Image[Abs[InverseFourier[ftimagedeblur3]]], {2.3, -.4, 1}]
```

The resulting image can be seen in figure (1.3).

When dealing with other image formats many of the same approaches can be used, though some care must be taken. For example, FITS stores values as reals with intensities ranging from 0 to 1, and are expressed as floating point numbers, whereas TIFF uses only integers. Some of the problems with FITS format input can be avoided by using IMAGEJ to import the FITS file and exporting it as TIFF. In general, one should avoid lossy formats to avoid the complications of a too vigorous image compression algorithm.

Exercises

1.1 The sun has an apparent magnitude of $m = -26.75$. Calculate its absolute magnitude.

1.2 The sun has a surface temperature of 5800 K. Assuming the sun is a blackbody, calculate its B-V index.

1.3 The star Rigel has a B-V index of -0.03. Calculate its surface temperature.

1.4 From the orbital properties of Alpha Centari A and B as given by Wikipedia, calculate and plot the orbit of B about A as seen from Earth.

1.5 Computationally explore the inverse Compton effect with 3K (cosmic background radiation) photons.

1.6 Show that ds^2 is unchanged under a Lorentz transformation. This behavior is known as Lorentz invariance.

1.7 Computationally explore the twin paradox using SR matrices:

1. By ignoring the turning process by simply reversing the direction of the velocity
2. By simulating the turnaround through a series of matrices that approximates a circular turning by a polygon

1.8 Image process a poor-quality astronomical image taken from the web.

Stellar atmospheres

The light we observe from a star is produced in its atmosphere. An understanding of stellar astrophysics must therefore begin with an understanding of stellar atmospheres. In this chapter we begin with a basic discussion of radiative thermodynamics and the radiative transfer of energy. From this foundation we derive the radiative transfer equation, which governs the behavior of light within a stellar atmosphere. Although this equation is in general quite complicated, it can be made more tractable through simplifying assumptions such as the gray body approximation. The theory developed here pertains for the most part to plane-parallel geometry. In Section 4.1 we not only explore some aspects of radiative transfer in spherical geometry, but also examine the problems of line profile formation in radially expanding atmospheres.

2.1 Radiative transfer and the flow of photons through matter

2.1.1 Specific intensity and photon transfer

Consider a surface of area dA through which an element of monochromatic light energy d^3E flows, going into solid angle $d\omega$, as seen in figure (2.1). This energy is proportional to

$$d^3E_\lambda = I_\lambda \, d\omega \, dA \, d\lambda \cos\theta, \tag{2.1}$$

where I_λ is the specific intensity and θ is the angle the light makes with dA.

As light travels through matter, it gains or loses energy through interaction with its surroundings. The main loss mechanism is atomic and molecular absorption or perhaps scattering. The details of the mechanism are not important, however, because the diminution is always proportional to the incident energy and the oblique path length ds.[1] Thus the absorption term is

$$\left(d^4E_\lambda\right)_a = -\kappa_\lambda \, d^3E_\lambda \, ds = -\kappa_\lambda \, ds I_\lambda \, d\omega \, d\lambda \, dA, \tag{2.2}$$

where κ_λ is the absorption coefficient. On the other hand, emission (and scattering) produce radiation into the surrounding volume $dV = dA \, ds$, giving an emission term

$$\left(d^4E_\lambda\right)_e = \varepsilon_\lambda \, d\omega \, d\lambda \, dV = \varepsilon_\lambda \, ds \, d\omega \, d\lambda \, dA, \tag{2.3}$$

where ε_λ is the emission coefficient. The result is

$$d^4E_\lambda = \left(d^4E_\lambda\right)_a + \left(d^4E_\lambda\right)_e = dI_\lambda \, dA \, d\omega \, d\lambda, \tag{2.4}$$

---

[1] Using ds eliminates the earlier $\cos\theta$ factor in the area because dA is perpendicular to ds.

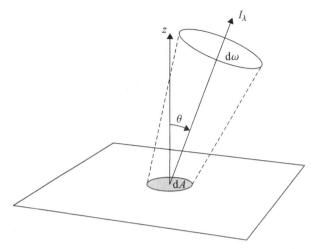

Fig. 2.1 Definitions of radiative transfer.

where dI_λ is the differential of specific intensity. Substituting gives

$$dI_\lambda \, dA \, d\omega \, d\lambda = -\kappa_\lambda \, ds I_\lambda \, d\omega \, d\lambda \, dA + \varepsilon_\lambda \, ds \, d\omega \, d\lambda \, dA. \qquad (2.5)$$

Now we can divide out many differentials to obtain the radiative transfer equation:

$$\frac{dI_\lambda}{ds} = -\kappa_\lambda I_\lambda + \varepsilon_\lambda. \qquad (2.6)$$

If we consider κ_λ and ε_λ as functions of the line-of-site distance s, this expression can be integrated directly. The sign of dI_λ/ds depends on ε_λ compared with $\kappa_\lambda I_\lambda$. It should be stressed that κ_λ and ε_λ are functions of λ, so the integration must be done at *each* wavelength. You *cannot* in general integrate over λ first.

In astronomy and other fields, traditionally it is customary to divide the radiative transfer equation by κ_λ. This enables a change of the variable of integration:

$$\frac{dI_\lambda}{\kappa_\lambda ds} = \frac{dI_\lambda}{d\tau_\lambda} = -I_\lambda + \frac{\varepsilon_\lambda}{\kappa_\lambda} = -I_\lambda + S_\lambda. \qquad (2.7)$$

Here $\tau_\lambda = \int \kappa \, ds$ is called the optical depth, and $S_\lambda = \varepsilon_\lambda/\kappa_\lambda$ is called the source function. The radiative transfer equation is then

$$\frac{dI_\lambda}{d\tau_\lambda} = -I_\lambda + S_\lambda. \qquad (2.8)$$

Although this equation looks simple enough, all depends on the nature of S_λ, which can be quite complex.

2.1.2 Thermodynamics of a black body gas

A black body is said to be in thermodynamic equilibrium. For equation (2.8) this implies that within the body,

$$\frac{dI_\lambda}{d\tau_\lambda} = 0 \qquad \text{or} \qquad S_\lambda = I_\lambda. \qquad (2.9)$$

The specific intensity for a black body is

$$I_\lambda = B_\lambda = \frac{2hc^2}{\lambda^5} \frac{1}{e^{hc/\lambda kT} - 1}, \tag{2.10}$$

where T is the temperature of the body, and B_λ is the wavelength form of the black body formula. Hence for a uniform black-body gas,

$$S_\lambda = B_\lambda, \tag{2.11}$$

which is known as Kirchoff's law of thermal radiation.

For stellar atmospheres we cannot assume a constant uniform temperature; however, we can assume that for small zones of the atmosphere $T = const$. That is, locally $S_\lambda = B_\lambda$. Such a situation is called Local Thermodynamic Equilibrium (LTE). What this implies is that at a particular λ the average photon path where it is "free" (no photon absorption or emission) is small compared to the distance where the temperature changes enough to shift the black body spectrum in a "significant" way, say by 10^{-6} K or less at that wavelength. This can be quantified, but we do not explore that here.

2.1.3 The effect of spectral lines

The temperature in the Planck formula is taken to be the kinetic temperature of the gas. In the laboratory, it has been observed (first by Kirchoff) that when a hot body is viewed through a cooler gas (atomic or molecular but not "too" low or "too" high density) a dark line or dark band spectrum is observed superposed on the hot body's black-body spectrum. On the other hand, if a hot thin gas of the same composition is viewed alone, a bright line spectrum is seen in the places where the dark lines were in the dark line spectrum.

To see the effect of spectral lines, let us return to the radiative transfer equation:

$$\frac{dI_\lambda}{d\tau_\lambda} = -I_\lambda + S_\lambda, \quad \text{where } d\tau_\lambda = \kappa_\lambda \, ds. \tag{2.12}$$

We want to find a solution to $I_\lambda (\tau_\lambda)$. The simplest solution is found by letting $S_\lambda = 0$, which yields e^{τ_λ}. For a more general solution, we take e^{τ_λ} as the integrating factor; thus

$$\frac{dI_\lambda}{d\tau_\lambda} e^{\tau_\lambda} + I_\lambda e^{\tau_\lambda} = +S_\lambda e^{\tau_\lambda}, \tag{2.13}$$

or

$$\frac{d}{d\tau_\lambda} (I_\lambda e^{\tau_\lambda}) = S_\lambda e^{\tau_\lambda}. \tag{2.14}$$

If S_λ is constant throughout the volume, then integrating:

$$I_\lambda e^{\tau_\lambda} \big|_0^{\tau_\lambda} = S_\lambda e^{\tau_\lambda} \big|_0^{\tau_\lambda}. \tag{2.15}$$

If I_{λ_0} is I_λ at $\tau_\lambda = 0$ this becomes

$$I_\lambda = I_{\lambda_0} e^{-\tau_\lambda} + S_\lambda \left(1 - e^{-\tau_\lambda}\right). \tag{2.16}$$

This is the radiative transfer through a uniform slab of gas. The first term on the left is the dimming of any incident radiation and the second term is the contribution from the gas itself.

Let us first consider the case where light is emitted by the gas alone. For this we assume no background source, i.e., $I_{\lambda_0} = 0$. Then

$$I_\lambda = S_\lambda \left(1 - e^{-\tau_\lambda}\right), \tag{2.17}$$

where, assuming LTE, the source function is a black-body distribution, and $S_\lambda = B_\lambda(T)$. If $\tau_\lambda << 1$, the gas is said to be optically thin. This can be the result of s, the real depth, being small or κ_λ being small, or both. Because τ_λ is small, $e^{-\tau_\lambda}$ may be expanded as $e^{-\tau_\lambda} \simeq 1 - \tau_\lambda$ and we have

$$I_\lambda = \tau_\lambda S_\lambda, \tag{2.18}$$

or for LTE

$$I_\lambda = \tau_\lambda B_\lambda. \tag{2.19}$$

Note that within a spectral feature κ_λ can be large compared with the κ_λ between spectral features. If this were not the case, a slab of gas, would not absorb light to form a dark feature when radiation from a hot source illuminates the slab. This is what makes a spectral line appear.

If κ_λ is constant throughout the slab, then $\tau_\lambda = \kappa_\lambda x$, where x is the physical thickness of the slab of gas, and

$$I_\lambda = \kappa_\lambda x S_\lambda, \tag{2.20}$$

or for LTE

$$I_\lambda = \kappa_\lambda x B_\lambda. \tag{2.21}$$

This shows that for a fixed x, the emergent intensity in the spectral feature (emission in this case) is proportional to the absorption coefficient κ_λ and the Planck function B_λ. The hotter the gas, the higher the intensity. This is why emission spectra have the same form as absorption spectra. If a slab of gas is optically thick, there are no emission lines. For the optically thick case, $\tau_\lambda \to \infty$; thus $e^{-\tau_\lambda} \to 0$. As a result, $I_\lambda = S_\lambda$, or in LTE $I_\lambda = B_\lambda$.

The second case to consider is when a slab of gas is illuminated by a background source, that is, the view through a gas when $I_{\lambda_0} \neq 0$.

Starting again with equation (2.16),

$$I_\lambda = I_{\lambda_0} e^{-\tau_\lambda} + S_\lambda \left(1 - e^{-\tau_\lambda}\right), \tag{2.22}$$

suppose we now let $I_{\lambda_0} = B_\lambda(T_2)$ and $S_\lambda = B_\lambda(T_1)$, such that

$$I_\lambda = B_\lambda(T_2) e^{-\tau_\lambda} + B_\lambda(T_1) \left(1 - e^{-\tau_\lambda}\right). \tag{2.23}$$

In the optically thin case $e^{-\tau_\lambda} \simeq 1 - \tau_\lambda$; thus

$$I_\lambda = B_\lambda(T_2)(1 - \tau_\lambda) + B_\lambda(T_1)\tau_\lambda, \tag{2.24}$$

or

$$I_\lambda = B_\lambda(T_2) + \tau_\lambda [B_\lambda(T_1) - B_\lambda(T_2)]. \tag{2.25}$$

The resulting effect depends on the temperature of the background source T_2 relative to that of the foreground gas T_1:

1. If $T_2 > T_1$ then the second term is negative and absorption occurs proportional to τ_λ.
2. If $T_2 < T_1$ then the second term is positive and emission occurs proportional to τ_λ.
3. If $T_2 = T_1$ then $I_\lambda = B_\lambda$, with neither absorption or emission.

In the optically thick case, only the B_λ of the foreground gas is seen regardless of temperature.

A good rough model for a stellar atmosphere can be obtained by treating the atmosphere as a series of slabs of different temperatures. From the preceding possible outcomes, it is clear that a decreasing temperature through the atmosphere produces absorption lines, whereas an increasing temperature through the gas produces emission lines. An atmosphere with a constant temperature would produce neither. Observationally we find that most stars show absorption lines (a few show emission lines), while few stars show a pure Planck spectrum. It is clear then that the temperature of a star typically changes with atmosphere depth.[2]

2.1.4 Determining stellar temperature

Outside of spectral features (i.e., in the continuum) κ_λ is small, but it is not 0. As a result, the effective background spectrum of a stellar atmosphere is essentially where $\tau_{\lambda(\text{continuum})} \to \infty$. This continuum background can be used to determine stellar temperature in several ways:

1. Because the continuous spectrum resembles a black body, the effective temperature for the star can be obtained from its luminosity (and with the assumption that the star is spherical)[3]

$$L = 4\pi R^2 \sigma T^4. \tag{2.26}$$

2. Similarly, one can determine the peak wavelength λ_{peak} of the continuous spectrum, and thus determine the Wien temperature[4]:

$$T = \frac{b}{\lambda_{\text{peak}}}. \tag{2.27}$$

3. The apparent magnitude of a star can be determined through two calibrated color filters. Comparison of these two color magnitudes allows one to determine a star's color temperature.
4. The continuous spectrum can be fit to the full black-body function to determine the Planck temperature.

These various methods do not all have to agree, but the closer they are the closer the stellar atmosphere is to Local Thermal Equilibrium (LTE). Each of these temperatures are radiation temperatures. They correspond to the actual gas temperatures only when they are in good agreement.

If one can measure the absolute amount of radiation at particular wavelength, the temperature can be determined from B_λ and is called the brightness temperature. This is used mainly in radio astronomy, where the long-wavelength approximation of the Planck formula holds.

To see how this works, let us start with the Planck formula:

$$B_\lambda = \frac{2hc^2}{\lambda^5} \frac{1}{e^{hc/\lambda kT} - 1}. \tag{2.28}$$

Because $e^{hc/\lambda kT} \simeq 1 + hc/\lambda kT$ when λT is large,

$$B_\lambda = \frac{2ckT}{\lambda^4}. \tag{2.29}$$

This is the Rayleigh–Jeans equation, which defines the brightness temperature. It is also often expressed in terms of the frequency ν,

$$B_\nu = \frac{2kT\nu^2}{c^2}. \tag{2.30}$$

---

[2] The simple slab model is also useful in modeling radio emission lines from low-temperature dust clouds.
[3] The Stefan–Boltzmann constant $\sigma = 5.670400(40) \times 10^{-8}\,\text{W} \cdot \text{m}^{-2} \cdot \text{K}$.
[4] Here b is known as Wien's displacement constant, with a value of $b = 2.8977685(51) \times 10^{-3}\text{m} \cdot \text{K}$.

In radio astronomy, if the temperature obtained in this way is in conflict with the other data about the source, then one can conclude that it is due to a nonthermal effect.

2.1.5 Some qualitative considerations

The absorption coefficient of the background continuum is much smaller than that of spectral features. That is, κ_λ (continuum) $\ll \kappa_\lambda$ (feature). As a result, it takes more physical depth (a larger optical path) to reach $\tau_\lambda \to \infty$. Thus the continuum comes from a deeper and hence hotter part of a star. The zone where τ_λ is large is called the photosphere. On the other hand, inside weak spectrum lines κ_λ is larger than in the continuum and the physical depth is small. This means that spectral features are formed above the continuum level in the photosphere where the temperature is lower and the gas appears darker because of its temperature. At the edge of the sun, for example, τ_λ is larger because the path length s is longer. Hence the atmosphere seen at $\tau_\lambda \to \infty$ is higher and cooler (darker). This effect is known as limb darkening.

In a very strong spectrum line κ_λ is very large, particularly at the line center. Often, residual emission lines can be seen at their line centers within the broader absorption line. This indicates that well above the photosphere (where continuum and weak absorption lines are formed), the local temperature rises again as you go outward. Such regions are called chromospheres and coronae. In the sun, viewing in Balmer α (Hα, 6563 Å) shows the chromosphere, while He 504 Å shows the corona.[5]

2.2 Formal solution of radiative transfer

2.2.1 Transfer equation for different geometries

In the previous section we utilized the wavelength form of the radiative transfer equation,

$$\frac{dI_\lambda}{\kappa_\lambda ds} = -I_\lambda + S_\lambda. \tag{2.31}$$

In this section we utilize the frequency form,

$$\frac{dI_\nu}{\kappa_\nu \rho \, dz} = -I_\nu + S_\nu. \tag{2.32}$$

Here κ_ν is the mass absorption per unit mass, ρ is the mass density, and z is an axis toward the observer. For spherical coordinates, I_ν is a function of r and θ; thus in general

$$\frac{dI_\nu}{ds} = \frac{\partial I_\nu}{\partial r}\left(\frac{dr}{dz}\right) + \frac{\partial I_\nu}{\partial \theta}\left(\frac{d\theta}{dz}\right), \tag{2.33}$$

where θ is the angle from the z-axis, with $dr = \cos\theta \, dz$ and $r\, d\theta = -\sin\theta \, dz$. Hence,

$$\frac{\partial I_\nu}{\partial \nu}\frac{\cos\theta}{\kappa_\nu \rho} - \frac{\partial I_\nu}{\partial \theta}\frac{\sin\theta}{\kappa_\nu \rho r} = -I_\nu + S_\nu. \tag{2.34}$$

---

[5] Recall that 1 Å $= 10^{-10}$ m.

Needless to say, this is a very tough equation to solve. We explore this in more detail in Section 4.1. For now we assume the atmosphere is very thin so that we can use the plane parallel approximation (thus I_ν is θ independent) such that

$$\cos\theta \frac{\mathrm{d}I_\nu}{\kappa_\nu \rho \, \mathrm{d}r} = -I_\nu + S_\nu. \tag{2.35}$$

In equation (2.35) $\mathrm{d}r$ increases outward. If we want depth inward, $\mathrm{d}\tau_\nu = -\kappa_\nu \rho \, \mathrm{d}r$, and

$$\cos\theta \frac{\mathrm{d}I_\nu}{\mathrm{d}\tau_\nu} = I_\nu - S_\nu, \tag{2.36}$$

where τ_ν is measured radially, not along the line of sight. If I_ν and S_ν are known, numerical solutions to the differential equation can be obtained fairly easily.

2.2.2 Integral equation description of radiative transfer

Formally, every differential equation has a corresponding integral equation form.[6] To determine the integral form of equation (2.36), we take its integral transform (similar to a Laplace transform). For the purely radial case ($\cos\theta = 1$), we multiply equation (2.36) by $\mathrm{e}^{-(\tau_\nu - t_\nu)}$, where τ_ν is some fixed optical depth. Integrating we then find

$$I_\nu(\tau_\nu) = \int_0^{\tau_\nu} S_\nu(t_\nu)\, \mathrm{e}^{-(\tau_\nu - t_\nu)} \mathrm{d}t_\nu + I_\nu(0)\, \mathrm{e}^{-\tau_\nu}. \tag{2.37}$$

For integral equations such as this, $S_\nu(t_\nu)$ is known as the kernel, and $\mathrm{e}^{-(\tau_\nu - t_\nu)}$ is known as the weighting function. For a line-of-sight integration, the $\cos\theta$ factor must be included in the weighting function; thus the solution becomes

$$I_\nu(\tau_\nu) = -\int_0^{\tau_\nu} S_\nu(t_\nu)\, \mathrm{e}^{-(\tau_\nu - t_\nu)\sec\theta} \sec\theta \, \mathrm{d}t_\nu + c. \tag{2.38}$$

Because $\sec\theta$ is an even function, we must be careful to distinguish between incoming light rays ($\theta \geq 90°$) and outgoing light rays ($\theta < 90°$). Normally, $I_\nu = 0$ at $\tau_\nu = 0$ for inward rays; hence

$$I_\nu^{in} = -\int_0^{\tau_\nu} S_\nu(t_\nu)\, \mathrm{e}^{-(\tau_\nu - t_\nu)\sec\theta} \sec\theta \, \mathrm{d}t_\nu. \tag{2.39}$$

For outgoing rays in the case of the $\tau_\nu = \infty$ limit,

$$I_\nu^{out} = \int_{\tau_\nu}^{\infty} S_\nu(t_\nu)\, \mathrm{e}^{-(\tau_\nu - t_\nu)\sec\theta} \sec\theta \, \mathrm{d}t_\nu. \tag{2.40}$$

Because the total specific intensity is

$$I_\nu(\tau_\nu) = I_\nu^{out}(\tau_\nu) + I_\nu^{in}(\tau_\nu), \tag{2.41}$$

[6] It should be noted that *Mathematica* has no explicit way of directly solving integral equations except through conversion to the equivalent differential equation. Then either DSolve[] or NDSolve[] can be used. Sometimes Solve[] or NSolve[] can be used with integrals but we have not consistently tested these on the "standard" integral equations. In notebook **9-5galaxy2D** we discuss this capability further. The best reference for such things is Davis (1962). The Davis book was written in the "pre-Chaos" days and yet covers many topics rediscovered 20–30 years later.

we find

$$
\begin{aligned}
I_\nu(\tau_\nu) = & \int_{\tau_\nu}^{\infty} S_\nu(t_\nu)\,e^{-(\tau_\nu - t_\nu)\sec\theta}\,\sec\theta\,dt_\nu \\
& - \int_{0}^{\tau_\nu} S_\nu(t_\nu)\,e^{-(\tau_\nu - t_\nu)\sec\theta}\,\sec\theta\,dt.
\end{aligned}
\tag{2.42}
$$

This is the integral equation form of the radiative transfer equation with the condition that $S_\nu \exp(-\tau_\nu) \to 0$ as $\tau_\nu \to \infty$. The conditions at $\tau_\nu = 0$ are

$$
I_\nu^{in}(0) = 0,
\tag{2.43}
$$

$$
I_\nu^{out}(0) = \int_{0}^{\tau_\nu} S_\nu e^{-(\tau_\nu - t_\nu)\sec\theta}\,\sec\theta\,dt_\nu.
\tag{2.44}
$$

These are called the emergent intensities, and they are the boundary conditions for the integral equation solution.

2.2.3 Mean intensity, radiative flux, and radiative pressure

The integral solution of the specific intensity is not much better than its differential form. However, some physical insights can be gained by looking at the moments of I_ν rather than the full solution. The zero-order moment of I_ν is known as the mean intensity and is simply the average of I_ν over the solid angle $d\omega$:

$$
J_\nu = \frac{1}{4\pi} \oint I_\nu\,d\omega.
\tag{2.45}
$$

The first-order moment is the radiative flux.[7] Flux is a measure of the total energy flow through an area dA, per time dt, per unit frequency (or per unit wavelength):

$$
F_\nu = \lim \frac{\sum \Delta E_\nu}{\Delta A\,\Delta t\,\Delta \nu} = \oint \frac{dE_\nu}{dA\,dt\,d\nu}
\tag{2.46}
$$

Because $dE_\nu = I_\nu\,dA\,dt\,d\nu\,d\omega \cos\theta$, which is the frequency form of equation (2.1), we find

$$
F_\nu = \oint I_\nu \cos\theta\,d\omega.
\tag{2.47}
$$

The flux is a measure of the anisotropy of the radiation field. If the field is isotropic, then $F_\nu = 0$.

If the specific intensity I_ν is isotropic, integration of equation (2.47) is straightforward. Formally,

$$
F_\nu = \int_{0}^{2\pi} d\phi \int_{0}^{\pi} I_\nu \cos\theta \sin\theta\,d\theta
\tag{2.48}
$$

for spherical coordinates. This can be split into two integrals,

$$
F_\nu = \int_{0}^{2\pi} d\phi \int_{0}^{\pi/2} I_\nu \cos\theta \sin\theta\,d\theta + \int_{0}^{2\pi} d\phi \int_{\pi/2}^{\pi} I_\nu \cos\theta \sin\theta\,d\theta,
\tag{2.49}
$$

---

[7] The first-order moment is sometimes expressed as $H_\nu = (1/4\pi) \oint I_\nu \cos\theta\,d\omega$, with the $1/4\pi$ factor consistent with J_ν and K_ν.

where the first term is the emergent flux leaving the enclosed surface, while the latter is the flux entering the surface. We can consider a star to be an isolated radiating sphere, and if we further suppose that we are on the outer surface of the star, the emergent flux over one hemisphere is

$$F_\nu = \int_0^{2\pi} d\phi \int_0^{\pi/2} I_\nu \cos\theta \sin\theta \, d\theta. \tag{2.50}$$

If I_ν is isotropic, there is no ϕ or θ dependence of I_ν, and

$$F_\nu = 2\pi I_\nu \int_0^{\pi/2} \cos\theta \sin\theta \, d\theta = \pi I_\nu. \tag{2.51}$$

In astrophysics, the flux is often normalized by a factor of $1/\pi$, so that $F_\nu = I_\nu$. The two are not, however, the same quantity. The specific intensity is independent of distance from the source as long as the area of the object is resolved and the solid angle $d\omega$ is defined. Flux is defined to be energy with no ω dependence and therefore varies according to the inverse square law.

The second-order moment of I_ν is also known as the K integral. It is defined as

$$K_\nu = \oint I_\nu \cos^2\theta \, d\omega \tag{2.52}$$

and represents the radiative pressure of the light energy.

Photons have a momentum $p_\nu = E_\nu/c$; thus the momentum flux is

$$dp_\nu = \frac{1}{c} dE_\nu \cos\theta = \frac{I_\nu}{c} \cos^2\theta \, dt \, dA \, d\nu \, d\omega. \tag{2.53}$$

The monochromatic radiative pressure is then

$$P_\nu \, d\nu = \frac{1}{c} \oint I_\nu \cos^2\theta \, d\nu \, d\omega. \tag{2.54}$$

Hence

$$P_\nu = \frac{K_\nu}{c}. \tag{2.55}$$

If I_ν is isotropic, this becomes

$$P_\nu = \frac{4\pi}{3c} I_\nu, \tag{2.56}$$

which can be integrated to determine the total radiation pressure:

$$P_R = \frac{4\pi}{3c} \int_0^\infty I_\nu \, d\nu. \tag{2.57}$$

For an ideal black body, $I_\nu = B_\nu$ and

$$\pi \int_0^\infty B_\nu \, d\nu = \sigma T^4, \tag{2.58}$$

so that

$$P_R = \frac{4\sigma}{3c} T^4. \tag{2.59}$$

2.2.4 Simplifying the source function

The source function $S_\nu = \varepsilon_\nu / \kappa_\nu$ is in general quite complex. However, a reasonable approximation can be made by considering two extreme cases, that of pure isotropic scattering and that of pure absorption. For pure isotropic scattering, all emitted energy is due to scattered photons. Because the scattering is uniform in all directions, the emission coefficient becomes

$$\varepsilon_\nu = \frac{1}{4\pi} \oint \kappa_\nu I_\nu \, d\omega. \tag{2.60}$$

The absorption coefficient κ_ν does not normally depend on ω so

$$S_\nu = \frac{\varepsilon_\nu}{\kappa_\nu} = \frac{1}{\pi} \oint I_\nu \, d\omega. \tag{2.61}$$

In other words,

$$S_\nu = J_\nu. \tag{2.62}$$

For pure absorption, the stellar atmosphere becomes a perfect black-body gas; thus the source function is that of a black body:

$$S_\nu = \frac{2h\nu^3}{c^2} \frac{1}{e^{h\nu/kt} - 1} = B_\nu. \tag{2.63}$$

Let us then consider the mixed case, such that

$$\varepsilon_\nu = \kappa_\nu^S I_\nu + \kappa_\nu^A B_\nu (T), \tag{2.64}$$

where $\kappa_\nu = \kappa_\nu^S + \kappa_\nu^A$. Thus,

$$S_\nu = \frac{\kappa_\nu^S}{\kappa_\nu^S + \kappa_\nu^A} I_\nu + \frac{\kappa_\nu^A}{\kappa_\nu^S + \kappa_\nu^A} B_\nu (T). \tag{2.65}$$

The assumption of isotropy also makes calculation of the mean intensity, flux, and radiative pressure somewhat more manageable. For the flux, recall from equation (2.47),

$$F_\nu = \oint I_\nu \cos\theta \, d\omega = 2\pi \int_0^\pi I_\nu \sin\theta \cos\theta \, d\theta, \tag{2.66}$$

or more explicitly,

$$F_\nu = 2\pi \int_0^{\pi/2} I_\nu^{in} \sin\theta \cos\theta \, d\theta + 2\pi \int_{\pi/2}^\pi I_\nu^{out} \sin\theta \cos\theta \, d\theta. \tag{2.67}$$

Substituting equations (2.39) and (2.40) for I_ν^{in} and I_ν^{out},

$$\begin{aligned}
F_\nu = \; & 2\pi \int_0^{\pi/2} \int_{\tau_\nu}^\infty S_\nu (t_\nu) \, e^{-(\tau_\nu - t_\nu)\sec\theta} \sin\theta \, dt_\nu \, d\theta \\
& - 2\pi \int_{\pi/2}^\pi \int_0^{\tau_\nu} S_\nu (t_\nu) \, e^{-(\tau_\nu - t_\nu)\sec\theta} \sin\theta \, dt_\nu \, d\theta.
\end{aligned} \tag{2.68}$$

Assuming S_ν is isotropic (pure scattering and pure absorption),

$$F_\nu = 2\pi \int_{\tau_\nu}^{\infty} S_\nu \int_0^{\pi/2} e^{-(\tau_\nu - t_\nu)\sec\theta} \sin\theta \, d\theta \, dt_\nu$$
$$- 2\pi \int_0^{\tau_\nu} S_\nu \int_{\pi/2}^{\pi} e^{-(\tau_\nu - t_\nu)\sec\theta} \sin\theta \, d\theta \, dt_\nu. \tag{2.69}$$

The form of the inner integral can be simplified by letting $\psi = \sec\theta$, and $\zeta = (\tau_\nu - t_\nu)$. Thus,

$$\int_0^{\pi/2} e^{-(\tau_\nu - t_\nu)\sec\theta} \sin\theta \, d\theta = \int_1^{\infty} \frac{e^{-\psi\zeta}}{\psi^2} \, d\psi. \tag{2.70}$$

This is an example of an exponential integral, which in general is defined as

$$E_n(\zeta) = \int_1^{\infty} \frac{e^{-\psi\zeta}}{\psi^n} d\psi \tag{2.71}$$

and can be computed numerically. We now can use the exponential integral notation to simplify the equations, and we obtain

$$F_\nu(\tau_\nu) = 2\pi \int_0^{\tau_\nu} S_\nu E_2(t_\nu - \tau_\nu) \, dt_\nu - 2\pi \int_{\tau_\nu}^{\infty} S_\nu E_2(t_\nu - \tau_\nu) \, dt_\nu. \tag{2.72}$$

For the emergent flux,

$$F_\nu(0) = 2\pi \int_0^{\infty} S_\nu(t_\nu) E_2(t_\nu) \, dt_\nu. \tag{2.73}$$

This same E_n formalism can be used for the mean intensity and radiative pressure as well, such that

$$J_\nu(\tau_\nu) = \frac{1}{2} \int_{\tau_\nu}^{\infty} S_\nu E_1(t_\nu - \tau_\nu) \, dt_\nu + \frac{1}{2} \int_0^{\tau_\nu} S_\nu E_1(t_\nu - \tau_\nu) \, dt_\nu, \tag{2.74}$$

and

$$K_\nu(\tau_\nu) = \frac{1}{2} \int_{\tau_\nu}^{\infty} S_\nu E_3(t_\nu - \tau_\nu) \, dt_\nu + \frac{1}{2} \int_0^{\tau_\nu} S_\nu E_3(t_\nu - \tau_\nu) \, dt_\nu. \tag{2.75}$$

2.2.5 Radiative equilibrium and the Milne approximation

Given the lifetimes of stars, it is reasonable to assume that temperatures within a stellar atmosphere are effectively constant. Conservation of energy therefore requires that the integrated flux F be a constant at all depths z. Thus

$$\frac{d}{dz} F(z) = 0, \tag{2.76}$$

where

$$F(z) = \int_0^{\infty} F_\nu(z) \, d\nu = \int_0^{\infty} F_\nu(0) \, d\nu. \tag{2.77}$$

When this is true, a stellar atmosphere is in radiative equilibrium. The flux is then that of a black body:

$$F(z) = F_0 = \sigma T_{eff}^4. \tag{2.78}$$

For a thin atmosphere, we can again use the plane-parallel approximation for radiative transfer. Thus, from equation (2.36),

$$\cos\theta \frac{\mathrm{d}I_\nu}{\mathrm{d}z} = \kappa_\nu \rho I_\nu - \kappa_\nu \rho S_\nu. \tag{2.79}$$

Assuming κ_ν is isotropic, and integrating over $\mathrm{d}\omega$,

$$\frac{\mathrm{d}}{\mathrm{d}z} \int I_\nu \cos\theta \, \mathrm{d}\omega = \kappa_\nu \rho \int I_\nu \, \mathrm{d}\omega - \kappa_\nu \rho \int S_\nu \, \mathrm{d}\omega. \tag{2.80}$$

Then

$$\frac{\mathrm{d}F_\nu}{\mathrm{d}z} = 4\pi \rho \int_0^\infty \kappa_\nu I_\nu \, \mathrm{d}\omega - 4\pi \rho \int_0^\infty \kappa_\nu S_\nu \, \mathrm{d}\omega, \tag{2.81}$$

or because $\mathrm{d}F_\nu/\mathrm{d}z = 0$,

$$\int_0^\infty \kappa_\nu I_\nu \, \mathrm{d}\omega = \int_0^\infty \kappa_\nu S_\nu \, \mathrm{d}\omega. \tag{2.82}$$

Another useful condition is obtained by multiplying equation (2.79) by $\cos\theta$ before integrating, with the result

$$\int_0^\infty \frac{\mathrm{d}K_\nu}{\mathrm{d}\tau_\nu} \mathrm{d}\nu = \frac{F_0}{4\pi}. \tag{2.83}$$

Note that deep in the photosphere $\mathrm{d}F_\nu/\mathrm{d}z \to 0$, and we may always assume that $I_\nu = S_\nu$. This is called the Milne approximation.

2.3 The gray-body approximation

2.3.1 The gray body

A great simplification occurs when the atmosphere is assumed to be "gray." That is, κ_ν or κ_λ is constant. This is not a realistic assumption, but it has a simple solution. It is also a useful stepping stone for more rigorous models. If numerical techniques are used, then a zero-order starting solution (even if not too accurate) is a great convenience. The gray body provides such a solution.

In the gray case the radiative transfer equation is again

$$\cos\theta \frac{\mathrm{d}}{\mathrm{d}z} \int_0^\infty I_\nu \, \mathrm{d}\nu = \rho \int_0^\infty \kappa_\nu I_\nu \, \mathrm{d}\nu - \rho \int_0^\infty \kappa_\nu S_\nu \, \mathrm{d}\nu, \tag{2.84}$$

where, with

$$I = \int_0^\infty I_\nu \, \mathrm{d}\nu, \qquad S = \int_0^\infty S_\nu \, \mathrm{d}\nu, \tag{2.85}$$

and letting $\kappa_\nu = \kappa$, then

$$\cos\theta \frac{\mathrm{d}I}{\mathrm{d}z} = \kappa \rho I - \kappa \rho S. \tag{2.86}$$

Applying the equations of radiative equilibrium to the gray case, we find

$$F = F_0, \qquad J = S, \qquad \frac{\mathrm{d}K}{\mathrm{d}\tau} = \frac{F_0}{4\pi}. \tag{2.87}$$

2.3.2   The Eddington approximation to the gray case

Arthur Eddington first proposed the gray body approximation in 1926. With it he derived a basic relation between atmospheric temperature and optical depth. Eddington began with a linear variation with depth, as

$$\frac{dK}{d\tau} = \frac{F_0}{4\pi} \rightarrow K(\tau) = \frac{F_0 \tau}{4\pi} + \text{const.} \tag{2.88}$$

He then assumed that the linear relationship between κ and F occurred because of a constant $I(\tau)$ for the inward and outward intensities:

$$I(\tau) = I_{in}(\tau) \qquad \theta > \frac{\pi}{2}, \tag{2.89}$$

$$I(\tau) = I_{out}(\tau) \qquad \theta \leq \frac{\pi}{2}. \tag{2.90}$$

This meant further that at a given τ, I is constant over a hemisphere (i.e., isotropic over a hemisphere). Thus the mean intensity becomes

$$J(\tau) = \oint I \frac{d\omega}{4\pi} = \frac{1}{2} \int_0^{\pi/2} I_{out} \sin\theta \, d\theta + \frac{1}{2} \int_{\pi/2}^{\pi} I_{in} \sin\theta \, d\theta, \tag{2.91}$$

$$= \frac{1}{2} I_{out} \int_0^{\pi/2} \sin\theta \, d\theta + \frac{1}{2} I_{in} \int_{\pi/2}^{\pi} \sin\theta \, d\theta, \tag{2.92}$$

or

$$J(\tau) = \frac{1}{2} \left[I_{out}(\tau) + I_{in}(\tau) \right]. \tag{2.93}$$

Likewise, the radiative flux becomes

$$F(\tau) = \pi \left[I_{out}(\tau) - I_{in}(\tau) \right], \tag{2.94}$$

and the K integral

$$K(\tau) = \frac{1}{6} \left[I_{out}(\tau) + I_{in}(\tau) \right]. \tag{2.95}$$

From equations (2.93) and (2.95), it is clear

$$K(\tau) = \frac{1}{3} J(\tau), \tag{2.96}$$

thus, from Eddington's linear form of K,

$$\frac{3F_0 \tau}{4\pi} + \text{const.} \tag{2.97}$$

At the outer boundary,

$$2\pi J(\tau) = F(0) = F_0. \tag{2.98}$$

Thus

$$J(\tau) = \frac{3}{4\pi} \left(\tau + \frac{2}{3} \right) F_0. \tag{2.99}$$

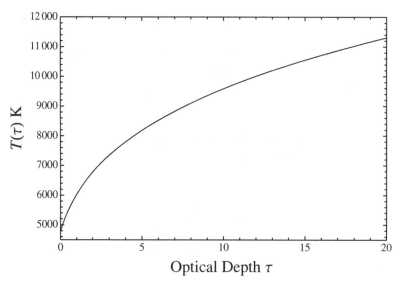

Fig. 2.2 Temperature variation in a gray body atmosphere.

Because $J(\tau) = S(\tau)$,

$$S(\tau) = \frac{3}{4\pi} \left(\tau + \frac{2}{3} \right) F_0. \tag{2.100}$$

In the gray case the source function varies linearly with optical depth.

Because the gray body is in local thermodynamic equilibrium, we also know that $S(\tau) = B(\tau)$, or

$$S(\tau) = \frac{\sigma}{\pi} T^4(\tau). \tag{2.101}$$

Thus

$$T^4(\tau) = \frac{3}{4} \left(\tau + \frac{2}{3} \right) T_{eff}^4, \tag{2.102}$$

and

$$T(\tau) = \left[\frac{3}{4} \left(\tau + \frac{2}{3} \right) \right]^{1/4} T_{eff}, \tag{2.103}$$

shown in figure (2.2). A more rigorous solution by Chandrasekhar (1960) gives

$$J(\tau) = \frac{3}{4\pi} [\tau + q(\tau)] F_0, \tag{2.104}$$

where $0.577 < q(\tau) < 0.710$, and

$$T(\tau) = \left\{ \frac{3}{4} [\tau + q(\tau)] \right\}^{1/4} T_{eff}. \tag{2.105}$$

This is still a simple result. It is not, however, strictly valid, because most of the time the atomic absorptions are not very smooth. Still, this relation gives a start to other methods of solution.

2.3.3 The limits of Eddington's approximation

A better calculation of the gray approximation is to assume κ_ν or κ_λ is constant, but to evaluate the mean intensity, flux, and K integrals directly. This requires us to evaluate the exponential integrals

$$E_n\,(\tau - t) = \int_1^\infty \frac{e^{-w(\tau - t)}}{w^n}\mathrm{d}w \tag{2.106}$$

and

$$E_n\,(t - \tau) = \int_1^\infty \frac{e^{-w(t - \tau)}}{w^n}\mathrm{d}w. \tag{2.107}$$

These are easily done in *Mathematica* with a bit of care.[8] Directly entering the integral, one finds

$$\int_1^\infty y^{-n}\,e^{-x\,y}\,dy$$

`ConditionalExpression[ExpIntegralE[n, x], Re[x] > 0 && Re[n] < 1]`

Mathematica states the result as a conditional expression, as this is the case only if x is positive. Note that in *Mathematica*, if assumptions are required as indicated in the `ConditionalExpression[]` answer, they must be clearly specified.[9] Thus

`y1 = Integrate[y⁻ⁿ e⁻ˣ ʸ, {y, 1, ∞}, Assumptions → Re[x] > 0]`

`ConditionalExpression[ExpIntegralE[n, x], Re[n] < 1]`

The result is left as $E_n(x)$ or `ExpIntegralE[n,x]` when n is unspecified. Here we have placed the needed assumptions inside the verbal command form of the integral and so the solution expression is isolated for further use. Other ways of isolating the solution expression exist, including using component notation or the function `Part`.[10] When a numerical exponent is given in the functional expression of `Integrate`, the recursion version of the exponential integral is given as reduced to a gamma function of order 0. This means we do not want to use the implicit function `ExpIntegralE[n,x]` in the integrals taken over the exponential integrals. To get the gamma function versions, additional assumptions must be made on τ and t when functions of $(t - \tau)$ or $(\tau - t)$ are used.[11]

For example, calculation of the flux, equation (2.72), requires one to determine $E_2\,(t - \tau)$ for both inward ($\tau > t$) and outward ($t > \tau$) flow. Thus

$$\texttt{expplus2 = Integrate}\left[\frac{e^{-(t-\tau)\times w}}{w^2}, \{w, 1, \infty\}, \texttt{Assumptions} \to \texttt{Re}[\tau] < \texttt{Re}[t]\right]$$

$$\texttt{expminus2 = Integrate}\left[\frac{e^{(t-\tau)\times w}}{w^2}, \{w, 1, \infty\}, \texttt{Assumptions} \to \texttt{Re}[t] < \texttt{Re}[\tau]\right]$$

---

[8] This is covered in greater detail in **2-2Moments**.

[9] In version 9 of Mathematica it is possible to 'calculate' with Conditional Expression[] by using them directly. We won't do that.

[10] See the *Mathematica* documentation for details.

[11] For the "in" solution, $Re[\tau] > Re[t]$, and for the "out" solution $Re[\tau] < Re[t]$.

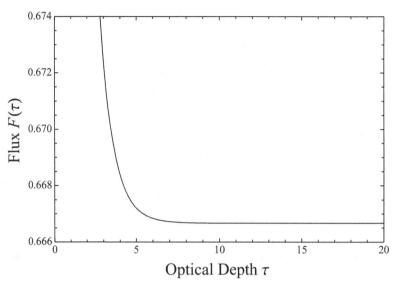

Fig. 2.3 Calculated flux $F(\tau)$ with $s_0 = 3/4$ and $s_1 = 1/2$.

yields

$$E_2^{\text{in}} = e^{(t-\tau)} + (t-\tau)\,\Gamma\,(0, -t+\tau) \tag{2.108}$$

and

$$E_2^{\text{out}} = e^{(-t+\tau)} + (-t+\tau)\,\Gamma\,(0, t-\tau) \tag{2.109}$$

respectively. For the gray-body approximation the source function, equation (2.100), is linear with optical depth. We can therefore let $S(\tau) = s_0 + s_1\tau$ in equation (2.72) and determine the flux. This yields

$$F(\tau) = F^{\text{in}}(\tau) - F^{\text{out}}(\tau), \tag{2.110}$$

where

$$
\begin{aligned}
F^{\text{in}}(\tau) = {} & \frac{1}{3}e^{-\tau}\left[e^{\tau}\tau^2 Ei(-\tau)\,(3s_0 + s_1\tau)\right. \\
& \left. + 3s_0\,(\tau + e^{\tau} - 1) + \left(\tau^2 - \tau + e^{\tau}\,(3\tau - 2) + 2\right)\right], \tag{2.111}
\end{aligned}
$$

$$F^{\text{out}}(\tau) = \frac{1}{3}\left[3s_0 + s_1\,(2 + 3\tau)\right]. \tag{2.112}$$

In the Eddington approximation, the flux $F(\tau) = F_0$ is taken to be constant. However, as seen in figure 2.3, the calculated flux diverges for small τ. The simple Eddington approximation is not valid in the top layers of the atmosphere. In real stars such as the sun, there is also convection in the upper layers and that alone renders a purely radiative approach inaccurate.

2.3.4 Solar temperature profile and limb darkening

For the sun it is possible to obtain the temperature–optical depth relationship because we can observe the radiation coming at different angles from the center, something we cannot do for most other stars. At shallower angles, the observed intensity originates from a region of the atmosphere that is higher, and therefore cooler, an effect known as limb darkening.

Given the specific intensity

$$I_v\left(0\right) = \int_0^\infty S_v \exp\left(-\tau_v \sec\theta\right)\sec\theta\,\mathrm{d}\tau_v, \tag{2.113}$$

and the gray-body approximation

$$S_v = a + b\tau_v, \tag{2.114}$$

we find

$$I_v\left(0\right) = a + b\cos\theta, \tag{2.115}$$

which observationally can be solved for a and b at some frequency or wavelength (usually 500 nm). From this a $T\left(\tau_o\right)$ scale can be found for the solar photosphere.[12]

Then this relationship can also be scaled to other stars. The relationship that we seek is

$$T\left(\tau_o\right) = QT_\odot\left(\tau_o\right), \tag{2.116}$$

where Q is the scaling factor. In the gray case,

$$T\left(\tau\right) = \left\{\frac{3}{4}\left[\tau + q\left(\tau\right)\right]\right\}^{1/4} T_{eff}, \tag{2.117}$$

or

$$T^*\left(\tau_o\right) = \frac{T_{eff}^*}{T_{eff}^\odot} T_\odot\left(\tau_o\right). \tag{2.118}$$

Equation (2.115) defines variation of intensity with the angle of incidence θ. When normalized to the intensity at $\theta = 0$ it becomes the limb darkening law, and is often expressed as

$$H(\mu) = \frac{3}{5}\left[\mu + q(\mu)\right], \tag{2.119}$$

where $\mu = \cos\theta$ and the function $q(\mu)$ is the same as that in equation (2.104), and is known as the Hopf function. In the Eddington approximation $q(\tau) = 2/3$.

The exact limb darkening formula (normalized at $\mu = 1$) is given by Mihalas (1978) as

$$H(\mu) = \frac{1}{1+\mu}\exp\left[\frac{1}{\pi}\int_0^{\pi/2}\frac{\phi\arctan\left(\mu\tan\phi\right)}{1-\phi\cot\phi}\mathrm{d}\phi\right]. \tag{2.120}$$

The exact equation does not have an analytical solution (even *Mathematica* fails to find one), so it must be determined numerically.[13]

---

[12] Here τ_o indicates the optical depth at a standard frequency or wavelength.

[13] This is covered in greater detail in **2-3LimbDark**.

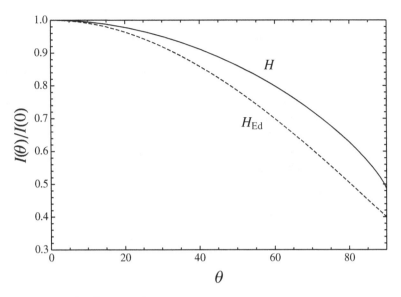

Fig. 2.4 Limb darkening of Eddington and Mihalas compared.

A comparison of Eddington's limb darkening approximation with the more general form (seen in figure 2.4) shows that Eddington's model predicts somewhat cooler temperatures near the sun's surface. This is due to the fact that $q(\tau) > 2/3$ for small τ, resulting in a higher surface temperature. As figure 2.5 shows, at larger depths the two become equivalent, and the difference is only a few percent at most, so it is typically reasonable to use the Eddington expression in the atmosphere calculations.

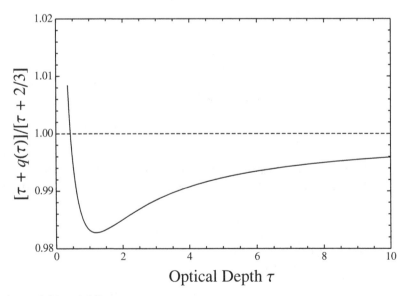

Fig. 2.5 Comparison of $\tau + q(\tau)$ to $\tau + 2/3$.

2.3.5 Pressure and density in a gray body

A simple model of atmosphere pressure can be derived by assuming the stellar atmosphere is an ideal gas. We also assume there are no atmospheric motions so that layers are in what is known as hydrostatic equilibrium.

We start with the weight of a small volume of gas:

$$dW = \rho g dA\, dz. \tag{2.121}$$

Over a short distance the pressure changes by

$$dP = \frac{dW}{dA} = \rho g\, dz, \tag{2.122}$$

where z goes inward (thus the usual minus sign giving a decrease with height is not present). But $d\tau_\nu = \kappa_\nu \rho\, dz$, so for the photosphere:

$$\frac{dP}{d\tau_\nu} = \frac{g}{\kappa_\nu}, \tag{2.123}$$

where

$$g = \frac{GM_x}{R_x^2}, \tag{2.124}$$

with M_x as the mass of the star and R_x as the radius of the star. If κ_ν is constant regardless of depth, then the pressure is linear with optical depth:

$$P(\tau_\nu) = \frac{g}{\kappa_\nu}\tau_\nu + P(0). \tag{2.125}$$

With temperature and pressure profiles known, the density profile is derived from the ideal gas law:

$$P(\tau_\nu) = \frac{\rho(\tau_\nu)\, k_b T(\tau_\nu)}{\mu}, \tag{2.126}$$

where k_b is the Boltzmann constant and μ is the mean molecular weight. With the assumption that the pressure $P(0) = 0$ at the surface, substituting equations (2.103) and (2.125) yields

$$\rho(\tau_\nu) = \frac{\mu g \tau_\nu}{\kappa_\nu k_b T_{\text{eff}} \left(\frac{1}{2} + \frac{3}{4}\tau\right)^{1/4}}. \tag{2.127}$$

Knowing the density as function of optical depth, we can also determine the relation between the optical depth and depth within the atmosphere. Because $d\tau = \rho \kappa_\nu\, dz$, we can multiply equation (2.127) by $\kappa_\nu\, dz$ and integrate; thus

$$\int dz = \frac{k_b}{\mu g T_{\text{eff}}} \int \frac{\left(\frac{1}{2} + \frac{3}{4}\tau\right)^{1/4}}{\tau_\nu}\, d\tau_\nu. \tag{2.128}$$

This is typically written in terms of the scaling factor,

$$z = \frac{k_b T_{\text{eff}}^*}{\mu g^*} \int_{(\tau_1)_\nu}^{(\tau_2)_\nu} \frac{\left(\frac{1}{2} + \frac{3}{4}\tau\right)^{1/4}}{T_{\text{eff}}^\odot} \frac{d\tau_\nu}{\tau_\nu}. \tag{2.129}$$

This expression gives the real depth–optical depth relationship between two optical depths. The constant in front of the integral is the atmospheric scale height and is a standard property of an isothermal atmosphere. The integral itself can be evaluated numerically, though it should be noted that the function diverges as $\tau \to 0$ or $\tau \to \infty$.

2.4 Continuous opacity in a real hydrogen star

2.4.1 Mechanisms of absorption

In a real star, the absorption coefficient κ_ν is not constant, but depends in a complicated fashion on the temperature, the chemical abundance of the absorbing gas, and the amount of ionization.[14] There are three main processes by which absorption can occur:

1. **Bound–bound transitions,** in which an atom absorbs (emits) a photon, resulting in a bound electron transitioning to a higher (lower) energy bound state. Such transitions are rare outside the line spectra wavelengths and do not greatly affect the absorption coefficient.
2. **Bound–free absorption,** or photoionization, in which an incident photon ionizes an electron. Because the now free electron can have any final kinetic energy, bound–free absorption can occur for a continuum of wavelengths.
3. **Free–free absorption,** in which a free electron near an ion absorbs a photon, changing its kinetic energy. This can also occur for a continuum of wavelengths.

For solar-type stars, the dominant factors of continuous absorption are the bound–free and free–free transitions of the negative ion of hydrogen (H^-) as well as the transitions of hydrogen. For lower temperature stars, diatomic and in some cases triatomic molecules (H_2O, CO_2, etc.) are important. Of course, in molecular clouds more complex absorbing molecules are found.

2.4.2 Continuous opacity for atomic hydrogen

For continuous opacity, the bound–free and free–free absorptions are generally calculated for an effective wavelength, typically $\lambda_0 = 5000\,\text{Å}$ for solar-type stars. A bound–free transition requires a minimum energy to overcome the binding energy of the atom. For a neutral hydrogen atom, this binding energy is[15]

$$E_n = -\frac{hRc}{n^2} = -\frac{13.6\,\text{eV}}{n^2};$$
(2.130)

thus the minimum n for a bound–free transition is $n_0 > \sqrt{R\lambda}$.

The bound–free absorption coefficient for an atom with energy E_n is

$$\alpha_n = \frac{\alpha_0 g_n^{bf} \lambda^3}{n^5},$$
(2.131)

where g_n^{bf} is a quantum corrective term known as the Gaunt factor, and

$$\alpha_0 = \frac{32\pi^2 e^6 R}{3\sqrt{3}h^3 c^3}.$$
(2.132)

---

[14] See **2-4Absorbtion** and **2-5Atmosphere** for details.
[15] We ignore the effect of orbital angular momentum here.

Table 2.1 Constants for H$^-$ Absorption Coefficient	
Constant	Value
a_0	1.99654
a_1	-1.18267×10^{-5}
a_2	2.64243×10^{-6}
a_3	-4.40524×10^{-10}
a_4	3.23992×10^{-14}
a_5	-1.39568×10^{-18}
a_6	2.78701×10^{-23}

To calculate the bound–free absorption for an "average" atom, we must determine what fraction of hydrogen atoms are in each E_n energy state. This is given by the Saha equation for atomic hydrogen:

$$\frac{N_n}{N} = n^2 e^{hc/\lambda kT}. \tag{2.133}$$

The bound–free absorption coefficient (in cm^2/atom) is then

$$\alpha\left(H_{bf}\right) = \sum_{n_0}^{\infty} \frac{\alpha_n N_n}{N} = \alpha_0 \sum_{n_0}^{\infty} \frac{\lambda^3}{n^3} g_n^{bf} e^{hc/\lambda kT}. \tag{2.134}$$

When calculating this absorption, typically only the first few terms of the sum are included.[16]

The free–free absorption is a thermal Bremsstrahlung effect and thus relies on the gas temperature. For a single atom the absorption coefficient is

$$\alpha_{ff} = \frac{2h^2 e^2 R}{3\sqrt{3}\pi m^3 c^3} g_\lambda^{ff} \sqrt{\frac{2m}{\pi kT}} \lambda^3, \tag{2.135}$$

Applying the Saha equation, the free–free absorption coefficient per atom is

$$\alpha\left(H_{ff}\right) = \alpha_{ff} \frac{2\pi mkT^{3/2}}{h^3} e^{hc/\lambda kT}. \tag{2.136}$$

A hydrogen atom can also hold an extra electron and be negatively ionized (H$^-$). Near solar temperatures, H$^-$ contributes to the absorption as well and must also be included. At somewhat higher temperatures the hydrogen is ionized to such an extent that it is no longer a strong absorber. For cooler stars it becomes the dominant absorber, but drops off for the coolest stars due to the lack of free electrons.

The bound–free absorption coefficient has been calculated by Wishart (1979) and others, and it is typically expressed as a polynomial fit to Wishart's data. Specifically, the polynomial fit

$$\alpha_{bf} = a_0 + a_1\lambda + a_2\lambda^2 + a_3\lambda^3 + a_4\lambda^4 + a_5\lambda^5 + a_6\lambda^6, \tag{2.137}$$

where the constants are given in table (2.1), gives the absorption in units of 10^{-18} cm^2 per H$^-$ ion and fits Wishart's data to within 0.2% for 2250 Å $< \lambda <$ 15 000 Å.

[16] See **2-4Absorption** for the calculation of this coefficient.

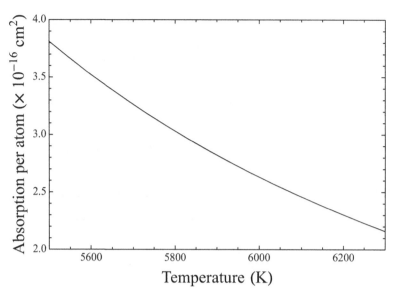

Fig. 2.6 Absorption coefficient for atomic hydrogen.

This is easily calculated numerically for a given wavelength. For the absorption of a typical atom, we must take into account the electron density. In *Mathematica* this can be determined by[17]

```
κbfion[θ_, eden_] := 4.158 × 10⁻¹⁰ × alph1 × eden × 1.38046 × 10⁻¹⁶ ×
    (5040/θ) × θ^(5/2) × 10^(0.754×θ)
```

where alph1 = α_{bf}, eden is the electron density and $\theta = 5040/T$.

The free–free absorption for H^- is again due to the thermal Bremsstrahlung. Its abundance depends on the electron pressure. Here we assume the electron pressure follows that of an ideal gas, and then calculate the ion fraction through the Saha equation.[18] The total absorption per atom can be obtained by summing all four of these coefficients, the result of which is seen in figure (2.6).

2.4.3 Evaluation of the continuous opacity

Because our absorption coefficient depends on wavelength, temperature, pressure, and chemical composition, we must make some assumptions as to their nature. We will take $\lambda_0 = 5000$ Å as the effective wavelength for our calculations. We have seen for the gray-body approximation the relation between temperature and optical depth is reasonably modeled by Eddington's approximation, equation (2.103):

$$T(\tau) = \left[\frac{3}{4}\left(\tau + \frac{2}{3}\right)\right]^{1/4} T_{eff}. \qquad (2.138)$$

---

[17] Here we use × explicitly for multiplication. One can also use a space between variables. Both forms are used in the text.
[18] See **2-4Absorption** for the numerical calculation of this value.

We also continue to assume the atmosphere is an ideal gas in hydrostatic equilibrium. For the gray body the optical depth is related to the real depth by $d\tau = \kappa\rho\, dx$; thus the pressure equation becomes explicitly independent of the gas density

$$\frac{dP}{d\tau} = \frac{GM}{\kappa R^2},\tag{2.139}$$

where M is the stellar mass and R the stellar radius.

For the composition, we follow the values given Aller (1963). The total absorption κ can then be calculated as a function of pressure and optical depth[19]:

```
κtotal[θ_, eden_] :=  ────────── ×
                       abundatwts

  ((κ_H[θ] + κbfion[θ, eden]) × (1 - 10^(-χ_λ×θ)) + κffion[θ, eden]) / (1 + ───Φ[θ]───);
                                                                            epress[θ, eden]
```

The absorption κtotal depends on both pressure and optical depth. Because the pressure is given by equation (2.139), we can substitute κtotal and calculate the function in *Mathematica* with NDSolve[]. The initial pressure at $\tau = 0$ cannot be zero, but must be some small initial value for which good convergence is obtained.

```
diff1 = NDSolve[{D[pgas[τ], τ] == gstar / κtotal[pgas[τ], τ], pgas[0] == 1. × 10^-15},
   pgas, {τ, 0., 100}, MaxSteps → 10 000]
```

This gives the pressure pgas as an interpolating function of optical depth, but this is sufficient for our needs. We can now determine the relationship between the optical depth and the real depth within a stellar atmosphere by integrating the relation $d\tau = \kappa\rho\, dx$. Because our pressure function is not an analytical solution, we must use NIntegrate[] in to determine a numerical solution

```
realdepth[τ_] := NIntegrate[

  (     1.38 × 10^-16 × 5040     ) /. diff1 // Evaluate,
   ──────────────────────────────────────────────
   κtotal[pgas[tt], tt] × pgas[tt] × θ[tt] × abundatwts

  {tt, .1, τ}]
```

The result is seen in figure (2.7).

2.5 The case of spectrum lines

2.5.1 The line transfer equation

In the region of a spectral line, the absorption and emission coefficients differ significantly from that of the continuum background. As a result, the radiative transfer equation in a line region takes a slightly

---

[19] See **2-5Atmosphere**.

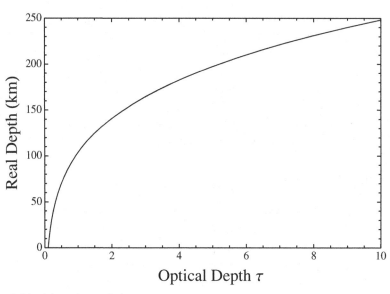

Fig. 2.7 Real depth vs. optical depth for a solar-type hydrogen star.

different form. If we denote l_ν and ε_ν^l as the line absorption and emission coefficients, and κ_ν and ε_ν^c as the continuum absorption and emission coefficients, then

$$d\tau_\nu = (l_\nu + \kappa_\nu)\,\rho\,d\nu, \tag{2.140}$$

$$S_\nu = \frac{\varepsilon_\nu^l + \varepsilon_\nu^c}{l_\nu + \kappa_\nu}, \tag{2.141}$$

with

$$\frac{dI_\nu}{d\tau_\nu} = -I_\nu + S_\nu. \tag{2.142}$$

If we separate the line source function from the continuum one,

$$S_l = \frac{\varepsilon_\nu^l}{l_\nu} \qquad S_c = \frac{\varepsilon_\nu^c}{\kappa_\nu}. \tag{2.143}$$

Then the total becomes

$$S_\nu = \frac{(l_\nu/\kappa_\nu)\,S_l + S_c}{1 + l_\nu/\kappa_\nu} = \frac{S_l + (\kappa_\nu/l_\nu)\,S_c}{1 + \kappa_\nu/l_\nu}. \tag{2.144}$$

The integral equations are as before,

$$I_\nu(0) = \int_0^\infty S_\nu(\tau_\nu)\,e^{-\tau_\nu \sec\theta}\,\sec\theta\,d\tau_\nu, \tag{2.145}$$

$$F_\nu(0) = 2\pi \int_0^\infty S_\nu(\tau_\nu)\,E_2(\tau_\nu)\,d\tau_\nu. \tag{2.146}$$

2.5.2　The line source function

Outside of a line region, we can still use Eddington's gray-body approximation. Thus, the source function is

$$S_\nu(\tau_\nu) \simeq \frac{3}{4\pi} F_\nu(0) \left[\tau_\nu + \frac{2}{3} \right].$$ (2.147)

When the optical depth $\tau_\nu = (4\pi - 2)/3 \simeq 3.5 = \tau_1$, we find

$$S_\nu(\tau_1) = F_\nu(0).$$ (2.148)

Because l_ν changes across the line with the largest value at the line center, the condition $\tau_\nu = \tau_1$ occurs higher in the atmosphere at the line center than in the wings (the part nearest the continuum). Because S_ν decreases with height (i.e., cooler above the photosphere), an absorption line is formed.

Without going into detail, we will assume that the line source function is the same as the continuum for LTE. For a weak line, we expect then that the flux profile will follow the shape of $l_\nu(\nu)$. In fact, it can be shown that in the weak line approximation

$$\frac{F_c - F_\nu}{F_c} \simeq \left(\frac{d \ln S_\nu}{d\tau_c} \right) \frac{l_\nu}{\kappa_o} \tau_o = \left(\frac{d \ln S_\nu}{d\tau_c} \right) \frac{l_\nu}{\kappa_o} \tau_c,$$ (2.149)

or

$$\frac{F_c - F_\nu}{F_c} \simeq \tau_1 \left(\frac{d \ln S_\nu}{d\tau_c} \right)_{\tau_1} \frac{l_\nu}{\kappa_\nu} = \text{const.} \left(\frac{l_\nu}{\kappa_\nu} \right).$$ (2.150)

Across the line κ_ν is fairly constant, so the weak line shape is to a first approximation the shape of l_ν. In other words, the shape of an absorption line is purely a matter of atomic physics.

Although this means absorption lines seen in stars are similar to emission lines created in the lab, there are several processes within a stellar atmosphere that can affect both line width and line strength:

1. Physical factors affecting spectral line shape:
   1. Natural line broadening – quantum effect (uncertainty principle)
   2. Doppler broadening – Doppler effect due to temperature
   3. Pressure broadening – Stark effect (electric fields during collision)
   4. Stellar rotation – Doppler effect
   5. Turbulence in the atmosphere – Doppler effect
   6. Magnetic broadening – Zeeman effect
2. Factors affecting line strengths:
   1. Temperature
   2. Abundance
   3. Turbulence

2.5.3　Line broadening mechanisms

1. Natural line broadening
   This is due to the Heisenberg uncertainty principle:

$$\Delta E \Delta t \leq \hbar.$$ (2.151)

Because $E = hc/\lambda$, the uncertainty in energy corresponds to an uncertainty in line wavelength (or frequency). Thus,

$$\Delta\lambda \approx \frac{\lambda^2}{2\pi c}\left(\frac{1}{\Delta t_i} + \frac{1}{\Delta t_f}\right), \tag{2.152}$$

where Δt_i is the lifetime of the initial state, and Δt_f is the lifetime of the final state. The statistical distribution of frequencies is then given by the Lorentzian profile; thus with $\gamma \approx \Delta t$, the line profile becomes

$$l_\nu\rho = N\frac{\pi e^2}{mc}\left(\frac{\gamma}{\Delta\omega^2 + (\gamma/2)^2}\right), \tag{2.153}$$

where e and m are the electron charge and mass, c is the speed of light, N is the number of atoms, and ω is the angular frequency.

Some atoms make their transitions rapidly, so Δt is short and ΔE is large. A short Δt has a higher probability of transiting and is called a permitted transition. The spectrum line has a large natural width. At the other extreme, there are transitions where Δt is large. The corresponding ΔE is small and the line narrow. Such transitions are highly unlikely (i.e., low probability) and called forbidden. This does not mean, however, that such a transition is impossible. Lasers, for example, use forbidden transitions that are artificially stimulated to have a high probability.

2. Doppler broadening

This is due to atomic motions in a gas. If we assume a stellar atmosphere is in kinetic equilibrium,

$$\frac{1}{2}m\bar{v}^2 = \frac{3}{2}kT, \tag{2.154}$$

where $\bar{v}^2$ is the mean square velocity. Such a gas has a Maxwellian distribution of speeds along a line of sight:

$$\frac{dN(v)}{dN} = \left(\frac{2m}{\pi kT}\right)^{3/2} v^2 \exp\left(-\frac{mv^2}{2kT}\right) dv. \tag{2.155}$$

Using this with the Doppler formula and the absorption coefficient,

$$\frac{v_0 - v}{v_0} = \frac{v}{c} = \frac{\lambda - \lambda_0}{\lambda_0} \tag{2.156}$$

gives the line shape called the Doppler profile.

3. Pressure broadening

Pressure broadening occurs mainly in the spectral lines of the light elements such as hydrogen and helium. Essentially it is due to light-element atomic electrons "feeling" the strong electric fields of other atoms and ions during collisions. This is called the Stark effect and is very complicated to calculate, even using advanced quantum mechanics. However, as we have shown, the pressure is a straightforward calculation for a star whose mass is known and whose atmosphere is static.

2.5.4 Line strengths

Observationally, the strength of a line is defined by the equivalent width. If the continuum level is equal to unity, then the equivalent width is

$$W = \int_{-\infty}^{\infty} \left(\frac{F_c - F_\nu}{F_c}\right) d\nu, \tag{2.157}$$

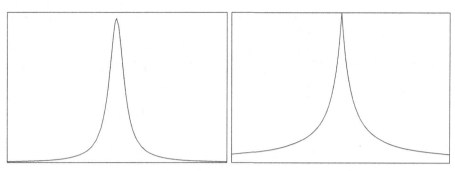

Fig. 2.8   The Voigt (left) and Hjerting (right) functions.

where $F_v = F_c e^{-\tau_v}$. But,

$$\tau_v = \int_0^L l_v \rho \, dz = A \int_0^L \left(\frac{N}{N_o}\right) N_H \alpha \, dz, \tag{2.158}$$

where A is the elemental abundance relative to hydrogen, N/N_o is the fraction of atoms available for absorbing, N_H is the number of hydrogen atoms, and α is the atomic absorption coefficient. For weak lines, $F_v \simeq F_c (1 - \tau_v)$, and $W \propto A$. For strong lines,

$$\frac{F_c - F_v}{F_c} = 1 - e^{-\tau_v}, \tag{2.159}$$

and $W \propto A^{1/2}$.

In addition to the dependence on A, there is a dependence on temperature through the Boltzmann equation for atoms

$$\frac{N}{N_o} = \frac{g_n}{g_l} e^{-(E_n - E_l)/kT(\tau_v)}. \tag{2.160}$$

In very hot stars lines may be weakened or absent because all the atoms have been converted into ions. The equation that describes this is called the Saha Equation, and is derived from the law of mass action from chemistry for the reaction

$$H \rightleftharpoons H^+ + e^- \tag{2.161}$$

$$\frac{N_1}{N_o} P_e = \frac{(2\pi m)^{3/2} (kT)^{5/2}}{h^3} \frac{2U_1(T)}{U_o(T)} e^{-\chi/kT}, \tag{2.162}$$

where P_e is the electron pressure (derived from gas pressure), U_1 is the ionic partition function, U_o is the atomic partition function, and χ is the ionization energy for the atom. Note that partition functions behave like statistical weights (g's) in the Boltzmann equation. Solving for the electron pressure requires iteratively using all the Saha equations (one for each element) together, which is not an easy task.

2.5.5 Computation of a line profile

A realistic line profile must combine both natural (Lorentzian) and Doppler (Gaussian) broadening. This can be obtained by taking a convolution of these two profiles, known as the Voigt profile (figure 2.8, left):

$$V(x, \sigma, \gamma) = \frac{\gamma}{\sigma \pi \sqrt{2\pi}} \int_{-\infty}^{\infty} \frac{e^{-t^2/2\sigma^2}}{(x - t)^2 + \gamma^2} dt. \tag{2.163}$$

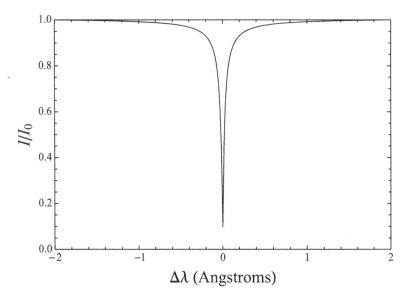

Fig. 2.9   A calculated "weak" line profile.

This must be determined numerically.[20] Here *Mathematica* needs some guidance in the numerical integration of the function. Singularities are suspected at 0 and ±1 as well as the end points, so they must be explicitly mentioned in the limits list for the integrating variable. We also force recursion with explicit limits to the process. Thus

$$\texttt{voigt[u\_, a\_, ΔvD1\_] :=} \frac{1}{\sqrt{\pi} \times \Delta vD1} \times \frac{a}{\pi} \times$$

$$\texttt{NIntegrate}\left[\frac{e^{-u1^2}}{(u - u1)^2 + a^2}, \{u1, -\infty, -1, 0, 1, \infty\}, \texttt{MinRecursion} \to 3, \texttt{MaxRecursion} \to 15\right]$$

Here $\Delta vD1 = (2kT/m)(v/c)$ is the Doppler width, $u = \Delta v/\Delta vD1$, and $a = (\gamma/4\pi)/\Delta vD1$.

The wavelength version of the profile is known as the Hjerting function (figure 2.8, right), where $u = \Delta\lambda/\Delta\lambda_D$ and $a = (\gamma\lambda^2/4\pi c)/\Delta\lambda_D$. It can be calculated from the Voigt function; however, there is an implied minus sign in the $\Delta\lambda_D$ that makes the function flip to negative values when a negative u is used. Hence the absolute value of the Voigt function must be taken.

In the Eddington approximation it can be shown for a weak spectral line that

$$\frac{F_c - F_v}{F_c} \simeq \tau_{\text{eff}} \left(\frac{d \ln S_v}{d\tau_c}\right)_{\tau_{\text{eff}}} \frac{l_v}{\kappa_v}, \tag{2.164}$$

where τ_{eff} is the level where $S_v = F(0)$; that is,

$$\tau_{\text{eff}} = \frac{4\pi - 2}{3} \simeq 3.5. \tag{2.165}$$

---

[20] See **2-6Lineabs** for details.

Because the derivative, κ_ν, and τ_{eff} are all constants, the line shape follows that of l_ν. This can be calculated in *Mathematica*; however, the resulting `lineabs` function (figure 2.9) contains a singularity at the line center and does not go to 0 at either end of the two wavelength extremes. These problems can be overcome by sampling the function at several (in this case 256) values and plotting with `ListPlot[]`:

```
depthlist = Table[lineabs[z], {z, -2., 2., 0.01563}];
linelist = Table[{z, lineabs[z]}, {z, -2., 2., 0.01563}];
plotlist = Table[{z, 1 + lineabs[2] - lineabs[z]}, {z, -2., 2., 0.01563}];
ListPlot[plotlist, PlotRange → All, AxesOrigin → {0, 0}, Joined → True]
```

Elimination of the singularity in `lineabs` also makes it possible to explore additional line broadening effects, such as rotational broadening, where it is necessary to take a numerical Fourier transform of our line function.[21]

Exercises

2.1 A certain star has a surface temperature of 3000 K. How much energy does it emit per second per area?

2.2 If the sun at 5500 K has a peak energy at a wavelength of 500 nm, at what wavelength would the 3000 K star peak?

2.3 Suppose there is a star $10\times$ larger than the sun, with a luminosity $500\times$ that of the sun. What is the temperature of the star?

2.4 Given

$$S_\nu = a + b\tau_\nu,$$

and

$$I_\nu(0) = \int_0^\infty S_\nu e^{-\tau_\nu \sec\theta} \sec\theta \, d\tau_\nu,$$

show that

$$I_\nu(0) = a + b\cos\theta.$$

2.5 In the gray body approximation, the constants in the temperature optical depth relationship are $a = 1/2\pi$ and $b = 3/4\pi$. If you do a least squares fit of solar limb darkening at 500 nm versus the cosine of the angle, you get $0.29 + 0.71\cos\theta$.
1. At what optical depth is $T = T_{\text{eff}}$ in the two cases?
2. What is the temperature at the edge of the sun according to a linear fit of the data?
3. If hotter means deeper, then which model probes deeper into the sun?

2.6 Compare the solar limb darkening data with the H formula in **2-3LimbDark**. Hint: Normalize the theoretical intensity at the disk center to unity, and then let the data be obtained from $\arcsin(r/R_o)$, where r is the radial distance from the center and R_o is the apparent solar radius in the same units.

---

[21] See **2-6Lineabs**.

2.7 A medium with an optical depth (at 300 nm) of 1.5 and a surface intensity of $2.1 \times 10^{-6} W/m^2 \cdot ster \cdot s \cdot$ Å is placed in front of a black-body source. What source function is needed to just produce dark lines in the spectrum? To what black-body temperature does this energy correspond?

2.8 At a certain wavelength, a medium with a specific intensity of $6.7 \times 10^{-5} W/m^2 \cdot ster \cdot s \cdot$ Å is seen in front of a body with a source function $2.5 \times 10^{-4} W/m^2 \cdot ster \cdot s \cdot$ Å. The observed intensity is $1.88 \times 10^{-4} W/m^2 \cdot ster \cdot s \cdot$ Å. What is the optical depth of the medium at this wavelength?

2.9 At a certain frequency, the mass absorption coefficient of dust filled air is 0.25 per kg/m^3 of air. For the air, assume that the density decreases exponentially with height according to the formula

$$\rho = 1.2 e^{-h/10,000},$$

where ρ is in kg/m^3, and h is in meters. What is the optical depth for vertical incidence?

2.10 The scale height for an atmosphere is

$$H = \frac{k_b T_{eff}}{\mu g^*}, \qquad \text{where} \qquad g^* = \frac{GM^*}{R_*^2}.$$

1. For Earth's atmosphere, $H = 10$ km. Compare this to the sun's photospheric scale height. The μ for the sun is m_p, the proton mass, while μ for Earth is $32 \times m_p$. The temperature for Earth is 300 K.
2. Compare the scale height of a hot white dwarf ($T = 50\,000$ K, $M = 1.0 M_\odot, R = 0.000001 R_\odot$) to that of a red supergiant ($T = 1500$ K, $M = 20 M_\odot$, $R = 1000 R_\odot$). Assume the mean molecular weight for the supergiant to be $1.2 m_p$ and for that of the white dwarf to be $4.8 m_p$.
3. The Stark effect is stronger for stars with higher gravity than lower gravity. From your value of g used in the calculation of the scale height, predict which star will have the broadest helium lines.

2.11 The real depth corresponding to two different optical depths is the product of the scale height and the integral

$$\int_{\tau_1}^{\tau_2} \left[\frac{1}{2\pi} + \frac{3}{4\pi} \tau \right]^{1/4} \frac{d\tau}{\tau}.$$

Find a solution to this integral. If no analytic solution exists, then try using a numerical one using 0.01 to 10.0 for the optical depth limits. Adjust the optical depth values until you find a range of optical depths that will span at least one atmospheric scale height.

2.12 Repeat the model calculations of **2-5Atmosphere**
1. For a pure helium star.
2. For a 10 000 K main sequence star.
3. For a 3000 K main sequence star.

2.13 Most stars are made of hydrogen atoms. What is the root mean square H-atom velocity at 10 000 K? Now suppose you observe a spectral line of such atoms at 650 nm wavelength. Because some atoms are coming directly toward you and some going directly away, there will be a Doppler broadening of the spectrum line. If Doppler broadening is the only mechanism involved, what is the average width of the observed spectrum line?

2.14 Find the breakup velocity for a one solar mass star and then calculate the rotational broadened weak line shown in **2-6Lineabs**.

2.15 Include the first-order stark effect in the line absorption coefficient of **2-6Lineabs** and show how the line profile is changed from that of a weak line. Hint: See collisional (or pressure) broadening theory, for example, Aller (1963, p. 310) or Gray (1992, p. 209).

3 Stellar interiors

In the previous chapter we examined stellar atmospheres. Even with broad approximations, we were able to determine some general properties. We know, for example, that stellar atmospheres get hotter the deeper you go, and they roughly obey the ideal gas relationship. We now take these assumptions deeper, into the stellar interior.

In a typical star, the atmosphere is so small in comparison with the radius of the sun that we can essentially ignore it, and thus the entire star from center to rim will be taken as the "interior." What we would like to obtain is a general description of pressure, temperature, and density as a function of depth, as well as relations among mass, radius, and luminosity.

3.1 The hydrostatic model

3.1.1 A simple stellar interior

Observationally, we know (for a stable star like the sun) a star's size and temperature remain constant over long periods of time. Thus it is reasonable to presume our stars are in hydrostatic equilibrium. We will also assume for now that our stars are spherical and irrotational so that each of our properties depends only on the radius r from the center of the star. This means (as before) the pressure within the interior of the star is exactly balanced by the gravitational weight of the star itself. Thus, we have for a change in pressure P,

$$dP(r) = -g(r)\, dm, \tag{3.1}$$

or

$$dP(r) = -\rho(r)g(r)\, dr. \tag{3.2}$$

Here $\rho(r)$ is the local density as a function of radius, and $g(r)$ is the local strength of the gravitational field. We use negative sign here because we are measuring r as the distance from the center of the star (and not the distance from its surface). The gravitational field may be expressed in terms of its potential,

$$g(r) = \frac{d\phi(r)}{dr}. \tag{3.3}$$

Thus, we have an equation for hydrostatic equilibrium:

$$\frac{dP}{dr} = -\frac{d\phi}{dr}\rho. \tag{3.4}$$

The potential is a solution to the Poisson equation:

$$\nabla^2\phi = 4\pi G\rho, \tag{3.5}$$

and because we are assuming spherical symmetry, this can be written as

$$\frac{1}{r^2} \frac{d}{dr} \left(r^2 \frac{d\phi}{dr} \right) = 4\pi G \rho. \tag{3.6}$$

Equations (3.4) and (3.6) form a complete description for hydrostatic equilibrium.

In general, total pressure is a function of the pressure of the material and the pressure of the radiation streaming from the material. That is,

$$P(r) = p_m(r) + p_r(r), \tag{3.7}$$

and can be a very complicated function. However, if we exclude very hot stars, the pressure from the radiation is much less than the pressure of the material, and can be ignored. We can even take this further and say that radiative flux can be ignored entirely. In actuality, the loss of energy via radiation is replaced by the energy creation (fusion) in the central core. For now we will ignore this dynamic process, and assume our stellar interior is hydrostatic with minimal loss of heat (via radiation). As such, we may assume that the interior is locally adiabatic. This is not a realistic assumption, but as before it can give some basic results that are worth exploring.

As a first approximation, let us assume further that our interior consists of a uniform ideal gas. Thus, as for the atmosphere,

$$P(r) = \frac{k}{\mu} \rho(r) T(r), \tag{3.8}$$

where k is the Boltzmann constant and μ is the mean molecular weight, which we assume to be constant. From this we define a molar volume:

$$V(r) = \frac{\mu}{\rho(r)}. \tag{3.9}$$

(where μ is now the mean *molar* weight). Because the interior is locally adiabatic,

$$P(r) v^{\gamma}(r) = \text{constant}, \tag{3.10}$$

where

$$\gamma = \frac{\text{specific heat at constant pressure}}{\text{specific heat at constant volume}}. \tag{3.11}$$

Substitution of equation (3.9) into (3.10), we have[1]

$$P = c\rho^{\gamma}. \tag{3.12}$$

This relation is called polytropic, and stars that obey this relation are called polytropic stars of index n, where

$$n = \frac{1}{\gamma - 1}. \tag{3.13}$$

Such a model assumes adiabatic processes, which can hold in the convective regions of stars. For now, we are assuming this holds for the entire star (thus we ignore for now the nuclear furnace in the core).

---

[1] Here c is not the speed of light, merely a free parameter.

3.1.2 The Lane–Emden equation

From equation (3.4), we have

$$\frac{\mathrm{d}\phi}{\mathrm{d}r} = -\frac{1}{\rho}\frac{\mathrm{d}P}{\mathrm{d}r}. \tag{3.14}$$

Substituting equation (3.12) into (3.14),

$$\frac{\mathrm{d}\phi}{\mathrm{d}r} = -\frac{1}{\rho}\frac{\mathrm{d}}{\mathrm{d}r}\left(c\rho^{\gamma}\right) = -c\gamma\rho^{\gamma-2}\frac{\mathrm{d}\rho}{\mathrm{d}r}. \tag{3.15}$$

Writing this in terms of the index and integrating, we find

$$\int \mathrm{d}\phi = -c\,(n+1)\,\frac{1}{n}\int \rho^{\left(\frac{1}{n}-1\right)}\,\mathrm{d}\rho. \tag{3.16}$$

Thus, taking $\phi = 0$ at the surface of the star, we have

$$\phi = -\,(n+1)\,c\rho^{1/n}, \tag{3.17}$$

or

$$\rho = \left(\frac{-\phi}{c\,(n+1)}\right)^{n}. \tag{3.18}$$

This gives the density in terms of gravitational potential. This can be substituted into the differential equation for ϕ,

$$\frac{\mathrm{d}^2\phi}{\mathrm{d}r^2} + \frac{2}{r}\frac{\mathrm{d}\phi}{\mathrm{d}r} = 4\pi G\left(\frac{-\phi}{c\,(n+1)}\right)^{n}. \tag{3.19}$$

We can simplify this by converting this equation into a parameter free differential equation. That is, if we let $\phi = \phi_c$ and $\rho = \rho_c$ be the values of potential and density at the center of the star, then substituting

$$A^2 = \frac{4\pi G}{c\,(n+1)}\rho_c^{(n-1)/n}, \tag{3.20}$$

$$z = Ar, \tag{3.21}$$

$$\omega = \frac{\phi}{\phi_c} = \left(\frac{\rho}{\rho_c}\right)^{1/n}, \tag{3.22}$$

our equation becomes

$$\frac{\mathrm{d}^2\omega}{\mathrm{d}z^2} + \frac{2}{z}\frac{\mathrm{d}\omega}{\mathrm{d}z} + \omega^n = 0. \tag{3.23}$$

This is known as the Lane–Emden equation for polytropic stars. If we can solve this, we can calculate the functions for density pressure and temperature as

$$\rho\,(r) = \rho_c\omega\,(r), \tag{3.24}$$

$$P\,(r) = P_c\omega^{(n+1)/n}, \tag{3.25}$$

$$T\,(r) = T_c\omega^{1/n}, \tag{3.26}$$

as found from our defining relations. The radius of the star is the point where $\rho(R) = P(R) = T(R) = 0$, and the mass of the star will be

$$M = 4\pi \int_0^R \rho(r)\, r^2 \, dr. \tag{3.27}$$

Thus, for our simple model, we can determine everything except luminosity, which is technically 0 in this model given our adiabatic assumption.

3.1.3 Eddington's approximation

Given the usefulness of the Lane–Emden equation, we would like to allow for radiation yet still preserve the polytropic nature of our stellar model. This can be done using an approximation of Eddington. The basic idea of Eddington is that stellar pressure consists of both matter and radiative pressure,

$$P = p_m + p_r, \tag{3.28}$$

with the additional assumption that radiation pressure is proportional to the total pressure throughout the star. That is,

$$\frac{p_r}{P} = 1 - \beta, \tag{3.29}$$

where β is a constant less than 1. This constraint allows the star to remain polytropic, because

$$P = \frac{p_m}{\beta} = \frac{kT}{\mu\beta}\rho. \tag{3.30}$$

Thus

$$p_r = \frac{(1 - \beta)}{\beta} p_m, \tag{3.31}$$

which means (assuming blackbody luminosity)

$$\frac{1}{3}\sigma T^4 = (1 - \beta)\frac{k}{\mu\beta}\rho T. \tag{3.32}$$

Solving for T, we have

$$T = \left[\frac{3k(1 - \beta)}{\sigma\mu\beta}\right]^{1/3} \rho^{1/3}. \tag{3.33}$$

Substituting this into our relation for total pressure:

$$P = \frac{k}{\mu\beta}\left[\frac{3k(1 - \beta)}{\sigma\mu\beta}\right]^{1/3} \rho^{4/3} = c\rho^{4/3}. \tag{3.34}$$

Thus, the Eddington model is polytropic, with $\gamma = 4/3$, or $n = 3$.

3.2 The polytropic star

3.2.1 Exact polytrope solutions

As we have seen, a polytropic star of index n is described by the Lane–Emden equation:

$$\frac{d^2\omega}{dz^2} + \frac{2}{z}\frac{d\omega}{dz} + \omega^n = 0. \tag{3.35}$$

To solve our equation, we must first impose boundary conditions:

$$\omega(0) = 1, \qquad \frac{d\omega(0)}{dz} = 0. \tag{3.36}$$

The first is simply the requirement that our solution be finite at the origin, whereas the second requires that the origin not be discontinuous. This is necessary because we are dealing with a radial equation.

The good news is exact solutions to the Lane–Emden equation are known for $n = 0, 1, 5$. The bad news is none of them are a good match to a real physical star. This means we will have to take a computational approach for other n values. However, a quick examination of these solutions will tell us a few things about polytropes and provide a starting place for a computational formulation.

Solution for $n = 0$

This is a special case because $n = 0$ means $\gamma = \infty$. The density for this solution is constant, $\rho(r) = \rho_c$. In other words, this is the case of an incompressible fluid. The solution is

$$\omega_0(z) = 1 - \frac{z^2}{6}, \tag{3.37}$$

with

$$P = P_c\left[1 - z^2/6\right], \tag{3.38}$$

$$T = T_c\left[1 - z^2/6\right]. \tag{3.39}$$

This solution is important in that it represents all solutions to order z^2, when expanded as a series.

Solution for $n = 1$

This reduces the Lane–Emden equation to a spherical Bessel equation. The solution can be found through a simple change of variables, $\omega(z) = \chi(z)/z$, which reduces equation (3.23) in this case to

$$\frac{d^2\chi}{dz^2} + \chi = 0, \tag{3.40}$$

which is the equation for a simple harmonic oscillator. Thus,

$$\omega(z) = \frac{A\sin(z)}{z} + \frac{B\cos(z)}{z}. \tag{3.41}$$

Applying the boundary conditions the solution in this case is

$$\omega_1(z) = \frac{\sin z}{z}. \tag{3.42}$$

The variances of pressure and temperature with depth are then

$$P = P_c \frac{\sin^2 z}{z^2},$$

(3.43)

$$T = T_c \frac{\sin z}{z}.$$

(3.44)

This is a better solution, but its n value is still too low. For the simplest gas model, that of a monatomic gas, $\gamma = 5/2$, and $n = 3/2$. If our gas is relativistic, and largely ionized, then $n = 3$ is more appropriate.

Solution for $n = 5$

Derivation of the $n = 5$ solution is rather involved. Here we begin with a slightly different form of the Lane–Emden equation,

$$\frac{1}{z^2} \frac{d}{dz} \left(z^2 \frac{d\omega}{dz} \right) = -\omega^n.$$

(3.45)

This is transformed with the substitution $x = 1/z$; thus

$$\frac{d}{dz} = -x^2 \frac{d}{dx},$$

(3.46)

and

$$x^4 \frac{d^2\omega}{d\omega^2} = -\omega^n.$$

(3.47)

We then guess a solution

$$\omega = ax^b,$$

(3.48)

where a and b are constants to be determined. Substitution into equation (3.21) yields

$$ab(b-1)x^{b+2} = a^n x^{nb}.$$

(3.49)

This must be valid for all values of x; thus it must be that

$$b = \frac{2}{n-1}, \qquad a = \left[\frac{2(n-3)}{(n-1)^2} \right]^{1/(n-1)}.$$

(3.50)

This is a real solution only for $n > 3$. Although this is a valid solution to the Lane–Emden equation, it diverges at $z = 0$. To eliminate this singularity, consider the solution

$$\omega(x) = ax^b \zeta(x),$$

(3.51)

where $\zeta(x)$ is sufficient to keep $\omega(x)$ finite. The resulting differential equation is further simplified by setting $x = e^t$; thus

$$\frac{d^2\zeta}{dt^2} + \frac{5-n}{n-1} \frac{d\zeta}{dt} - \frac{2(n-3)}{(n-1)^2} \zeta \left(1 - \zeta^{n-1} \right) = 0.$$

(3.52)

For $n = 5$, this equation simplifies to

$$\frac{d^2\zeta}{dt^2} = \frac{1}{4} \zeta \left(1 - \zeta^4 \right).$$

(3.53)

Multiplying by $d\zeta / dt$,

$$\frac{1}{2} \frac{d}{dt} \left(\frac{d\zeta}{dt} \right)^2 = \frac{1}{4} \zeta \left(1 - \zeta^4 \right) \frac{d\zeta}{dt}, \tag{3.54}$$

which can be integrated to yield

$$\frac{1}{2} \left(\frac{d\zeta}{dt} \right)^2 = \frac{1}{8} \zeta^2 - \frac{1}{24} \zeta^6 + D, \tag{3.55}$$

and our boundary conditions require $D = 0$. Because the derivative must be real, ζ is bounded by the roots of the right-hand side polynomial, but this is not a problem here. Thus,

$$\frac{d\zeta}{\zeta \sqrt{1 - \frac{1}{3} \zeta^4}} = -\frac{1}{2} dt. \tag{3.56}$$

Substituting

$$\frac{1}{3} \zeta^4 = \sin^2 \theta, \tag{3.57}$$

then

$$\frac{d\theta}{\sin \theta} = -dt, \tag{3.58}$$

which integrates to

$$\tan \left(\frac{\theta}{2} \right) = C e^{-t}, \tag{3.59}$$

where C is a constant of integration. Reverting to our original variables and applying the boundary conditions, the solution for $n = 5$ is then

$$\omega_5 (z) = \frac{1}{\sqrt{1 + z^2/3}}. \tag{3.60}$$

Unlike the other two solutions, this solution does not have a finite radius (although it does have a finite mass). Higher values of n likewise have infinite radii. To have a star of finite size, it is necessary to require $n < 5$.

3.2.2 Approximate analytical polytrope solution

Because an analytic solution of a polytropic star is not known for general n, we are faced with two possibilities: obtain a solution computationally or determine an approximate analytic solution. Although we will obtain more accurate results computationally, it is worth pursuing the analytical approach a bit further.

Let us consider the Lane–Emden equation for small values of z. As mentioned earlier, all solutions to second order have the form

$$\omega (z) = 1 - \frac{1}{6} z^2 + \cdots \tag{3.61}$$

This means that near the center of the star,

$$\frac{1}{z} \frac{d\omega}{dz} \simeq -\frac{1}{3}, \tag{3.62}$$

and

$$\frac{d^2\omega}{dz^2} \simeq -\frac{1}{3} = \frac{1}{z}\frac{d\omega}{dz}. \tag{3.63}$$

Let us therefore assume that near the origin, the Lane–Emden equation (3.23) reduces to

$$\frac{3}{z}\frac{d\omega}{dz} + \omega^n = 0. \tag{3.64}$$

This can be integrated as

$$\int \frac{d\omega}{\omega^n} = \frac{1}{3}\int z\,dz; \tag{3.65}$$

thus

$$\frac{1}{n-1}\omega^{1-n} = \frac{z^2}{6} + c, \tag{3.66}$$

(so long as $n \neq 1$). Rewriting this equation, and because $\omega(0) = 1$ means $c = 1/(n-1)$, we have

$$\omega_1(z) = \left(1 + \frac{n-1}{6}z^2\right)^{1/(1-n)}. \tag{3.67}$$

This we will take as the solution of the central region for our star.

For the region near the surface, we cannot make such an approximation. However, we can rewrite equation (3.23) as

$$\frac{d\omega}{dz} = -\frac{z}{2}\omega^n - \frac{z}{2}\frac{d^2\omega}{dz^2}. \tag{3.68}$$

By taking the approximate solution we know, and substituting, we have

$$\frac{d\omega}{dz} = -\frac{z}{2}\omega_1^n - \frac{z}{2}\frac{d^2\omega_1}{dz^2}. \tag{3.69}$$

Thus

$$\omega = -\int \frac{z}{2}\omega_1^n dz - \int \frac{z}{2}\left(\frac{d^2\omega_1}{dz^2}\right)dz. \tag{3.70}$$

Working this out, and with $\omega(0) = 1$ again, we have the solution

$$\omega_2(z) = -1 + 2\left(1 + \frac{n-1}{6}z^2\right)^{1/(1-n)} + \frac{z^2}{6}\left(1 + \frac{n-1}{6}z^2\right)^{n/(n+1)}. \tag{3.71}$$

We then have two solutions, one for the interior region and one for the surface region. Our complete solution is then a sum of the two,

$$\omega(z) = a\omega_1(z) + (1-a)\omega_2(z). \tag{3.72}$$

Typically, there are heuristic arguments to be made about the "right" value of a. However, in our case we consider it fair to weigh both equally; thus $a = 1/2$. Our approximate solution is then

$$\omega(z) = -\frac{1}{2} + \frac{3}{2}\left(1 + \frac{n-1}{6}z^2\right)^{1/(1-n)} + \frac{z^2}{12}\left(1 + \frac{n-1}{6}z^2\right)^{n/(n+1)}. \tag{3.73}$$

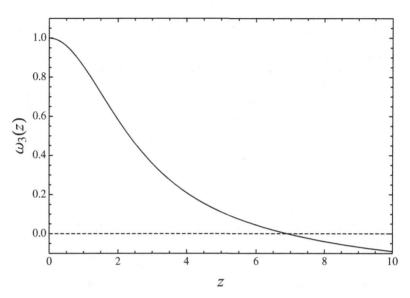

Fig. 3.1   Lane–Emden solution for $n = 3$.

3.2.3 Computational polytrope solutions

The alternative approach for polytropes of general n is to obtain the solution computationally.[2] In *Mathematica*, attempting to use DSolve to find an analytical solution to the Lane–Emden equation fails. This is not surprising given the convoluted path it took to solve $n = 5$. We must therefore resort to obtaining a numerical solution with NDSolve. Simple substitution of equation (3.23) into NDSolve fails as well due to a particularly difficult singularity at $z = 0$. To avoid the singularity we must start our solution at a small point $\epsilon > 0$, in this case $\epsilon = 10^{-44}$. Furthermore, to ensure proper behavior near the origin, we require the solution to follow the $n = 0$ solution for small z. Thus, for Eddington's approximation ($n = 3$), we have

```
nn = 3; eps = 10⁻⁴⁴;
diff = NDSolve[{ωω''[zz] + 2/zz × ωω'[zz] + ωω[zz]ⁿⁿ == 0,
    ωω[eps] == 1. - eps²/6, ωω'[eps] == -eps/3}, ωω, {zz, eps, 10}]
```

Mathematica gives the solution as an interpolating polynomial:

```
{{ωω → InterpolatingFunction[{{1.×10⁻⁴⁴, 10.}}, <>]}}
```

Although it appears *Mathematica* has failed to find a solution, the InterpolatingFunction[] can be treated in much the same way as an analytical function. For example, a plot of $\omega(z)$ can be seen in figure (3.1). From this it is clear the root of the function lies within $6 < z < 8$. The exact root can be

---

[2] For further details, see **3-1Polytrope**.

found by FindRoot[], starting with an initial "guess" of $z = 6$:

```
trial = 6;
z0 = FindRoot[ω[z1] /. diff[[1]], {z1, trial}]
```

which finds the root to be $Z_0 = 6.89685$. From this we can determine the mass function:

$$m_0 = \int_0^{Z_0} \omega(z) z^2 \, dz. \tag{3.74}$$

Thus

```
z3 = z1 /. z0[[1]];
m0 = Integrate[(ω[z2] /. diff[[1]]) × z2², {z2, 0, z3}];
N[m0]
```

which yields $m_0 = 14.1915$. From this we can determine stellar properties such as temperature, pressure, and density.

3.2.4 Determining the properties of a polytropic star

Having obtained a solution to the Lane–Emden equation, the properties of a polytropic star can be determined following a standard procedure:

1. Measure (or deduce) the radius (R) and mass (M) of the star in question.
2. Deduce (or guess) an appropriate value for n.
3. For that particular n, evaluate the Lane–Emden equation (3.23) to determine the first zero Z_0, where $\omega(Z_0) = 0$.
4. The parameter A is then be found from equation (3.21),

$$A = \frac{Z_0}{R}.$$

5. Calculate the integral

$$m_0 = \int_0^{Z_0} \omega(z) z^2 \, dz.$$

6. Because the mass of our star is calculated by equation (3.27), with $\rho = \rho_c \omega$ and $r = z/A$,

$$M = \frac{4\pi \rho_c}{A^3} m_0.$$

We can therefore use the measured mass M to determine the central density ρ_c:

$$\rho_c = \frac{MA^3}{4\pi m_0}.$$

7. The free constant c is then calculated from equation (3.20):

$$c = \frac{4\pi G}{A^2(n+1)}\rho_c^{(n-1)/n}.$$

8. The central pressure is then calculated by equation (3.12):

$$P_c = c\rho_c^\gamma = c\rho_c^{(1+n)/n}.$$

9. The central temperature is found from equation (3.8):

$$T_c = \frac{\mu P_c}{k\rho_c}.$$

10. The distributions are then known:

$$\rho(r) = \rho_c\omega(r),\tag{3.75}$$

$$P(r) = P_c\omega^{(n+1)/n},\tag{3.76}$$

$$T = T_c\omega^{1/n}.\tag{3.77}$$

As an example, consider the Eddington ($n = 3$) model for the sun.[3] The mass and radius of the sun are known: $R_\odot = 6.96 \times 10^{10}$ cm and $M_\odot = 1.99 \times 10^{33}$ g. In the previous section we found for $Z_0 = 6.89685$ and $m_0 = 14.1915$. Thus

$$\rho_c = 10.8632 \text{ g/cm}^3,$$

$$P_c = 2.51755 \times 10^{15} \text{ dyne/cm}^2,\tag{3.78}$$

$$T_c = 3.62404 \times 10^6 \text{ K}.$$

The central density here is orders of magnitude lower than that of a real star, which is a limitation of the polytropic model. Even here, however, the central temperature is millions of Kelvin, which implies that nuclear processes are present in the core.

The variation of density, pressure, and temperature with depth is easily found from $\omega(z)$. Because *Mathematica* has determined ω as an interpolating polynomial, plotting pressure and temperature requires use of Evaluate. For example, the pressure plot is obtained by

```
pressc = 2.51755 × 10^15; n = 3; end = 11;
Plot[Evaluate[pressc ω[t]^(n+1)/n /.diff], {t, eps, end}]
```

From the plot of pressure and temperature, figure (3.2), it is clear the polytropic model is almost isothermal (except near the star's surface). It also shows that these stars are what would be expected when the layers are so opaque that little radiant energy flows outward; thus the adiabatic condition $dQ = 0$ is well satisfied.

---

[3] See **3-2Interior**.

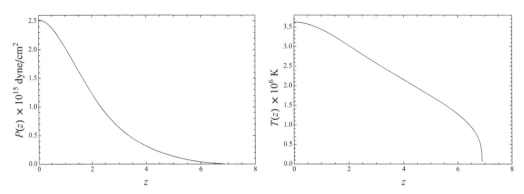

Fig. 3.2 Variance of pressure (left) and temperature (right) for an $n = 3$ polytropic star.

3.3 Stellar populations

3.3.1 Hertzsprung–Russell diagrams

For stars, two quantities can be measured with good accuracy: brightness (apparent magnitude) and color (from which the effective temperature, T_{eff}, may be determined). For nearby stars it is also possible to determine distance and therefore calculate its absolute magnitude (total luminosity, L). Then one can create a scatterplot of stars' temperature and luminosity (or related quantities) known as a Hertzsprung–Russell diagram, such as figure (3.3).

A Hertzsprung–Russell diagram allows us to determine the connection among a star's temperature, luminosity, and mass. If there were no connection among these quantities, one would expect such a scatterplot to be randomly distributed. However, it is clear that stars are clustered together, such as the roughly linear group in figure (3.3) known as the main sequence of stars.[4] Stars in the smaller cluster of the figure are known as giant stars. The nonrandom distribution of a Hertzsprung–Russell diagram provides information about the life cycle of stars. It also provides a point of comparison for the polytropic model. To compare our polytropic stellar model to real stars, we need a way to relate L and T_{eff} or L and M. This can be done by looking at the general properties of polytropes and making a few additional assumptions.

3.3.2 Polytropic mass–radius relation

Since all polytropic stars have the same form, we can define a new variable ξ, where

$$\xi = \frac{m_1(r)}{M_1} = \frac{m_2(r)}{M_2}, \tag{3.79}$$

which is the same for any two stars. Then we can relate any two stars by a set of values that are the same for all ξ, and are thus constants. In general these values are not constants, but this assumption

---

[4] Data for figure (3.3) is taken from the HYG-Database, github.com/astronexus/HYG-Database.

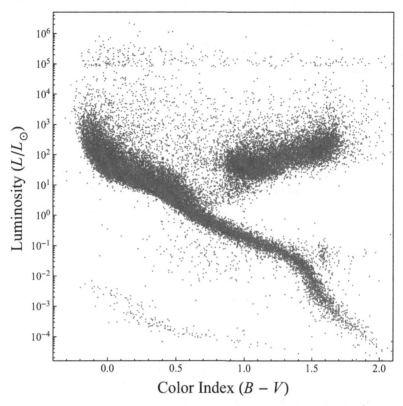

Fig. 3.3 Hertzsprung–Russell diagram of nearby stars.

will suffice for polytropic stars:

$$x = \frac{M_1}{M_2}, \tag{3.80}$$

$$z = \frac{r_1\,(\xi)}{r_2\,(\xi)} = \frac{R_1}{R_2}, \tag{3.81}$$

$$p = \frac{P_1\,(\xi)}{P_2\,(\xi)} = \frac{P_1^c}{P_2^c}, \tag{3.82}$$

$$d = \frac{\rho_1\,(\xi)}{\rho_2\,(\xi)} = p^{n/(n+1)} = p^{\alpha}. \tag{3.83}$$

Recall the equations for hydrostatic equilibrium, this time expressed in a slightly different way. Consider the mass of the star for a particular radius to be $m\,(r)$. If the local density is then $\rho\,(r)$, then a change in r gives a change in mass of

$$dm = 4\pi r^2 \rho \, dr. \tag{3.84}$$

This means

$$\frac{dm}{dr} = 4\pi r^2 \rho, \tag{3.85}$$

or

$$\frac{\partial r}{\partial m} = \frac{1}{4\pi r^2 \rho}. \tag{3.86}$$

Because for differential operators,

$$\frac{\partial}{\partial m} = \frac{\partial r}{\partial m}\frac{\partial}{\partial r}, \tag{3.87}$$

we have

$$\frac{\partial}{\partial m} = \frac{1}{4\pi r^2 \rho}\frac{\partial}{\partial r}. \tag{3.88}$$

This is useful, because we want to relate our equations to ξ, which is related to m and not r.
   Our equation for hydrostatic equilibrium was

$$\frac{\partial P}{\partial r} = -g\rho = -\frac{Gm}{r^2}\rho. \tag{3.89}$$

Thus, we can transform this to

$$\frac{\partial P}{\partial m} = \frac{1}{4\pi r^2 \rho}\frac{\partial P}{\partial r} = -\frac{Gm}{4\pi r^4}. \tag{3.90}$$

With the definition of $\xi = m(r)/M$, we have the set of equations as

$$\frac{dr}{d\xi} = \frac{1}{4\pi}\frac{M}{r^2 \rho}, \tag{3.91}$$

$$\frac{dP}{d\xi} = -\frac{G}{4\pi}\frac{\xi M}{r^4}, \tag{3.92}$$

These derivatives must be the same for each mass; thus

$$\frac{dr_1}{d\xi} = \frac{1}{4\pi}\frac{M_1}{r_1^2 \rho_1} \quad \text{and} \quad \frac{dr_2}{d\xi} = \frac{1}{4\pi}\frac{M_2}{r_2^2 \rho_2} \text{ etc.,} \tag{3.93}$$

which can be true only if

$$\frac{x}{z^3 d} = 1, \tag{3.94}$$

$$\frac{x^2}{z^4 p} = 1. \tag{3.95}$$

These can be solved by relating all constants to x, such that (with $\alpha = n/(n+1)$)

$$z = x^z, \quad p = x^{p'}, \quad d = x^{\alpha p'}. \tag{3.96}$$

These can be substituted into the preceding equations, which simplifies them to

$$1 - 3z' - \alpha p' = 0, \tag{3.97}$$

$$2 - 4z' - p' = 0. \tag{3.98}$$

These have the solutions

$$z' = \frac{2\alpha - 1}{4\alpha - 3} = \frac{n - 1}{n - 3}, \tag{3.99}$$

$$p' = -\frac{2}{4\alpha - 3} = -\frac{2n + 2}{n - 3}. \tag{3.100}$$

Thus, the relation between the masses and radii of polytropic stars is

$$\frac{R_1}{R_2} = \left(\frac{M_1}{M_2}\right)^{z'} = \left(\frac{M_1}{M_2}\right)^{(n-1)/(n-3)}. \tag{3.101}$$

It is interesting to note that for $n = 1$, the radius of all stars is the same, regardless of mass. For $n = 3$, the masses of all stars are the same, regardless of radius. Perhaps more interesting is the fact that for $1 < n < 3$, the larger the mass the smaller the radius. This will have important consequences for white dwarf stars, as we see in Chapter 4. For $n > 3$ the radius of a star increases with increasing mass.

3.3.3 Radiative diffusion and mass–luminosity

We now return to the idea of luminosity, which we so hastily discarded earlier. To consider luminosity, we assume again a gray body. That is, the absorption coefficient κ_ν is constant over all frequencies (and wavelengths). Thus $\kappa_\nu = \kappa$. If we assume this is constant for all depths, then the mean free path for a photon is

$$l_\gamma = \frac{1}{\kappa \rho}. \tag{3.102}$$

This represents the average distance a photon will travel before it interacts with something. So long as this is much shorter than the distances we are considering (and we will assume this is true), we can treat the transfer of energy to be conductive. Conduction of anything is driven by the diffusion equation

$$F = -D\frac{\partial n}{\partial r}, \tag{3.103}$$

where F is the particle flux, n is the particle density, and D is the coefficient of diffusion, given by

$$D = \frac{1}{3}vl_p. \tag{3.104}$$

where v is the RMS velocity. For photons, the particle density is given by

$$n = \sigma T^4, \tag{3.105}$$

and the diffusion coefficient becomes

$$D = \frac{c}{3\kappa\rho}. \tag{3.106}$$

where c is the speed of photon diffusion. The flux then becomes

$$F = -\frac{4\sigma c T^3}{3\kappa\rho}\frac{\partial T}{\partial r}. \tag{3.107}$$

Because the luminosity is given as $l = 4\pi r^2 F$, we have

$$\frac{\partial T}{\partial r} = -\frac{3}{16\pi\sigma c}\frac{\kappa\rho l}{r^2 T^3}. \tag{3.108}$$

This is then the transport equation for radiation within the star.

We can now derive the relation between mass and luminosity. To begin with, this equation can be expressed as

$$\frac{\partial T}{\partial m} = -\frac{3}{64\pi^2\sigma c}\frac{\kappa l}{\rho r^4 T^3}, \tag{3.109}$$

and from the definition for luminosity,

$$\frac{\partial l}{\partial m} = \varepsilon \sim \rho T^{\nu}. \tag{3.110}$$

These can be expressed as before,

$$\frac{dT}{d\xi} = -\frac{3}{64\pi^2 \sigma c} \frac{\kappa l M}{r^4 T^3}, \tag{3.111}$$

$$\frac{dl}{d\xi} = \varepsilon M. \tag{3.112}$$

If we then define the ratios

$$t = \frac{T_1(\xi)}{T_2(\xi)} = \frac{T_1^c}{T_2^c}, \tag{3.113}$$

$$s = \frac{l_1(\xi)}{l_2(\xi)} = \frac{l_1^c}{l_2^c}, \tag{3.114}$$

$$k = \frac{\kappa_1}{\kappa_2} = 1, \tag{3.115}$$

$$e = \frac{\varepsilon_1}{\varepsilon_2}, \tag{3.116}$$

then we have the conditions

$$\frac{ex}{s} = 1, \tag{3.117}$$

$$\frac{sx}{z^4 t^4} = 1. \tag{3.118}$$

Then relating each constant to x as before,

$$t = x^{t'}, \quad e = x^{e'}, \quad s = x^{s'}, \tag{3.119}$$

the equations become

$$1 + \alpha p' + \nu t' - s' = 0, \tag{3.120}$$

$$1 + s' - 4z' = 4t'. \tag{3.121}$$

These have the solutions

$$s' = \frac{3\nu n - \nu + 4n + 12}{(n-3)(\nu-4)}, \tag{3.122}$$

$$t' = \frac{4n + 2}{(n-3)(\nu-4)}. \tag{3.123}$$

Now, it can be shown that if you assume the star to be an ideal gas, you find

$$t' = \frac{\nu - 1}{\nu + 3} = \frac{n - 1}{n - 3}. \tag{3.124}$$

Thus

$$\nu = -2n + 3, \tag{3.125}$$

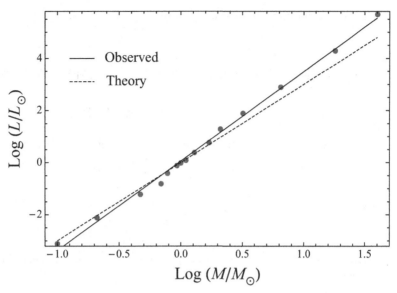

Fig. 3.4 Mass–luminosity for main sequence stars.

which yields

$$s' = 3. \tag{3.126}$$

The mass luminosity relation is then

$$\frac{l_1}{l_2} = \left(\frac{M_1}{M_2}\right)^3. \tag{3.127}$$

Comparison the observed mass–luminosity relationship for main-sequence stars (Allen, 2000) gives a value of $s' = 3.43$, as seen in figure 3.4.

3.3.4 Eddington's estimation of mass–luminosity

Long before nuclear processes were determined to be powering the stars, Sir Arthur Eddington estimated an upper limit to the luminosity of an object that is imposed on the process of radiative equilibrium. This result in turn implied an upper limit on the mass of a "stellar" object, because at this luminosity limit radiation pressure would be totally dominant and so would, if not limited, blow a massive object apart. Eddington's work first indicated that there would indeed be a mass–luminosity relationship at a time when no convincing observational existence for one existed.

As shown by Prialnik (2000), to obtain an expression for the so-called Eddington luminosity one starts with the equations for radiative equilibrium of stars. Although at the time (the 1920s), Eddington did not know the source of a star's luminosity, he was aware of the pressure of photons in a hot gas and its temperature dependence. Because the radiative pressure of photons is $P_{\text{rad}} = aT^4/3$, its derivative is

$$dP_{\text{rad}} = \frac{4}{3}aT^3 \, dT. \tag{3.128}$$

The radiative pressure can also be expressed in terms of the radiative flux F,

$$\frac{dP_{\text{rad}}}{dr} = -\frac{\kappa \rho}{c} \frac{F}{4\pi r^2}, \tag{3.129}$$

where κ is an effective opacity, ρ is the density, and c is the speed of light. The temperature gradient is then

$$\frac{dT}{dr} = -\frac{3}{4ac} \frac{\kappa \rho}{T^3} \frac{F}{4\pi r^2}. \tag{3.130}$$

If we assume a star is spherical and in hydrostatic equilibrium, from equation (3.2),

$$\rho = -\frac{r^2}{GM} \frac{dP}{dr}. \tag{3.131}$$

Thus

$$\frac{dP_{\text{rad}}}{dP} = \frac{\kappa F}{4\pi c GM}, \tag{3.132}$$

where the total pressure $P = P_{\text{gas}} + P_{\text{rad}}$. Because the left side of this equation is dimensionless, and F has units of luminosity, we can define

$$L_{\text{ED}} = \frac{4\pi c GM}{\kappa} \tag{3.133}$$

as the Eddington luminosity. The Eddington luminosity is a limit imposed by radiation pressure, because $\kappa F < 4\pi c Gm < 4\pi c GM$. An equation for L_{ED} in terms of the surface electron scattering opacity is given by Prialnik (2000) as

$$\frac{L_{\text{ED}}}{L_{\odot}} = 3.2 \times 10^4 \left(\frac{M}{M_{\odot}} \right) \left(\frac{\kappa_{\text{es}}}{\kappa} \right), \tag{3.134}$$

where κ_{es} is the reference opacity for electron scattering.

Eddington devised a "standard" model in which the flux/mass at any level was defined as

$$\frac{F}{m} = \eta \frac{L}{M}. \tag{3.135}$$

Assuming $\kappa \eta = \kappa_s$ is constant, then from equation (3.132) the luminosity is found to be $L = L_{\text{ED}}(1 - \beta)$, where β is the ratio of radiation pressure to total pressure. As shown earlier, from Eddington's approximation one derives a polytropic star of $n = 3$. This in turn leads to Eddington's quartic law:

$$(1 - \beta) = 0.003 \left(\frac{M}{M_{\odot}} \right)^2 \mu^4 \beta^4, \tag{3.136}$$

where μ is the mean molecular weight of the gas. The roots of this equation can be found analytically in *Mathematica*,[5]

```
βsol = Solve[ (1 - β1) == 0.003 (m / ms)² μ⁴ β1⁴, β1]
```

though they are not simple, and it is not readily apparent which root is needed.[6] Because β must be real and positive, a quick numerical test (setting $M_{\odot} = 1$ and $M = 2$) can be used to determine the

---

[5] The use of $==$ is correct in Mathematica, meaning "compare equals."
[6] See **3-3Masslum** for the full analytic roots.

correct root,

βsol[[1, 1, 2]] /. {ms -> 1, μ -> 1.5, m $\to$ 2}

0.93232 - 2.29852 i

βsol[[2, 1, 2]] /. {ms -> 1, μ -> 1.5, m $\to$ 2}

0.93232 + 2.29852 i

βsol[[3, 1, 2]] /. {ms -> 1, μ -> 1.5, m $\to$ 2}

-2.81507

βsol[[4, 1, 2]] /. {ms -> 1, μ -> 1.5, m $\to$ 2}

0.950429

which finds βsol[[4]] as the only positive real root.

It is now possible to construct a theoretical mass–luminosity diagram based on the Eddington standard model. At one solar mass, $L_{ED}/L_\odot = 3.2 \times 10^4$; thus for mass in terms of solar masses, and luminosity in terms of solar luminosities, the mass–luminosity function can be determined by

$$L = \frac{96\mu^4}{\kappa_s} M^3 \beta^4, \tag{3.137}$$

which in *Mathematica* becomes

lum[m1_, μ1_, κs_] := $\dfrac{96\ \mu1^4}{\kappa s}$ m1^3 βsol[[4, 1, 2]] /. {ms $\to$ 1, μ $\to$ μ1, m $\to$ m1}

In figure (3.5) this function (with $\mu = 1.5$ and photospheric opacity between 300 and 700) is compared to the observed mass–luminosity relation (Allen, 2000). Agreement is surprisingly good given the crudeness of Eddington's assumptions.

Although Eddington's theoretical model diverges as $m \to \infty$, a simple logistic fit of the actual data does not, as we can show using NonlinearModelFit[].[7] A square root mass dependence is expected from the exponents in the mass–luminosity theory; thus

nlm1 = NonlinearModelFit$\left[$masslum, -a + b e$^{-\sqrt{m}/m0}$, {a, b, m0}, m$\right]$

eqn1 = Normal[nlm1]

This fit gives the absolute visual magnitude[8] as a function of solar masses as

$$\Delta M = -5 + 25\,e^{-\sqrt{m}}. \tag{3.138}$$

The actual mass–luminosity relationship of main sequence stars is well represented by this empirical function. The observed main sequence mass limits are 0.07 to 100 solar masses, and so our empirical function does very well at setting the absolute magnitude limits in the visible range.

---

[7] For the details of NonlinearModelFit see the *Mathematica* online documentation.

[8] The notation can be a bit confusing here, as we use m and M for both mass quantities and magnitudes. Usually the two are not mixed, but here they are.

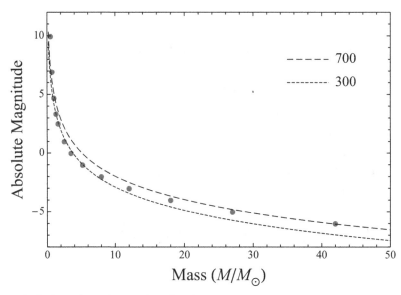

Fig. 3.5   Eddington's mass–luminosity vs. observed at two values of opacity.

3.4   The virial theorem and stellar structure

3.4.1   Virial theorem and the ergodic hypothesis

The virial theorem in a variety of forms was known as early as 1850. It has suffered over the years from a variety of misinterpretations that have tended to limit the extent of its applicability. Modern references that mention the virial theorem in any detail are fairly few in number and they are often classical mechanics texts (Goldstein, Poole, and Safko, 2002), celestial mechanics texts (Pollard, 1966a; Pollard 1966b; Danby, 1988), and stellar dynamics monographs (Contopoulos, 1966). These references, along with general texts such as Harwitt (1988), discuss the theorem only as it applies to the stellar N-body problem. On the other hand, the virial theorem discussion found in recent stellar interior books (Böhm-Vitense, 1992; Prialnik, 2000) stress the continuum mechanics version of the theorem, though not in the detail of Chandrasekhar (1961). A delightful exception to the specialization of modern accounts is the brief text by Collins (1978), devoted entirely to the virial theorem in stellar astrophysics, with emphasis on stellar structure applications.

As Collins points out, the virial theorem for a continuum density distribution has its conceptual roots in the Boltzmann transport equation.[9] The theorem had its mathematical origins in the mechanical theory of heat and in its simplest form describes an equilibrium situation with energies (scalars). Just as the integrals of the Boltzmann equation produce well-known continuity and conservation equations, so do integrals of equations of motion become the virial theorem in its various forms. The most familiar version is a scalar equation derived from Lagranges' identity by taking a time average. That is,

---

[9] For worked examples of the Boltzmann equation and its relatives, see **7-5Maxwell**, **7-6Boltzmann**, and **7-7CollisionB**.

starting with[10]

$$\frac{1}{2}\frac{d^2 I}{dt^2} = 2T + \Omega, \tag{3.139}$$

where I is the moment of intertia of the body, T is the kinetic energy, and Ω is the potential energy for a "gravitational" force. If the body is stable then the average over time of the LHS is 0, and we have the sum of the two time averages on the RHS is therefore

$$2\overline{T} + \overline{\Omega} = 0. \tag{3.140}$$

This is the most commonly seen virial form for both continuum and point mass situations. Except for the fact that in astrophysics it is very hard to obtain valid time averages, this form would be routinely used to solve many problems as is. The difficulty is that it is often necessary to substitute space averages for the required time averages. This raises issues embodied in the ergodic hypothesis, a principle that is not without controversy even today.[11]

In its original form, the ergodic hypothesis stated that given enough time, a system would visit every point in its available phase space. This was later modified to become the quasi-ergodic hypothesis. In statistical mechanics, where the relaxation or system crossing times are relatively short, the ergodic hypothesis is a reasonable assumption. Collins (p. 22) points out that for stellar structure applications, a similar ergodic assumption leading to equality of time and space averages is similarly reasonable. Where difficulty can arise is in dynamical systems where particle numbers are "too small" or the particles are subject to isolating integrals of motion (such as galactic dynamics) that restrict the phase space motions of particles to only certain regions within phase space.

3.4.2 Hydrostatic equilibrium and stellar age

In a discussion of hydrostatic equilibrium, Böhm-Vitense (1992) derives the equilibrium relationship between the thermal energy as measured by the ideal gas law and pressure integral over the radius of a spherical star,

$$E_{\text{thermal}} = \int_0^R \frac{3n}{2} kT 4\pi r^2 \, dr = \frac{1}{2} \int_0^R \rho \frac{GM_r}{r} 4\pi r^2 \, dr, \tag{3.141}$$

where R is the radius of the star, n is the local number density of the gas, T is the local gas temperature, ρ is the local mass density, r is the local radius, M_r is the mass below r, and G is the gravitational constant. The thermal energy E_{thermal} is the integral over the whole star; and the RHS is minus the integral of the potential energy over the whole star, thus this is actually a statement of the virial theorem, $2E + \Omega = 0$.

Computation of the LHS integral is achieved by converting nT to the equivalent ideal gas pressure, then integrating assuming hydrostatic equilibrium. The RHS integral is evaluated by building up the star by having successive spherical mass shells fall from infinity to the radius r. This gives the classical result

$$E_{\text{thermal}} = -\frac{1}{2} E_{\text{gravitation}}, \tag{3.142}$$

---

[10] Collins, p. 18.
[11] Collins, p. 18.

which holds regardless of $P(r)$, $T(r)$, and $M(r)$ as long as one assumes an ideal gas in hydrostatic equilibrium. For a constant density star,

$$E_{\text{gravitation}} = -\frac{3}{5}\frac{GM^2}{R}. \qquad (3.143)$$

From this one can determine the thermal energy produced by gravitational contraction. Dividing this by the star's luminosity, one finds the Kelvin–Helmholtz contraction time. For the sun, with $L_\odot = 3.8 \times 10^{26}\,\text{W}$,

$$t_{\text{KH}} = \frac{E_{\text{thermal}}}{L} = \frac{3}{10}\frac{GM_\odot^2}{R_\odot L_\odot} = 9.53788 \times 10^6 \text{ years}. \qquad (3.144)$$

A more general calculation[12] finds a slightly higher value of $t_{\text{KH}} = 2.6 \times 10^7$ years; thus the contraction time is the same order of magnitude whether or not one assumes uniform density. This result implies that the sun contracted from "infinity" to its present configuration in some 10 to 30 million years. Calculation of this number was considered an early triumph of astrophysical theory.

3.4.3 Hydrostatic equilibrium and core temperature

A common "rule of thumb" found in elementary astronomy textbooks states that 90% of the sun's mass is within 50% of the radius and the central density is 100 times the mean density. These numbers are based on modern numerical models. As a general approximation we can assume solar density is a Gaussian profile fit to this 50–90 rule. Thus defining a Gaussian density function and resulting mass

```
rho[r_, ρ_, σ_] := Module[{}, ρ e^(-(r/σ)^2)];

mass[ρ0_, σ0_, r0_] := ∫₀^r0 4 π r² rho[r, ρ0, σ0] dr
```

We then define $\texttt{rstarm} = 7 \times 10^8$ and $\texttt{mstarkg} = 2 \times 10^{30}$ as the solar radius (in meters) and mass (in kilograms), as well as the average density

```
avden = N[ mstarkg / (4/3 π rstarm³) ];
```

which yields $\bar{\rho} = 1392.03$. With $\texttt{Solve}$ we can then determine the central density ratio to the average density as a function of the ratio of the Gaussian standard deviation σ to the total radius R assuming the 50–90 rule

```
sol1 = Solve[{mass[rhoratio avden , 0.2 rstarm, 0.5 rstarm] == 0.9 mstarkg,
    rho[rstarm, rhoratio avden, 0.2 rstarm] == 10⁻⁷ avden}, rhoratio]
sol1[[1, 1, 2]]
mass1[r0_] := mass[rhoratio avden , 0.2 rstarm, r0] /. rhoratio → sol1[[1, 1, 2]];
rho1[r2_] := rho[r2, rhoratio avden, 0.2 rstarm] / avden /. rhoratio → sol1[[1, 1, 2]];
```

---

[12] See **3-4 Virial**.

The central pressure can then be estimated by assuming hydrostatic equilibrium and integrating from the surface downward. The pressure integral (taking a nonzero upper limit to maintain convergence) is then

$$P_c = -G\bar{\rho} \int_{R_\odot}^{0.0001R_\odot} \frac{\varrho(r)m(r)}{r^2} \, \mathrm{d}r.$$ (3.145)

To find the central temperature, the central mass density must be converted into a particle density, which in turn requires an assumption of the mean molecular weight. The actual sun's interior is a mixture of mostly hydrogen and helium. Cox (2000) puts the central content of hydrogen by weight as $X = 0.355$, and assuming no other elements but helium, $Y = 0.645$. The mean molecular weight is then $\mu = (X + 4Y)m_p = 4.9895 \times 10^{-27}$ kg. The number density of atoms can then be found by ρ_c/μ. Because the hydrogen and helium within the core are ionized, we must multiply this by a factor of 4 to give the actual value of electrons plus ions. Thus $n = 4\rho_c/\mu = 9.49983 \times 10^{31}$.

The central pressure must include both gas pressure and radiation pressure; therefore

$$P_c = nkT_c + \frac{a}{3}T^4,$$ (3.146)

where k is the Boltzmann constant and a is the radiation constant. In *Mathematica* Solve finds four roots for T_c, only one of which is real and positive, giving the central temperature as $T_c = 9.0185 \times 10^7$ K. This is nearly 30 times higher than the temperature derived from the simple polytropic model, and too high compared to contemporary models that place $T_c = 1.571 \times 10^7$ K. The reason for this discrepancy is that the model does not allow the "core" to cool down through radiative flux to the outside. A more realistic model would need to account for both radiative and hydrostatic equilibrium.

Although early in the 20th century virial applications to stellar structure seemed to be quite reasonable, both central temperature estimates and contraction time estimates were actually very far from the mark.[13] Formation of stars and other matters of stellar evolution (not just structure) are beyond the scope of this present work and we do not elaborate further on this. Instead we pass to a more successful modern virial theorem application that has held up very well.

3.4.4 Stellar pulsation

Perhaps the most important modern application of the virial theorem is its application to stellar pulsation. Collins (pp. 61–102) spends considerable effort on this aspect of the virial theorem, building on the earlier work of Ledoux (1945).

Collins' approach is to apply a small variation to the virial theorem, equation (3.139):

$$\frac{1}{2}\frac{\mathrm{d}^2 I}{\mathrm{d}t^2} = 2T + \Omega.$$ (3.147)

Because we consider only radial pulsations, we retain spherical symmetry. Thus we can define mass within a radius r as $m(r)$. By conservation of mass, the variation of $m(r)$ must satisfy

$$m(r_0 + \delta r) = m(r_0).$$ (3.148)

Thus

$$\frac{1}{2}\frac{\mathrm{d}^2 \delta I}{\mathrm{d}t^2} = 2\delta T + \delta\Omega.$$ (3.149)

---

[13] For an explanation of why this was so, see Bohm-Vitense (1992), chapters 9, 10, and 11.

From the definition of moment of inertia,

$$I = \int_0^M r^2 \, dm, \tag{3.150}$$

and so

$$\delta I = \int_0^M 2r\delta r \, dm + \int_0^M r^2 d\,(\delta m). \tag{3.151}$$

By conservation of mass, $\delta m = 0$, so the second term vanishes.

We consider radial pulsation about an equilibrium radius r_0. Because $dI = r^2 dm$, at r_0

$$dm = \frac{dI_0}{r_0^2}, \tag{3.152}$$

we have to first order in r,

$$\delta I = 2 \int_0^{I_0} \frac{\delta r}{r_0} dI_0. \tag{3.153}$$

For the potential energy

$$\Omega = -\int_0^M \frac{Gm(r)}{r} dm, \tag{3.154}$$

and to first order

$$\delta\Omega = 2 \int_0^{\Omega_0} \frac{\delta r}{r_0} d\Omega. \tag{3.155}$$

The kinetic energy consists of the bulk motion kinetic energy, T_1, and the thermal kinetic energy of the gas, T_2. Thus

$$T = T_1 + T_2 = \frac{1}{2} \int_0^M \left(\frac{dr}{dt}\right)^2 dm + \int_0^V \frac{3}{2} NkT \, dV. \tag{3.156}$$

For the first term,

$$\delta T_1 = \int_0^M \frac{dm \, d(\delta r^2)}{dt^2}, \tag{3.157}$$

which is of second order in r and can therefore be ignored. For the second term, since we assume an ideal gas, $NkT = PV$, and noting $\rho dV = dm$,

$$T_2 = \frac{3}{2} \int_0^M \frac{P}{\rho} dm. \tag{3.158}$$

If we assume the oscillations are adiabatic, then $P\rho^{-\gamma} = $ constant, where the constant γ is the ratio of specific heats initially defined in equation (3.11). To first order we then have

$$2\delta T = 3 \int_0^M \frac{P_0}{\rho_0} (\gamma - 1) \frac{\delta\rho}{\rho_0} dm. \tag{3.159}$$

Because we have assumed hydrostatic equilibrium, stellar pulsations must be driven by some mechanism in the interior. We do not explore the details of this mechanism here, but rather simply assume such a mechanism exists to explore what oscillatory frequencies might be. For simplicity let us assume simple harmonic motion radially. That is

$$\frac{\delta r}{r_0} = \xi = \xi_0 e^{i\omega_p t}, \tag{3.160}$$

where ξ_0 is the amplitude and ω_p is the angular frequency of the pulsation.

Substituting this into equations (3.153) and (3.155),

$$\delta I = 2e^{i\omega_p t} \int_0^{I_0} \xi_0 \, dI_0, \tag{3.161}$$

$$\delta \Omega = -e^{i\omega_p t} \int_0^{\Omega_0} \xi_0 \, d\Omega_0. \tag{3.162}$$

The kinetic energy term depends not on $\delta r / r_0$, but rather $\delta \rho / \rho_0$, so we must invoke conservation of mass to determine its variation. Because mass is conserved, its variation is 0; thus

$$\delta m = \delta \left(4\pi r^2 \rho dr \right) = 0. \tag{3.163}$$

Applying the variation and substituting equation (3.160),

$$\frac{\delta \rho}{\rho_0} = - \left(3\xi + r_0 \frac{d\xi}{dr_0} \right). \tag{3.164}$$

Substitution into equation (3.159) yields[14]

$$2\delta T = 3e^{i\omega_p t} \int_0^{\Omega_0} \xi_0 \left(\gamma - 1 \right) d\Omega_0. \tag{3.165}$$

Substituting these into equation (3.149), the virial equation then becomes

$$- \omega_p^2 e^{i\omega_p t} \int_0^{I_0} \xi_0 \, dI_0 = 3e^{i\omega_p t} \int_0^{\Omega} \xi_0 \left(\gamma - 1 \right) d\Omega_0 - e^{i\omega_p t} \int_0^{\Omega_0} \xi_0 \, d\Omega_0. \tag{3.166}$$

Evaluating the integrals, this reduces to

$$\omega_p^2 = - \frac{(3\gamma - 4) \, \Omega_0}{I_0}. \tag{3.167}$$

The moment of inertia for a sphere about an axis is $3/2$ that about its center; thus

$$I_z = \frac{2}{5} M R_0^2 = \frac{3}{2} T_0, \tag{3.168}$$

or

$$I_0 = \frac{4 M R_0^2}{15}, \tag{3.169}$$

and

$$\Omega_0 = -\frac{3}{5} \frac{G M^2}{R_0}. \tag{3.170}$$

The oscillation frequency then becomes

$$\omega_p = \frac{2}{3} \sqrt{\frac{G M \left(3\gamma - 4 \right)}{R_0^3}}, \tag{3.171}$$

---

[14] In Collins' original derivation there is an additional integral term containing a $d\gamma$ factor. Because we have assumed our star to be polytropic, here γ is constant and thus this second term vanishes. See **3-4Virial** for more details.

and the period of oscillation is then

$$P = 2\pi \sqrt{\frac{R_0^3}{GM(3\gamma - 4)}}.$$ (3.172)

Because ω_p must be real for oscillatory motion, it must be that $\gamma > 4/3$, or $n < 3$.

Cox (2000, p. 400) lists data on five different types of pulsating stars. These are listed in $\log_{10}$ form, so we must take log of the period for comparison:

$$\log P = \log\left(\frac{2\pi}{\sqrt{G(3\gamma - 4)}}\right) + 1.5 \log R - 0.5 \log M.$$ (3.173)

The Cox data can be compared to the foregoing relation by applying `LinearModelFit[]` to the data. For example, fitting the classical Cepheids,

`lm1 = LinearModelFit[classceph, {x, y}, {x, y}]`

FittedModel$\left[\,\middle|\;\boxed{-1.38377 + 0.94539\,x + 0.844539\,y}\;\middle|\,\right]$

where $x = R/R_\odot$ and $y = M/M_\odot$. The result does not compare well with the expected values of $1.5x + 0.5y$.[15] This is not unexpected, as Cepheid oscillations are not adiabatic. This lack of good agreement is seen for other star types as well. However, if we take the overall fit of various pulsating stars (Cepheids, Pop II Cepheids, RR Lyr, δ Scuti, β Cep)[16] we find

$$\log P = A + Bx + Cy,$$ (3.174)

where

$$\begin{aligned} A &= -1.7371 \pm 0.6017 \\ B &= 1.5345 \pm 0.3952 \\ C &= 0.4535 \pm 0.3411 \end{aligned}$$ (3.175)

which is in reasonable agreement with our model.

The constant A term can be used to fit a value for $\gamma = 2.66$, which can be approximated by $\gamma = 8/3$ or $n = 3/5$. This does not follow the expectation for an ideal gas, but radial pulsations are fairly restricted to special groups of stars so that the theory is not applicable to main-sequence stars. Nonradial pulsations can be excited in many types of stars. The observation of such oscillations has led to helioseismology and astroseismology, but these are beyond the scope of this text.

3.4.5 Stellar rotation

The rotation of a star can be approximated by considering a homogeneous rotating fluid. The properties of such a rotating fluid can be found in several places (Grossman, 1996; Danby, 1988; O'Keefe, 1966), usually following earlier investigations of Maclauren and Jacobi. In general there are an infinite number of possible shapes for a rotating fluid, but here we assume that of an ellipsoid where the surface is expressed by

$$\frac{x^2}{a^2} + \frac{y^2}{b^2} + \frac{z^2}{c^2} = 1.$$ (3.176)

---

[15] The last term has an expected value of 0.5, not -0.5, because the log values are plotted directly.
[16] See **3-4 Virial**.

For simplicity we take the z-axis as the axis of rotation, and by symmetry $a = b$.

In a frame co-rotating with the star, the potential energy for a point particle of unit mass becomes

$$\Omega = V - \frac{1}{2}\omega^2 r_0^2, \tag{3.177}$$

where V is the gravitational potential energy of the star in the nonrotating frame, ω is the angular velocity, and $r_0 = x^2 + y^2$ the distance from the axis of rotation. If we assume our star is a homogeneous ellipsoid of constant density ρ, then the gravitational potential is

$$V = -\frac{1}{2}G\rho\left(D - Ax^2 - By^2 - Cz^2\right), \tag{3.178}$$

where

$$D = 2\pi abc \int_0^\infty \frac{du}{\sqrt{(a^2 + u)(b^2 + u)(c^2 + u)}}, \tag{3.179}$$

and

$$A = 2\pi abc \int_0^\infty \frac{du}{\sqrt{(a^2 + u)^3(b^2 + u)(c^2 + u)}}. \tag{3.180}$$

The integrals for B and C follow that of A for b and c respectively. These integrals are easily evaluated in *Mathematica*.[17]

The surface of the rotating star can be assumed to be at constant pressure. Because pressure depends on the gravitational potential, this is equivalent to the condition of constant Ω. That is,

$$\left(\omega^2 - AG\rho\right)x^2 + \left(\omega^2 - BG\rho\right)y^2 + CG\rho z^2 = \text{constant}, \tag{3.181}$$

which defines the star's shape in terms of its angular velocity. Because z is the axis of rotation, $r_0^2 = x^2 + y^2$ is the equatorial radius and we can define $c = r_0\sqrt{1 - \varepsilon^2}$, where ε is the eccentricity of the ellipse. Evaluating the integrals one finds

$$\frac{\omega^2}{G\rho} = \frac{2\pi}{\varepsilon^3}\sqrt{1 - \varepsilon^2}\left[\left(3 - 2\varepsilon^2\right)\arcsin\varepsilon - 3\varepsilon\sqrt{1 - \varepsilon^2}\right], \tag{3.182}$$

which relates eccentricity and angular velocity.

In figure (3.6) it is clear there is a maximum rate of rotation at an eccentricity of about 0.95. The exact point can be found with FindMaximum[],

FindMaximum[ω2norm, {ecc, 0.95}]

where ω2norm is $\omega^2/G\rho$ expressed in terms of the eccentricity ecc, and 0.95 is the initial test value for the maximum. This finds the maximum rotation to be $\omega^2/G\rho = 1.4116$ at an eccentricity $\varepsilon = 0.92996$. Beyond this maximum rotation limit, a self-gravitating body of uniform density would break apart. This limit does not apply to a polytropic star for which the density is allowed to change with depth; however, it can be taken as a zero-order limit for white dwarfs and neutron stars, which are largely solids.

---

[17] Details can be found in **3-4Virial**.

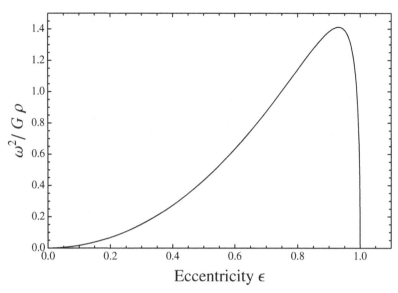

Fig. 3.6   Eccentricity related to rotational velocity.

We can now apply the virial theorem to our rotating star model. For simplicity we will assume a constant moment of inertia (no pulsation) where the rotation is the only mass motion. In a real star Coriolis forces cause meridian circulation, but these are slow compared to the rotational motion. We also assume the potential energy is for Newtonian gravity only.

If we express the virial theorem terms purely as volume integrals, we find

$$\frac{3}{2}\int \frac{\rho}{\mu}kT\,dV + \int \rho\left(\vec{r}\times\vec{\omega}\right)\cdot\vec{v}\,dV - \int \frac{GM}{r}\rho\,dV = 0. \tag{3.183}$$

Keeping only the kernels of the integrals give a "local" solution in which density cancels out. We can take this as the outer equipotential surface of the rotating object:

$$\frac{3}{2}\frac{1}{\mu}kT + \omega^2 r\sin\theta - \frac{GM}{r} = 0, \tag{3.184}$$

where θ is the angle from the axis of rotation.

At the equator, the local equation becomes

$$\frac{3}{2}\frac{1}{\mu}kT_{eq} + \omega^2 r_{eq} - \frac{GM}{r_{eq}} = 0, \tag{3.185}$$

while at the pole it is

$$\frac{3}{2}\frac{1}{\mu}kT_{pol} - \frac{GM}{r_{pol}} = 0. \tag{3.186}$$

Because $r_{pol} < r_{eq}$, this means $T_{eq} < T_{pol}$. That is, the equatorial temperature is less than that of a nonrotating star of the same mass and radius. This is also true for the luminous flux ($F_{eq} < F_{pol}$).

When hydrostatic equilibrium is combined with radiative equilibrium, the result for rotation is called von Zeipel's theorem. This can be stated fairly simply. The emergent flux is proportional to the local

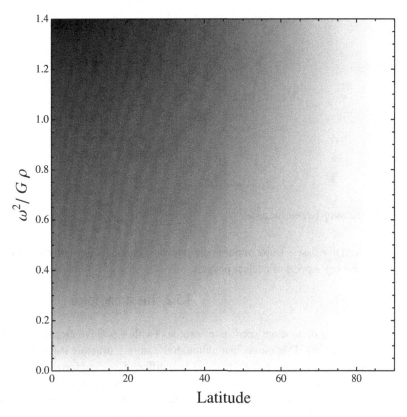

Fig. 3.7   Emergent flux as a function of rotation and latitude.

value of gravity as measured by g when centriptal acceleration is included. Because g is high for the polar regions, the poles are "bright" whereas the opposite is true for the equator.

This relation can be approximated by using our ellipsoidal model of a rotating star and graphing $T^4(\theta)/T^4_{\mathrm{pol}}$ as a function of latitude and $\omega^2/G\rho$ as seen in figure (3.7).[18]

3.5   Fusion in the stellar core

3.5.1   The case for fusion

In our discussion of stellar interiors we estimated the central temperature of a solar-type star to be in the range of 3.6×10^6 K under the polytropic assumption of hydrostatic equilibrium to 9.0×10^7 K when applying the virial theorem. This tells us that the temperature of the central region is high enough for nuclear fusion to occur. This provides a power source for stars; however, as we will see, the exact nature of this power source is quite temperature sensitive. The addition of a power source also means

---

[18] See **3-4Virial** for details.

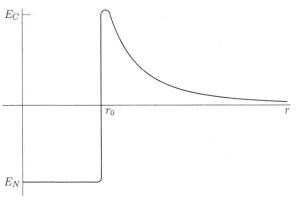

Fig. 3.8   Potential energy between two nuclei.

we need to reassess some of our assumptions when modeling stellar interiors. We must therefore look at some key aspects of nuclear physics.

3.5.2 The strong force

The nucleus of an atom consists of protons (with a positive electric charge) and neutrons (with no electric charge). This means that although the atomic structure itself is driven by the Coulomb force, the nucleus itself must be held together by a different, nuclear force. Because this force must be able to overpower the electric repulsion of the protons, it is known as the strong force.

The strong force is actually very complicated, particularly when talking about the interior of a nucleus itself. For our purposes, we consider only the force of attraction between two nucleons (protons or neutrons), which is given by the Yukawa potential:

$$V\,(r) = -\frac{g^2}{r}\mathrm{e}^{-r/R},\tag{3.187}$$

where g is the coupling constant and

$$R = \frac{\hbar}{mc},\tag{3.188}$$

known as the range of the potential. The total potential between two protons would then be

$$V\,(r) = \frac{\mathrm{e}^2}{r} - \frac{g^2}{r}\mathrm{e}^{-r/R}.\tag{3.189}$$

Even though the strong force is larger and dominates when nucleons are close, the electric force dominates when they are farther away. This means that two protons repel each other more and more strongly until they get within a certain distance, whereupon they are strongly attracted.[19]

The actual potential between protons is shown in figure (3.8). Classically, two nucleons would never bind together. To get close to each other they would need a combined kinetic energy greater than the maximum potential, but then they would have too much energy to stay bound. However, quantum mechanics allows for a nucleon to tunnel through the potential and become bound.

---

[19] A second nuclear force known as the weak force causes free neutrons to decay into protons, so we need not worry about them for now.

3.5.3 Gamow peak

To undergo a nuclear reaction, two nucleons must be within a distance $d \approx 10^{-11}$ cm. Thus, the potential energy between two nuclei that would need to be overcome is

$$V_0 \approx \frac{Z_1 Z_2 e^2}{d}. \tag{3.190}$$

Here, eZ_1 and eZ_2 are the charges of the two particles. The temperature required to do this is then

$$3kT \geq V_0, \tag{3.191}$$

or in the case of two protons

$$T \geq \frac{e^2}{3kd} \approx 56 T_6 = 5.6 \times 10^7 \text{K}. \tag{3.192}$$

(Note: $T_n \equiv 10^n$K is a shorthand for temperature.) This is within the range of our calculated central temperature for the sun, but it is higher than the contemporary value of 1.6×10^7K. This value is the requirement on the central temperature for most of the nucleons to engage in nuclear reactions. However, kinetic theory shows that even at the calculated T_c some nucleons might have nuclear reactions. Classically they would not form new bound states, but bound states can occur through quantum tunneling.

In quantum mechanics, the uncertainty principle allows a particle to spontaneously gain enough energy to overcome the barrier and then spontaneously lose it, returning to its original energy. If the energy of a particle is small compared with the height of the barrier (as it typically is for nuclear reactions), then the probability of this occurring is given by the Gamow factor:

$$G(E) = \exp\left[-\pi \sqrt{2m/E} Z_1 Z_2 e^2 / \hbar \right]. \tag{3.193}$$

Thus, the higher the energy, the more probable the reaction. From the Maxwell–Boltzmann distribution[20] the number of particles with kinetic energy between E and $E + dE$ is

$$n_E dE = \frac{2n}{\sqrt{\pi}} \frac{1}{(kT)^{3/2}} \sqrt{E} e^{-E/kT} dE. \tag{3.194}$$

Thus for any given temperature the higher the energy, the less likely a particle is to have it. The chance of forming a bound state is then given by the product of these two probabilities. The result is a window of opportunity known as the Gamow peak, seen (not to scale) in figure (3.9).

Quantum theory allows for the fusion of nuclei well below the "critical" temperature. However, it should be noted that the higher the atomic number, the greater the potential barrier that must be overcome. Thus, successively higher temperatures must be reached for the fusion of higher elements to occur.

3.5.4 Binding energy and nuclear abundances

The amount of energy that can be released (or absorbed) depends on the relative strength of the nuclear bound state. This is typically expressed in terms of the binding energy of a nucleus. The binding energy

---

[20] See chapter 1 and **7-5Maxwell**.

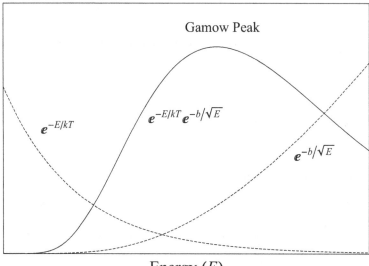

Fig. 3.9 Demonstration of the Gamow peak.

of a nucleus is defined as the energy required to break up and disperse to infinity all the constituent nucleons (protons and neutrons) in that nucleus. Thus

$$B = (M_N - M) c^2, \tag{3.195}$$

where B is the binding energy, M_N is the mass of all the constituent nucleons, and M is the mass of the bound nucleus. Intuitively, one might think that the larger the nucleus the greater the binding energy (because there are more nucleons). However, the exponential decay of the strong force means that beyond a critical diameter nucleons on opposite sides do not really attract each other. This means that (as a general rule) binding energy increases with atomic number up to a critical number (^{56}Fe), beyond which the binding energy per nucleon actually decreases, seen in figure (3.10). Thus in general, energy is released by fusion up to the iron peak at ^{56}Fe, and by fission beyond that.

For light elements there are some exceptions to the general trend of increasing binding energy. In particular, ^{7}Li, ^{7}Be, and ^{8}B have binding energies less than that of helium. This means that before the interior of a star gets hot enough for helium production, it becomes hot enough (around 3 million Kelvin) to photodisintegrate elements such as lithium. This actually has cosmological consequences. Standard models predict that hydrogen and helium ($\sim$70%/30%) make up the primordial elements of the universe. However, they also predict a lithium ratio of

$$1 \times 10^{-9} \lesssim \frac{^7\text{Li}}{^1\text{H}} \lesssim 8 \times 10^{-9}. \tag{3.196}$$

In the surface of the sun, we find that

$$\frac{^7\text{Li}}{^1\text{H}} \sim 10^{-11}. \tag{3.197}$$

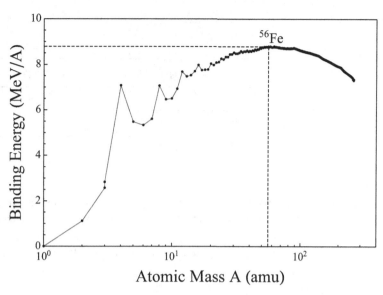

Fig. 3.10 Binding energy per nucleon.

This means that the sun must be formed from the remnants of long dead stars, which depleted the lithium in their burning. As we will see, the various fusion cycles help to explain not only the absence of certain elements, but also the relative commonality of other elements.

3.5.5 P–P chains (hydrogen burning)

The simplest nuclear cycle happens to be the first (lowest temperature) cycle and also the most energy producing. Given its simple nature, we will look at it in great detail. Commonly known as the p–p (proton–proton) chain, this reaction has the end result of taking four ^{1}H nuclei (protons) and forming a ^{4}He nucleus. However, this does not happen directly, but rather in stages.

The first stage occurs $\sim T_6$ and is simply a proton–proton collision:

$$^1\text{H} + {}^1\text{H} \rightarrow {}^2\text{H} + e^+ + \nu_e; \tag{3.198}$$

that is, the two protons bond, causing one proton to decay into a neutron, forming deuterium, and releasing a positron and a (electron) neutrino. The neutrino has such a small cross section for interaction that it moves straight out of the star, taking its energy with it. The positron typically is reabsorbed by some nuclei later in the cycle; however, if there is a free electron in the area, it annihilates it as

$$e^+ + e^- \rightarrow 2\gamma. \tag{3.199}$$

The net result of this is that over time there are no electrons in great quantities in a nuclear soup. In the core we have only the nuclei of atoms.

In the second stage, our deuterium bonds to another proton, forming our first helium nuclei:

$$^2\text{H} + {}^1\text{H} \rightarrow {}^3\text{He} + \gamma. \tag{3.200}$$

Table 3.1 Stages of the PP-I Cycle	
Stage	Energy (MeV)[a]
${}^1\text{H} + {}^1\text{H} \rightarrow {}^2\text{H} + e^+ + \nu_e$	1.442
${}^2\text{H} + {}^1\text{H} \rightarrow {}^3\text{He} + \gamma$	5.493
${}^3\text{He} + {}^3\text{He} \rightarrow {}^4\text{He} + {}^1\text{H} + {}^1\text{H}$	12.859

[a] Energy data are taken from Shore (2003).

The cycle is not complete, however, as ${}^3\text{He}$ is not stable. The next stage can go several ways (as we will see) but the simplest (and initially most common) is for two of these nuclei to bond

$$
{}^3\text{He} + {}^3\text{He} \rightarrow {}^4\text{He} + {}^1\text{H} + {}^1\text{H}, \tag{3.201}
$$

releasing two hydrogen nuclei for other cycles.

This set of reactions is known as PP-I. Its energy production can be seen in table (3.1). Because the third stage requires two ${}^3\text{He}$ nuclei, the energies of the first two stages must be counted twice. Thus, the total energy released is 26.729 MeV. The energy of the neutrinos does not contribute to the heating of the sun (because they just leave town), so we must subtract their energy (about 0.26 MeV each); thus the net energy gain is 26.2 MeV.

The first stage of PP-I is actually a very slow stage, much slower than the second stage. Thus, almost immediately after the first stage occurs, it is followed by the second. The third stage is somewhat slower than the second. As a result, the second stage is left waiting for a bit, which gives an opportunity for processes to finish the job. The first of these alternatives is called PP-II, seen in table (3.2). This cycle loses an additional 0.80 MeV via neutrinos, and has a net energy release of 25.67 MeV. The production of beryllium can also follow a different cycle, known as PP-III and seen in table (3.3), which has a net energy release of 19.20 MeV.

Power output of the PP cycle can be calculated by the formula:

$$
\varepsilon_{pp} = \left(2.32 \times 10^6\right) \frac{\rho X^2}{T_6^{2/3}} e^{-\frac{33.81}{T_6^{1/3}}} \text{ erg/s}, \tag{3.202}
$$

Table 3.2 Stages of the PP-II Cycle	
Stage	Energy (MeV)[a]
${}^3\text{He} + {}^4\text{He} \rightarrow {}^7\text{Be} + \gamma$	1.586
${}^7\text{Be} + e^- \rightarrow {}^7\text{Li} + \nu_e + \gamma$	0.861
${}^7\text{Li} + {}^1\text{H} \rightarrow {}^4\text{He} + {}^4\text{He}$	17.347

[a] Energy data are taken from Shore (2003).

Table 3.3 Stages of the PP-III Cycle	
Stage	Energy (MeV)^a
$^3\text{He} + {}^4\text{He} \rightarrow {}^7\text{Be} + \gamma$	1.586
$^7\text{Be} + {}^1\text{H} \rightarrow {}^8\text{B} + \gamma$	0.135
$^8\text{B} \rightarrow {}^8\text{Be} + e^+ + \nu_e$	17.98
$^8\text{Be} \rightarrow {}^4\text{He} + {}^4\text{He}$	0.095

^{a} Energy data are taken from Shore (2003).

where X is the hydrogen mass fraction and T_6 is the temperature in units of 10^6 K.[21] It is clear that as the temperature increases, the total output increases almost exponentially. In a stellar core, as the initial PP-I cycle begins, it increases the core temperature. This in turn increases the rate of nuclear production so that over time the core temperature of a star increases. Although the initial relative frequency of these three cycles depends on the initial chemical composition, each cycle dominates in turn with increasing temperature. In general, as temperature increases, PP-II overtakes PP-I in relative energy production and becomes the dominant cycle. PP-II is in turn overtaken by PP-III at $T_6 \sim 24$. This means that as we move deeper into the star, each gains dominance. However, the total relative energy production of all PP cycles is overtaken by the much more powerful CNO cycle at $T_6 \sim 15$, as seen in figure (3.11). The temperature of the sun's core is such that the CNO cycle greatly dominates in the center of the core. This significantly more energetic region actually induces a convective region within the sun.

3.5.6 CNO cycle

The CNO or Carbon–Nitrogen–Oxygen cycle is another way in which hydrogen can be consumed. The typical cycles are seen in table (3.4). Here carbon acts as a catalyst for hydrogen to helium fusion, moving through to nitrogen and oxygen to do so. By itself it is sometimes called the CN cycle. Any of the included reactions can be taken as the starting point for this cycle, as it forms a kind of closed loop. The addition of secondary reaction chain, table (3.5), forms the complete CNO cycle or tricycle. The ^{14}N produced at the end of the tricycle then enters the CN cycle.

The energy output for such a cycle is about 26.734 MeV. The power output is given by

$$\varepsilon_{CNO} = \left(8 \times 10^{27}\right) \frac{\rho X Z}{T_6^{2/3}} e^{-\frac{152.31}{T_6^{1/3}}} \text{ erg/s}, \tag{3.203}$$

where Z is the mass fraction of carbon and nitrogen.

At this point we have reached the limit of energy production for a stable sun-sized star. The concept of a CNO core with a convective region of PP is consistent with observations of the sun. We can even make a rough calculation for how long the sun will remain in this stage. That is,

$$t = (4m_H - m_{He}) c^2 \frac{XM}{L} \sim 11 \times 10^9 \text{ y}. \tag{3.204}$$

---

[21] In real models of stars such power equations are not used directly. Rather the curves are fit to power law functions. See **3-5Nuclear** for details.

Table 3.4 Stages of the CN Cycle	
Stage	Energy (MeV)[a]
$^{12}C + {}^{1}H \rightarrow {}^{13}N + \gamma$	1.954
$^{13}N \rightarrow {}^{13}C + e^+ + \nu_e$	2.221
$^{13}C + {}^{1}H \rightarrow {}^{14}N + \gamma$	7.550
$^{14}N + {}^{1}H \rightarrow {}^{15}O + \gamma$	7.293
$^{15}O \rightarrow {}^{15}N + e^+ + \nu_e$	2.761
$^{15}N + {}^{1}H \rightarrow {}^{12}C + {}^{4}He$	4.965

[a] Energy data are taken from Shore (2003).

As we move beyond hydrogen burning, we move into the late stages of a star. While the production of higher elements is crucial to our own existence, it is also an indication that a star's days are numbered.

3.5.7 Helium burning and higher elements

The basic helium burning cycle is seen in table (3.6). Note that in the first reaction energy is lost, only to be gained in the second stage. This cycle begins at temperatures $\sim T_8$.

When the hydrogen within the core of a star is depleted, the temperature and pressure of the core is insufficient to fuse helium. Thus in the core, nuclear reactions cease while a thin shell of hydrogen

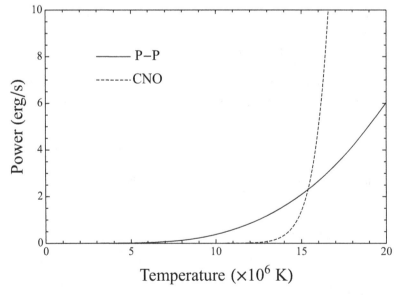

Fig. 3.11 Temperature dependence of P-P and CNO power generation.

Table 3.5 Stages of the NO Cycle	
Stage	Energy (MeV)[a]
$^{15}N + {}^{1}H \rightarrow {}^{16}O + \gamma$	12.1
$^{16}O + {}^{1}H \rightarrow {}^{17}F + \gamma$	0.6
$^{17}F \rightarrow {}^{17}O + e^{+} + \nu_{e}$	2.76
$^{17}O + {}^{1}H \rightarrow {}^{14}N + {}^{4}He$	1.19

[a] Energy data are taken from Harwit (1988).

Table 3.6 Stages of Hydrogen Burning	
Stage	Energy (MeV)[a]
$^{4}He + {}^{4}He \rightarrow {}^{8}Be - \gamma$	−0.092
$^{8}Be + {}^{4}He \rightarrow {}^{12}C + \gamma$	7.366

[a] Energy data are taken from Shore (2003).

Table 3.7 Oxygen-Producing Reactions
$^{12}C + {}^{4}He \rightarrow {}^{16}O + \gamma$
$^{13}C + {}^{4}He \rightarrow {}^{16}O + n$
$^{14}N + {}^{4}He \rightarrow {}^{18}F + \gamma \rightarrow {}^{18}O + e^{-} + \bar{\nu}_{e}$

Table 3.8 Helium Burning to Higher Elements
$^{16}O + {}^{4}He \rightarrow {}^{20}Ne + \gamma$
$^{18}O + {}^{4}He \rightarrow {}^{21}Ne + n$
$^{18}O + {}^{4}He \rightarrow {}^{22}Ne + \gamma$
$^{22}Ne + {}^{4}He \rightarrow {}^{25}Mg + n$
$^{22}Ne + {}^{4}He \rightarrow {}^{26}Mg + \gamma$
$^{25}Mg + {}^{4}He \rightarrow {}^{28}Si + n$
$^{25}Mg + {}^{4}He \rightarrow {}^{29}Si + \gamma$

Table 3.9 Burning of Higher Elements

$^{12}C + {}^{12}C \rightarrow {}^{20}Ne + \alpha$
$^{12}C + {}^{12}C \rightarrow {}^{23}Na + p$
$^{16}O + {}^{16}O \rightarrow {}^{28}Si + \alpha$
$^{16}O + {}^{16}O \rightarrow {}^{31}P + p$
$^{16}O + {}^{16}O \rightarrow {}^{31}S + n$

fusion remains around the core. In the absence of fusion, the core cannot fight against gravity and starts contracting. Part of this gravitational energy feeds the hydrogen shell, which causes the hydrogen to burn faster. This has a twofold effect. The helium produced in the hydrogen shell falls to the core, accelerating the collapse of the core, while the energy produced by the shell swells the outer layers of the star. Thus, the star begins to expand into a red giant, while the core collapses. Finally, when the temperature of the star reaches about T_8, helium fusion can occur.

For sun-like stars (with $m \lesssim 2M_\odot$), the core actually reaches a state of degeneracy–for which pressure depends on density alone–before reaching the critical temperature for helium fusion. This means that when burning starts, the first increase in temperature does not influence pressure, and the burning has a runaway effect. Over the span of a few seconds, the temperature can reach $\sim T_9$, and core luminosity can reach $10^{11} L_\odot$ (about that of an entire galaxy). This effect is known as the helium flash (though it is not seen from outside the star). After this, the core loses degeneracy, and expands until cooling to an equilibrium temperature for helium burning.

Once helium burning begins, the by-products of the CNO cycle can also fuse with helium, generating a hierarchy of elements. These reactions produce ^{18}O, from which a second hierarchy is generated. These cycles continue until all the ^4He is depleted. The free neutrons generated in these cycles can be absorbed by heavy elements to build elements heavier than ^{56}Fe.

After helium depletion, if the core of the star reaches temperatures $\sim T_9$, then carbon and oxygen can begin to fuse. At this point things become quite complex. However, some of the more common reactions can be seen in table (3.9). Finally, at $\sim T_9 = 3$, silicon interacts with other elements. Given time, this terminates at ^{56}Fe. After that, the star has breathed its last breath.

3.6 Beyond the polytropic model

3.6.1 Fully convective stars

Building a more realistic model of a star requires leaving behind Eddington's polytropic model and its scaling relationships to solve the equations of stellar structure numerically. Here we consider a fully convective star and integrate down from the surface. Using the approximate expressions for the energy yields from both the proton–proton reaction and the CNO cycle outlined in the previous section, we can construct stellar models that are a bit more realistic than the polytropic model. This improvement is largely due to the fact that we allow for energy production and transport, which is not considered in the polytropic form.

The radiation process alters stellar structures and are governed by the conditions of radiative or convective energy transport equilibrium in addition to the normal hydrostatic equilibrium and ideal gas conditions.[22] We also assume the model to be homologous, i.e., that the star is of uniform chemical composition and that the mass and radius functions scale in simple proportion. This approach was worked out by Martin Schwarzschild before digital computers were widely available, though here we follow the method of Bohm-Vitense (1992).

For a fully convective model there are four equations to be solved. As with our polytropic model we assume hydrostatic equilibrium, equation (3.4):

$$\frac{dP}{dr} = -\frac{d\phi}{dr}\rho, \tag{3.205}$$

where

$$\frac{1}{r^2}\frac{d}{dr}\left(r^2\frac{d\phi}{dr}\right) = 4\pi G\rho, \tag{3.206}$$

and the ideal gas relation, equation (3.8):

$$P(r) = \frac{k}{\mu}\rho(r)\,T(r). \tag{3.207}$$

To this we add the condition of thermal equilibrium:

$$\frac{dL}{dr} = 4\pi r^2 \rho(r)\epsilon(r), \tag{3.208}$$

where $\epsilon = \epsilon_0 \rho T^{\nu}$ is the equation of nuclear energy generation and ν is an exponent depending on whether energy is generated by proton–proton ($\nu = 4$) or CNO ($\nu = 16$) cycles, and the condition of convective equilibrium:

$$\frac{dT}{dr} = \left(1 - \frac{1}{\gamma}\right)\frac{T}{P}\frac{dP}{dr}, \tag{3.209}$$

which describes an adiabatic temperature gradient.

To solve these equations numerically we must first set the necessary star parameters and physical constants. For our example we will take values known for a solar-type star.[23] Our equations have a singularity at the core, just as the polytropic equation had. Physically this means a purely convective star is not stable and will either collapse or develop a radiative core as more sophisticated models indicate.

Computationally, this means *Mathematica* must be kept away from the central singularity. Thus, we must start some small distance epsy $= 10^{-4}$ from the singular point. In addition, *Mathematica* can be run only with the independent variable from the minimum to the maximum, i.e., not through negative steps. Because our known boundary conditions are at the surface of the star, the computation runs over $y = 1 - x$ so that our solution is computed from the surface downward. The solution to our equations

---

[22] Because we still assume an ideal gas equation of state, this does not apply to extreme stars such as white dwarfs.
[23] See **3-6Convective** for details.

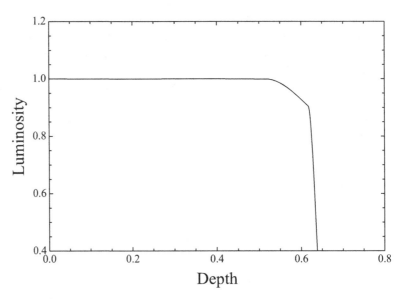

Fig. 3.12 Luminosity vs. depth for a fully convective star.

can then be computed via NDSolve[].

```
epsy = 10⁻⁴;
stardown = NDSolve[{p'[y] == p[y] × q[y] / (t[y] × (1 - y)²), . . .},
  {p, q, t, fl}, {y, epsy, 1 - epsy}, AccuracyGoal → 10⁻¹⁶,
  MaxSteps → 100 000, StartingStepSize → 0.000001]
```

Here x, q, fl, t, p represent the variables of radius, mass, luminosity, temperature, and pressure respectively. The results are interpolating functions, but these can be graphed and analyzed as regular functions. For example,

```
Plot[fl[z] /. stardown // Evaluate, {z, 0.001, 0.9}]
```

seen in figure (3.12) shows that outside the core the luminosity is constant, as would be expected in the parts of the star where there is no energy generation.

Although the solution gives a better model than the polytrope model, one must remember it is still not a good model for a realistic star. Rather it is a class of model that is in pure convective equilibrium and homologous, which no real star can be. A fully convective star is a good model for a contracting protostar that is fully mixed, so the chemical composition is uniform through the star. Once nuclear reactions have gone on for a while, the chemical composition of the core changes and the star is no longer homologous and cannot be described by such a model.

3.6.2 Radiative stars

Another way to extend the polytropic model is to consider a fully radiative stellar model. Here we again assume hydrostatic and thermal equilibrium for an ideal gas, but rather than convective equilibrium we impose radiative equilibrium. This condition is given by

$$\frac{dT}{dr} = \frac{L(r)}{4\pi r^2} \frac{\kappa_g}{T^3} \frac{3}{16\sigma}, \tag{3.210}$$

where κ_g is the Rosseland mean opacity of the material. We continue to assume a gray-body situation in which the opacity is uniform across all wavelengths. The opacity will, however, vary with depth. Here we assume a Kramer's type opacity:

$$\kappa_g = \kappa_0 \frac{\rho^\alpha}{T^\beta}, \tag{3.211}$$

where κ_0 is a parameter proportional to the metallic abundance of the gas. Following Kramer's classical computation we will assume $\alpha = 1$ and $\beta = 3.5$. More recent calculations of real stars favor $\alpha = 0.5$ and $\beta = 2.5$ for nonhomologous stars, but the classical values give more stable solutions at our level of sophistication.

As with the convective model, the stellar parameters and physical constants are based upon a solar-type star.[24] For the interior of the sun, the core is surrounded by a radiative envelope with a convective atmosphere. The bottom of the convective zone is at a real depth of $0.008R_\odot$. From the convective model in the previous section, the temperature at this depth is $T \sim 115\,000$ K and pressure $P \sim 5 \times 10^8$ dynes/cm^2. These will be taken as the "surface" boundary for the solution. Thus via NDSolve[]

```
epsy = 8 × 10⁻³; tempzone = 115000; preszone = 3 × 10⁸;
stardown = NDSolve[{p'[y] == p[y] × q[y] / (t[y] × (1 - y)²), . . .}, {p, q, t, fl},
   {y, epsy, .7}, AccuracyGoal → 6, MaxSteps → 500000, StartingStepSize → 0.00001]
```

Mathematica has a difficult time keeping the solution stable even though the initial step size is small. A plot of the results to various depths shows the model begins to fail at about half the star's radius. This failure is due to our initial assumptions. In a real star the mean atomic weight μ and the opacity parameters α and β are not constants, but they are rather significant functions of radius. One can adjust the parameters to obtain solutions to greater depth, but this can be tedious and will still fail to provide a complete interior solution.

The alternative is to look at two solutions, a core solution calculated from the center outward and an envelope solution calculated from the surface inward. Each of these can have different physical parameters that can be adjusted so that the core and envelope solutions match at some intermediate point in the star. We must therefore determine a reasonable core model to match our radiative solution.

3.6.3 The hybrid core

The heat transport mechanism in a stellar core is controlled by whichever energy production mechanism is dominant. In low-mass stars such as red dwarfs the p–p cycle dominates. Energy generation is spread throughout the core, the radial temperature gradient is relatively shallow, and the core is radiative. If the star is very massive, then the CNO process dominates, even though the abundance of carbon and nitrogen is much lower than that of hydrogen. In this case the temperature gradient is large (because the CNO cycle is highly temperature sensitive) and the core is convective. In a solar mass star, both

[24] See **3-7Radiative**.

processes occur with comparable strengths. Thus for our model we must use a parameterization that blends the two cases.

To combine the two models we introduce a new parameter `fract` that represents the fraction of the gradient that is convective (CNO); thus 1 − `fract` represents the fraction that is radiative (p–p). A reasonable result can be found by setting `fract` = 0.15. The equations can then be solved numerically as before:

```
epsx = 10⁻⁴; rcore = 0.54; fract = 0.15;
starup = NDSolve[{px'[x] == -px[x] × qx[x] / (tx[x] × x²), qx'[x] == px[x] × x² / tx[x],
    tx'[x] == (1 - fract) (-bigsee × flx[x] × px[x]ᵅ⁺¹ / (x² × tx[x]ᵅ⁺ᵝ⁺⁴)) + . . . },
    {px, qx, tx, flx}, {x, epsx, rcore}, MaxSteps → 500 000, StartingStepSize → 0.001]
```

Because we start at the center rather than the surface, the boundary conditions must be determined by trial and error. This can be done by selecting a reasonable μ and then choosing a suitable starting distance from the center and adjusting the initial flux fl and mass q within that distance. The integration is very nonlinear and not very stable.[25] Again this is due to our assumption of constant μ, α, and β. A reasonable solution can be obtained with $\mu = 0.65$, which is close to the "cosmic" value of $\mu = 0.62$ while still providing a stable solution. This solution is still crude, as it assumes the gray body approximation and homogeneity, but it is sufficient to create a composite model for a sun-like star.

3.6.4 The composite star

We are now able to create a composite stellar model of a sun-like star. We will assume our star consists of a hybrid core with a radiative envelope and a convective atmosphere.

Because we defined the "surface" of the radiative envelope to be the bottom of the convective atmosphere, we need only match the envelope to the hybrid core. This can be done for any quantity (pressure, luminosity, etc.), but for demonstrative purposes we look at the interior temperature.[26]

The temperature functions of the hybrid core and radiative envelope can be done as before, with the core calculated outward from the center and the envelope inward from the atmospheric surface. Because we wish to compare these two functions in *Mathematica*, the two solutions must be expressed as separated functions; thus we use t for the radiative solution and tx for the core solution, following the same pattern for the other parameters.

A plot of these two temperatures, figure (3.13), shows they intersect at $z \sim 0.25$ and $z \sim 0.55$. The solution we seek is that of the core up to the first intersection and that of the envelope beyond that. The location of the first intersection can be found by FindRoot[]. Because we seek the first intersection, we set the initial guess to be 0.2. Thus

```
FindRoot[(Evaluate[t[1 - z] temp0] /. stardown) - (tx[z] temp0 /. starup) == 0, {z, 0.2}]
```

which determines the intersection to be at $z = 0.230403$. The interior temperature function can then be determined by matching our two solutions at that point, as seen in figure (3.14).

---

[25] You can see this by changing values in **3-8HybridCore**.
[26] Matching of other quantities can be seen in **3-9CompoundStar**.

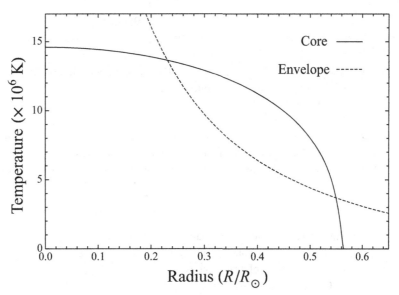

Fig. 3.13 Comparison of temperatures for the hybrid core and radiative envelope.

This approach reaches the limit of our gray-body model for stellar interiors. The preferred modern approach for computation is called Henyey's method, in which the differential equations and boundary conditions are replaced with linearized difference equations. The problem is then solved algebraically on a grid of points. Such a grid method is beyond the scope of this text; however, a good introduction to astrophysical grid methods can be found in Bowers and Wilson (1991).

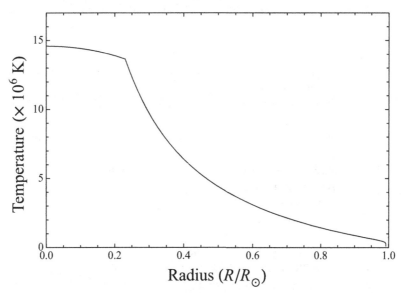

Fig. 3.14 Interior temperature for a solar-type star.

Exercises

3.1   If we assume the sun is a polytropic star of index 1, with $M_\odot = 2 \times 10^{33}$ kg and $R_\odot = 7 \times 10^{10}$ cm,
   1. What is its central temperature?
   2. What is its central pressure?
   3. What is its central density?

3.2   Plot a graph of the Lane–Emden solutions $\omega_n(z)$ for $n = 0, 1, 2, 3, 4, 5$. With the exception of the $n = 5$ solution, computationally determine the roots where $\omega_n(z) = 0$.

3.3   Suppose observation of several main-sequence stars yields a mass-radius relation of

$$\frac{R_1}{R_2} = \left(\frac{M_1}{M_2}\right)^{3.2}$$

while the mass-luminosity relation is that of an ideal gas. Given the aforementioned solar mass and radius, and the solar luminosity $L_\odot = 4 \times 10^{33}$ erg/s,
   1. Calculate the polytropic index n.
   2. Calculate the radius of a $15 M_\odot$ star.
   3. Calculate the luminosity of a $15 M_\odot$ star.
   4. From these, calculate such a star's effective surface temperature.

3.4   In Section 3.3 we derived the mass–radius and mass–luminosity relations for an ideal gas star. Following the same procedure, assume

$$\rho \sim \frac{\mu P}{T}, \quad \varepsilon \sim \rho T^\nu,$$

and derive the two relations. Note: For these relations you can show $\gamma = 1$, which is the case for an isothermal star. The polytropic index does not come into play here.

3.5   In Eddington's estimation of mass–luminosity we defined the quantity

$$\beta = \frac{P_{gas}}{P_{gas} + P_{rad}}.$$

   1. Assuming an ideal gas with black-body radiation pressure, determine β in terms of the gas density ρ and temperature T.
   2. With the same assumptions, derive β as a function of density ρ and pressure P.
   3. What does β approach in the limits $P \to 0$ and $P \to \infty$?
   4. What do these limits say about the conditions at which the pressure is radiation dominated or gas dominated?

3.6   Using **3-4Virial**, make a crude variational analysis of:
   1. A white dwarf star
   2. A neutron star

3.7   Using **3-6Convective**, make a convective analysis for:
   1. A giant star such as Arcturus
   2. A supergiant star such as Betelgeuse

3.8   Using **3-7Radiative**, make a radiative analysis of:
   1. A giant star such as Arcturus
   2. A supergiant star such as Betelgeuse

3.9   If we ignore the kinetic energy involved in a fusion process, the energy produced for each step of a fusion cycle can be calculated directly from the famous equation $E = mc^2$.
1. Calculate the energy for each step of CNO cycle.
2. From conservation of energy, how much energy must be "lost" from the neutrinos?
3. What is the net energy generated in the production of one helium atom?

3.10 Consider a simple $25M_\odot$ star. We will assume that it is initially composed entirely of hydrogen. We will also assume that its luminosity is constant.
1. Assuming the star is an ideal gas, calculate its luminosity.
2. Assuming the star can fuse all of its hydrogen into helium, what is the lifetime of its "main sequence" stage?
3. In the same way, how long could it fuse helium into carbon?
4. How long for carbon up to iron?
5. Do you think these results are good rough calculations for a star of this size? Why?

3.11 Evolutionary models indicate that there is a time dependence for the chemical composition of a stellar core (i.e., helium is made at the expense of hydrogen). For solar mass stars the core value of μ increases as the star ages. Using the hybrid core model, determine what happens to the core properties as the star ages.

3.12 In stellar evolution, the protostar stage is basically a fully convective star that has not reached "ignition." By altering the surface temperature and perhaps the surface radius of the convective model in **3-6Convective**, find what surface temperature produces a core temperature that effectively suppresses any thermonuclear reactions.

4 Extreme classical stars

In Chapters 2 and 3 we looked at atmospheres and interiors of main sequence stars. For such stars we may assume both local thermodynamic equilibrium and the physics of an ideal gas. In this chapter we move beyond these simple assumptions. We begin by looking at atmospheric models in which local thermodynamic equilibrium is not valid, such as the case of expanding atmospheres. We then examine extreme stars such as white dwarfs and neutron stars, in which electron degeneracy plays a central role. As we shall see, this will take us to the limits of classical physics.

4.1 Atmospheres beyond local thermodynamic equilibrium

4.1.1 Brief survey of non-LTE situations

In Chapter 2 we developed an atmospheric model that could be approximated by assuming local thermodynamic equilibrium (LTE). Such models work well to describe the photospheres of solar-type main sequence stars. These relatively simple models do not work at all for the chromospheric or coronal regions of a stellar atmosphere. For these regions we must look to non-LTE (NLTE) models.

The most comprehensive treatment of radiative transfer in LTE and NLTE situations is that of Mihalas (1978), and we follow the conventions adopted there in our treatment here. Our present use of NLTE concepts is toward atmospheres that have a systematic (and in some cases differential) macroscopic radial flow.[1] NLTE models are also described in Gray (1992), which outlines how to deal with "turbulence," which must be handled by direct atmospheric modeling.

Mihalas also considers techniques suitable for high temperature situations including hot massive main-sequence stellar photospheres where large-scale mass motions are well known from observations. For such stars the NLTE effects on spectral absorption lines are particularly detectable. Mihalas' review of radiative transfer is masterfully comprehensive and is recommended reading for those wanting to understand the subject in great depth. Overall the book is at a graduate level, so we do not delve into many of the complexities here. However, there are several key points concerning NLTE models that we restate here:

1. The radiative transfer equations are simply statements of the Boltzmann transport equations for photons.[2]
2. There are many shortcomings of the boundary conditions that were avoided in the previous analysis by assuming the source function was that of a black body and by assuming a linear dependence of

---

[1] This differs from the mass motions such as solid body, low-velocity stellar rotation considered in Section 2.5.
[2] See **7-5Maxwell** and **7-6Boltzmann** for elements of the Boltzmann transport equations and its derivative equations.

the source function on optical depth. In that case the emergent intensity becomes the Eddington–Barbier relation, which is linear in $\mu = \cos\theta$. It is this result that is used to justify the often stated relationship that images of the sun taken in the center of the strongest absorption spectrum lines arise higher in the atmosphere in terms of real depth than images from the continuum. The same reasoning leads to x-ray and radio images being of the solar chromosphere and corona and not the photosphere. The peak intensity in the continuum of the sun should be characteristic of the "lowest" photospheric levels.

3. Non-LTE situations basically occur when the Eddington–Barbier relationship fails, and these in turn most often are the result of some sort of scattering such as Thompson (electrons) or Rayleigh (molecules) equaling or dominating absorption processes within an atmosphere.

4. Scattering, unlike absorption, tends to simply change the direction of travel of the photon without much loss of energy. The distance a photon travels between scatterings is called the mean free path. Absorption, on the other hand, is an inelastic collision than can result in the entire energy of the photon being transferred into thermal motions after a single collision. At levels at which absorption dominates over scattering, the radiation field is said to be thermalized.

4.1.2 Departures from local thermodynamic equilibrium

One of the first requirements for LTE is that the radiative flux J_ν be that of a black body B_ν, which necessitates that the radiation field be isotropic. At the outer boundary of a star this cannot be true because the radiative flux streams outward into a nearly black sky (barring exceptions such as close binaries). This means the radiation solid angle must be less than the complete 4π steradians as one moves above the photosphere toward the stellar boundary. This is expressed by the so-called dilution factor:

$$W = \frac{\omega_*}{4\pi} = \frac{1}{2}\left(1 - \sqrt{1 - (r_*/r)^2}\right), \tag{4.1}$$

where ω_* is the solid angle of the photosphere as seen by the line forming region at a distance r from the center of the star, and r_* is the radius of the photosphere.

Dilution is not the only factor in the outer layers. The amount of line emission in many wavelength regions far exceeds that of the photospheric continuum. This, along with widely variable opacity in many regions, means that the radiation field is not representable by a Planck function at a single temperature. For the sun, dilution occurs in the chromosphere and corona, which are non-LTE for other reasons.

The ionization and excitation equilibria are another factor in non-LTE effects. Mihalas (p. 124) argues that in stellar photospheres, the ionization radiative rates dominates collisional rates, which makes them very sensitive to non-LTE effects where $J_\nu \neq B_\nu$. It should be noted that the radiative/collision ratio is proportional to the dilution factor W. This means that in the very outer layers (the stellar wind regions), ionization becomes collisionally dominated once again. The same comments apply to excitation. As a result, the statistical equilibrium equations in NLTE are somewhat more complicated than in LTE and must be solved simultaneously with the radiative transfer equation. The statistical mechanical requirements are specified by Mihalas (pp. 127–145) and readers are referred there for details. If the atmospheres are also moving, this adds a further complication, as we see later.

4.1.3 Scattering and the transfer equation

The main process that makes $J_\nu \neq B_\nu$ in a sun-like atmosphere is the scattering of photons. In hot stars the primary process is the Thompson scattering of electrons, which is gray scattering. In cool stars it is Rayleigh scattering of molecules, which has a λ^{-4} wavelength dependence. A detailed examination of photon scattering can be found in Mihalas (1978), while Chandrasekhar (1960) gives a comprehensive discussion of Rayleigh scattering.

A useful approach is to begin with the scattering transfer equation in a simple traditional form and then describe a possible iterative method of solution to transit from LTE to non-LTE. In general there are two issues that must be addressed. First, direct integration starting with the boundary condition $J_\nu = B_\nu$ at large optical depths does not produce the expected result. This means that there are really two boundary conditions required, something that cannot be specified with the integral solutions of the transfer equation. Thus with scattering, the transfer equation must be a differential one. Second, scattering can be coherent (no frequency change) or incoherent (having a frequency change). If the scattering particle is moving, even coherent scattering becomes incoherent due to the Doppler effect. In astrophysical terms it is usually the amount of frequency redistribution in scattering that is of concern.

As we have seen in Chapter 2, the source function for pure absorption is that of a black body, $S_\nu = B_\nu$, while for pure isotropic scattering the source function is the mean intensity, $S_\nu = J_\nu$. Thus we can define a mixed state

$$S_\nu = \lambda_\nu B_\nu + (1 - \lambda_\nu) J_\nu, \tag{4.2}$$

where

$$\lambda_\nu = \frac{\kappa_\nu}{\kappa_\nu + \sigma_\nu}, \tag{4.3}$$

and κ_ν is the absorption coefficient and σ_ν is the scattering coefficent. The ratio λ_ν represents the probability that a photon is destroyed (degraded into thermal energy) per scattering.

The transport equation for the mean intensity J_ν in differential form can then be written as

$$\frac{\partial H_\nu}{\partial \tau_\nu} = J_\nu - S_\nu = \lambda_\nu (J_\nu - B_\nu), \tag{4.4}$$

where $H_\nu = F_\nu / 4\pi$ is the Eddington form of the radiative flux. For the flux, the differential form is

$$\frac{\partial K_\nu}{\partial \tau_\nu} = H_\nu. \tag{4.5}$$

As before, we assume the Eddington approximation; thus the K integral $K_\nu = J_\nu / 3$ and

$$\frac{1}{3} \frac{\partial^2 J_\nu}{\partial \tau_\nu^2} = \lambda_\nu (J_\nu - B_\nu). \tag{4.6}$$

Following Mihalas, we consider the simple case that varies linearly with optical depth:

$$B_\nu (\tau_\nu) = a_\nu + b_\nu \tau_\nu. \tag{4.7}$$

In this case the second derivative of B_ν vanishes, so we may define a new variable $\Delta JB = J_\nu - B_\nu$ such that the equation becomes

$$\frac{1}{3} \frac{\partial^2 \Delta JB}{\partial \tau_\nu^2} = \lambda_\nu \Delta JB. \tag{4.8}$$

The solution is easily found by hand or through *Mathematica*[3], and takes the form

$$\Delta JB = c_1 e^{-\sqrt{3\lambda_\nu}\tau_\nu} + c_2 e^{\sqrt{3\lambda_\nu}\tau_\nu}. \tag{4.9}$$

At large optical depths $J_\nu = B_\nu$; thus $c_2 = 0$. For the surface boundary condition we assume the gray-body result:

$$J(0) = \frac{3}{4\pi}F(0)q(0) = 3H(0)q(0), \tag{4.10}$$

where $q(\tau)$ is the Hopf function, with $q(0) = 1/\sqrt{3}$. Substituting the equations for H_ν and K_ν, the surface boundary condition becomes

$$J_\nu(0) = \frac{1}{\sqrt{3}}\left[\frac{dJ_\nu}{d\tau_\nu}\right]_0. \tag{4.11}$$

Taking the derivative in the limit $\tau \to 0$, we find

$$c_1 = \frac{-3a + \sqrt{3}b}{3\left(1 + \sqrt{\lambda_\nu}\right)}, \tag{4.12}$$

The mean intensity is then

$$J_\nu = a + b\tau_\nu + \frac{-3a + \sqrt{3}b}{3\left(1 + \sqrt{\lambda_\nu}\right)}e^{-\sqrt{3\lambda_\nu}\tau_\nu}. \tag{4.13}$$

The constants a and b depend on the type of model being considered. Here we will assume a nearly gray atmosphere with electron scattering, such that

$$b \sim \frac{3}{4\pi}\left(\sigma_R T_{eff}^4\right), \tag{4.14}$$

and

$$a \sim b\overline{q(\tau)} \approx 0.7b, \tag{4.15}$$

where σ_R is the Stefan–Boltzmann radiation constant. A purely absorbing atmosphere has $\lambda = 1$ while a purely scattering atmosphere has $\lambda = 0$. These two extremes are seen in figure (4.1) as a log plot compared to the LTE model where $J = B$. It is clear that in the absorption extreme, LTE holds well except for very near the surface ($\tau \lesssim 3$). Scattering on the other hand deviates from LTE over a wide depth ($\tau > 30$).

Even without mass motions in the atmosphere, the failure of LTE at shallow optical depths can be considerable, particularly when the temperature is high and the pressure is low. It that case electron production is enhanced and scattering is comparable to absorption. Because resonance lines have high particle absorption coefficients, their cores originate at smaller optical depths (at greater real heights) than comparable weak lines. Thus non-LTE effects must often be considered for observed line profiles, which can make analysis and modeling difficult.

---

[3] See **4-1MovingAtm** for details.

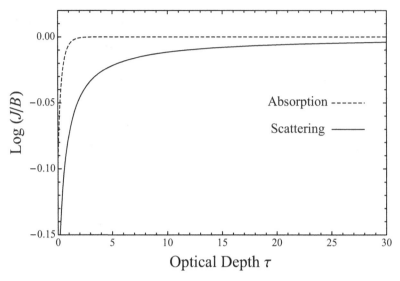

Fig. 4.1 Pure absorption vs. pure scattering.

4.1.4 Spherical gray-body atmospheres

It was mentioned earlier that systematic mass motions within the atmosphere can produce non-LTE situations due to the Doppler effect. However, there are situations in which static but extended atmospheres can, under non-LTE conditions, produce line profiles that would be interpreted as Doppler effects under LTE. To distinguish the two we must consider both cases, beginning with the extended atmosphere. In our modeling so far we have considered only plane-parallel atmospheres. This constraint is too limiting in the case of an extended atmosphere. We will, however, continue to assume our atmosphere is gray. In this way we may keep most of our simplifying assumptions, but relax the plane-parallel constraint.

In the plane parallel approximation the radiative transfer equation takes the simple form

$$\frac{\partial H_\nu}{\partial \tau_\nu} = J_\nu - S_\nu, \tag{4.16}$$

but in a spherical, static (time-independent) atmosphere a more general form of the equation must be used:

$$\frac{1}{r^2}\frac{\partial r^2 H_\nu}{\partial r} = \eta_\nu - \chi_\nu J_\nu, \tag{4.17}$$

where (following Mihalas) η_ν is the emission coefficient while χ_ν is the extinction coefficient. For the second moment we also have

$$\frac{\partial K_\nu}{\partial r} + \frac{1}{r}\left(3K_\nu - J_\nu\right) = -\chi_\nu H_\nu. \tag{4.18}$$

This is often expressed in terms of the Eddington factor f_ν. In the gray-body approximation, Eddington imposed a relationship between K and J such that $K = J/3$. In modern models this is considered too severe; thus one defines $f = K/J$ as a variable with the classical value of $1/3$, and equation (4.18) becomes

$$\frac{\partial f_\nu J_\nu}{\partial r} + \frac{1}{r}(3f_\nu - 1)J_\nu = -\chi_\nu H_\nu. \tag{4.19}$$

The gray-body form of these equations were obtained by Chandrasekhar (1934) and Kosirev (1934) by imposing radiative equilibrium in the form of

$$\int_0^\infty \eta_\nu \, d\nu = \int_0^\infty \chi_\nu J_\nu \, d\nu = \chi J, \tag{4.20}$$

which yields

$$\frac{1}{r^2}\frac{\partial r^2 H}{\partial r} = 0, \tag{4.21}$$

$$\frac{\partial}{\partial r}(fJ) + \frac{1}{r}(3f - 1)J = -\chi H. \tag{4.22}$$

Integration of equation (4.21) yields

$$H = \frac{H_0}{r^2} = \frac{L\pi^2}{16}, \tag{4.23}$$

where L is the standard luminosity of the star and H_0 is the Eddington flux at $r = 1$. This can be related to the optical depth by the integral

$$\tau(r) = \int_r^R \chi(r) \, dr, \tag{4.24}$$

where R is some outer radius and χ is the extinction coefficient.

For equation (4.22) two boundary conditions are needed. At large optical depths the radiation field is isotropic and the Eddington factor is $f = 1/3$. The equation thus becomes

$$\frac{\partial J}{\partial r} = -3\chi H = -3\chi \frac{H_0}{r^2}, \tag{4.25}$$

with the Eddington-type boundary condition at $\tau = 0$ as

$$J(0) = 2H(0) = \frac{2H_0}{R^2}. \tag{4.26}$$

Near the outer radius $r \sim R$ and $f \to 1$, and equation (4.22) becomes

$$\frac{\partial(J/r^2)}{\partial r} = -\chi r^2 H = -\chi H_0, \tag{4.27}$$

with the same boundary conditions at $\tau = 0$.

We must then adopt a radial dependence for the opacity $\chi(r)$. Following Kosirev (1934), we presume a power law dependence,

$$\chi(r) = \frac{c(n)}{r^n}. \tag{4.28}$$

Kosirev justifies this step on dynamical grounds. The extended atmosphere does not need to be in hydrostatic equilibrium. Instead it could be in hydrodynamical equilibrium, which would not be detected because the lack of frequency dependence means the Doppler effect would not be observed. If the hydrodynamic situation is assumed, then the law of continuity for radial motion requires that $\rho v r^2$ be constant. This means that the density ρ is proportional to $1/r^2$. If we assume the opacity is proportional to density, it should be radially dependent in a similar way. Thus a power-law opacity can allow for both static and moving atmospheres in the gray-body model. We can now derive a function of temperature vs. optical depth for an expanding atmosphere. Although this can be done analytically, it will be useful to derive the result in *Mathematica*.[4] Because our equations are expressed in terms of r, we must first find $\tau(r)$. Substituting the power function for χ in equation (4.24) the integral becomes

$$\tau\mathbf{1} = \int_{\mathbf{r}}^{\mathbf{bigr}} \mathbf{cn\ rp^{-n}\ drp}$$

where $\mathtt{cn} = c(n)$ and $\mathtt{bigr} = R$. This integration is quite slow due to the conditions required for a general solution. We can ignore the conditions and select for the answer by setting $\mathtt{\tau2 = \tau1[[1]]}$, which yields

$$\tau(r) = c(n)\frac{R^{1-n} - r^{1-n}}{1 - n}.$$

For our purposes we can assume $r << R$ and set $R \to \infty$; thus

$$\tau(r) = \frac{c(n)}{1 - n}r^{1-n}. \tag{4.29}$$

The mean intensity $J(r)$ can be obtained by noting $d\tau = -c(n)r^{-n}dr$; thus

$$\mathbf{jay1 = fluxH0}\left(3\int_{\infty}^{\mathbf{r}} - \left(1 / r\right)^2\ \mathbf{cn\ r^{-n}\ dr}\right)$$

where $\mathtt{fluxH0}$ is H_0. This is again a conditional expression, thus imposing $\mathtt{jay3 = jay1[[1]]}$ we find

$$J(r) = \frac{3c(n)L\pi^2r^{-(1+n)}}{16(1+n)}. \tag{4.30}$$

In the small r limit we can express $c(n)$ as function of optical depth; thus

$$J = \mathtt{jayd1} = \frac{3L(-1+n)\pi^2\tau}{16(1+n)r^2}, \tag{4.31}$$

which represents the mean intensity J at large optical depths. Near the surface J is found by direct integration of equation (4.27), which yields

$$J = \mathtt{jayu1} = \frac{L\pi^2(1+\tau)}{16r^2}. \tag{4.32}$$

To complete the solution we need an interpolation between these two formulae. Mihalas suggests a function that is asymptotic to them, so we define a linear function

$$\mathbf{jayf = slope\ (\tau r1 + intercept);}$$

---

[4] See **4-1MovingAtm** for complete details.

where $\tau r1 = \tau$. At large optical depths τ is the dominant term, so we may set $\mathtt{jayd2} = \mathtt{slope}\tau\mathtt{r1}$. At the surface τ is negligible, so $\mathtt{jayu2} = \mathtt{slope\ intercept}$. We can then use $\mathtt{Solve[]}$ to find the slope

slopeJ = Solve[jayd1 == jayd2, slope]
slope1 = slopeJ[[1, 1, 2]]

which gives the slope as

$$\mathtt{slope1} = \frac{3L(-1 + n)\pi^2}{16(1 + n)r^2}. \tag{4.33}$$

With the slope now known, we can solve for the intercept:

interceptJ = Solve$\left[\mathtt{fluxH0}\,/\,\mathtt{r^2} = \mathtt{slope1\ intercept},\ \mathtt{intercept}\right]$
intercept1 = interceptJ[[1, 1, 2]]

which gives

$$\mathtt{intercept1} = \frac{1 + n}{3(-1 + n)}. \tag{4.34}$$

The result is

$$J = \mathtt{jayf}\tau\mathtt{r} = \frac{3L(-1 + n)\pi^2}{16(1 + n)r^2}\left(\frac{1 + n}{3(-1 + n)} + \tau\right), \tag{4.35}$$

which is the expression given by Mihalas (p. 246). This equation still has r^2 in it, however, so we must convert it to optical depth via equation (4.29) to become

$$\frac{J(\tau)}{J(1)} = \mathtt{ratioJ} = \frac{1 + n + 3(-1 + n)\tau}{-2 + 4n}\tau^{2/(n-1)}. \tag{4.36}$$

If we impose the LTE condition that $J(\tau) = B(\tau) \sim T^4(\tau)$, we can solve for temperature T/T_{eff} as a function of optical depth for various n values by

temp3 $=$ Solve[ratioJ $==$ ratioT^4, ratioT]

which has only one positive root.

In the plane parallel derivation the temperature profile took the form

$$\frac{T}{T_{eff}} = \left[\frac{3}{4}\left(\tau + \frac{2}{3}\right)\right]^{1/4}. \tag{4.37}$$

This can be compared to the expanded atmosphere results as seen in figure (4.2).

It is clear that when $\tau > 1$ the temperature increases faster with depth the lower the n value. At depths $\tau < 1$ the plane-parallel case remains warmer than the extended atmospheres. Because resonance lines form mainly when $\tau < 1$, they form in a cooler gas than for the plane-parallel case. On the other hand, the continuum and weaker lines form in $\tau > 1$, and thus form in a hotter gas. This means that spectral lines of stars with extended atmospheres will be shifted relative to continuum colors/temperatures. This is a well known phenomenon seen in many giant and supergiant stars.

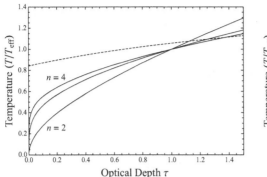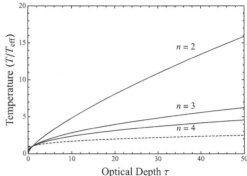

Fig. 4.2   Temperatures of an expanded atmosphere.

4.2 Expanding atmospheres and the Sobolev method

4.2.1 Line formation in an expanding atmosphere

Observationally, the existence of very large scale and rapid atmospheric expansion motions is well established. Often the spectrum lines are mixtures of emission and absorption features that defy unique interpretation. Although we focus on expansion because of its observational prevalence, contraction also occurs in star formation where interstellar and circumstellar dust tend to obscure view at optical wavelengths.

Doppler effects tend to complicate the radiative transfer, particularly when other non-LTE effects are present. Regardless of how computations are done, the problem is extremely complex. The early theories assumed that the atmospheres were optically thin, and many calculations were done with this assumption. The advance made by Sobolev (1960) and Ambartsumian (1958) was to show that a velocity gradient in an expanding medium actually simplifies the line transfer problem because it dominates the photon escape and thermalization process. The implication as stated by Mihalas (p. 474) is that there is geometric localization of the source function.

Because there is a complete breakdown of the $J = B$ relation in an expanding atmosphere we can no longer formulate an optical depth temperature relationship in the traditional way. It also means that if we are going to consider the line transfer problem in our calculations we will need to consider a matrix solution, which we have avoided thus far. We have avoided matrix methods not because *Mathematica* cannot do them (it can), but because matrix methods often require considerable abstraction of the mathematics with little enlightenment as to the underlying physics.

For our present purposes, we consider spectral line formation in an expanding spherical envelope that is very transparent. We assume that inside the envelope is a spherical stellar photosphere with a black-body continuum distribution. If the stellar disk size is vanishingly small compared with the size of the envelope, then an emission spectrum dominates. If the disk is relatively large, then an additional absorption spectrum will appear in front of the disk, where the envelope occults the photosphere from our vantage point. Because the envelope is expanding, the absorption lines will be shifted to the violet. Emission spectra near the absorption line region will also show a violet shift, gradually moving to a red shift at the edges of the envelope region.

Mihalas (p. 474) quantifies this model by considering only the emission coefficient integrated over the volume of the envelope, an approach that we will follow because the emission coefficient can be taken as a power law of the material density, it may be expressed as

$$\eta(r, v) = \eta_0 \left(\frac{\rho}{\rho_0}\right)^{\alpha} \phi(v - v_0). \tag{4.38}$$

Here $\phi(v - v_0)$ is the line profile without a Doppler effect. If we assume the velocity gradient is a power law,

$$V = V_0 \left(\frac{r}{r_0}\right)^{n}, \tag{4.39}$$

then with the mass continuity requirement

$$\rho V r^2 = \rho_0 V_0 r_0^2, \tag{4.40}$$

we find the emission without Doppler motion is

$$\eta(r, v) = \eta_0 \left(\frac{r}{r_0}\right)^{-(n+2)\alpha} \phi(v - v_0). \tag{4.41}$$

The initial values represent those at the photospheric surface. For simplicity we can assume our quantities are relative to the surface and set $V_0 = 1$ and $r_0 = 1$.

We can now consider the Doppler effect on the emission profile. Given a velocity $V(r)$, the velocity along the line of site is $V(r)\cos\theta$ and the line profile becomes

$$\phi\left[v - v_0\left(1 + \cos\theta\frac{V(r)}{c}\right)\right].$$

Setting $x = (v - v_0)/\Delta vd$, where $\Delta vd = v_0 V_0/c$, then the energy received at each value x is found by the integral over the volume of the envelope minus that occulted by the star. Thus

$$E(x) = \eta_0 \int_{vol} \phi(x - \cos\theta \, V(r)) \, r^{-(n+2)\alpha} dr 2\pi \sin\theta \, r^2 \, d\theta. \tag{4.42}$$

If the thermal velocity is small compared to the flow velocity, then the zero-order ϕ function will be a delta function such that

$$\phi(x - \cos\theta \, V(r)) = \delta(x - \cos\theta \, V(r)), \tag{4.43}$$

so the contours of $x = \cos\theta V(r)$ are the approximate contours of the emission lines in the optically thin case.

The radial velocity $V(r)$ can take on several forms. We consider three[5] cases here:

1. $V(r) = $ constant, in which our thin shell of atmosphere has reached escape velocity and is coasting.
2. $V(r) = V_0/\sqrt{r}$, where the shell has an initial high velocity V_0 at $r = 1$ and decelerates as it expands.
3. $V(r) = V_\infty\left(1 - \Psi e^{(1-r)}\right)$. This is an exponential version of the second case where there is a higher concentration of material near the star itself. Here $\Psi < 1$ is a small factor such that $V_0 = (1 - \Psi)V_\infty$.

To evaluate equation (4.42) for our models, we must first estimate the emissivity of the gas. We can do this by invoking Kirchoff's law (opacity = emissivity) and then finding the density dependence of the representative Rosseland mean oapacity. Cox (p. 111) provides the Rosseland mean opacity at

----

[5] For more examples, see **4-1MovingAtm**.

various densities in $\log_{10}$ format. Taking the data for 30 000 K and applying `LinearModelFit[]` one finds

$$\log \eta = \log \eta_0 + \alpha x = 9.22524 + 1.01286\, x, \tag{4.44}$$

where $x = \log(\rho/\rho_0)$. Thus we may set $\alpha = 1$ as a reasonable value. Because we are interested only in the shape of the emission lines, we may also choose an effective optical depth such that $\eta_0 = 1$. We also assume for simplicity that the non-Doppler line shape follows a Gaussian profile:

$$\phi(x) = e^{-x^2}. \tag{4.45}$$

Case 1: Constant velocity

If we assume the velocity is constant then the integrals are separable, and our integral equation becomes

$$E(x) = 2\pi \eta_0 \int_\theta e^{-(x - V_0 \cos\theta)^2} \sin\theta \; d\theta \int_r r^{-(n+2)\alpha+2} \; dr. \tag{4.46}$$

We can set η_0, α, and n to unity, which simplifies the integral further, however this does not have an analytical solution. To evaluate the integral in *Mathematica* we first introduce a normalization factor

```
normalize = NIntegrate[
    e^-x² 2 π NIntegrate[NIntegrate[ Sin[θ], {θ, 0, π}] , {r, 1, 100}], {x, -10, 10}] / 20;
```

Because we are using `NIntegrate[]`, we must choose a finite range for our variables r and x. The energy integral then becomes

```
energy[x_, n_, α_, vr1_] :=
    2 π NIntegrate[NIntegrate[ (e^-(x-Cos[θ] vr1)²)/normalize  Sin[θ], {θ, 0, π}] r^-(n+2) α+2, {r, 1, 100}]
```

Here $V_0 = $ vr1. One advantage of *Mathematica* here is that we can evaluate the integral numerically while keeping our variables (n, α, V_0) undefined. We do not have to substitute specific values to perform the evaluation. If we define our variables the result can be plotted, as seen in figure (4.3), where $n = \alpha = 1$. The result produces the well-known P-Cygni profile. The flat top of the profile is characteristic of the optically thin case.

While this calculation is straightforward, it assumes the central star subtends a negligible angle so that there is no absorption or occultation zone. This allowed us to perform the integration over all θ.

To account for the absorption/occultation zone, we define the observed radius of the star in terms of a radian angle θ_0. This means our angle integral must be evaluated as an absorption for $\theta < \theta_0$ and an emission otherwise. This can be handled by introducing an `If` condition directly into `NIntegrate[]`. For the sake of generality, we also introduce the absorption to emission ration as a new variable χ to η.

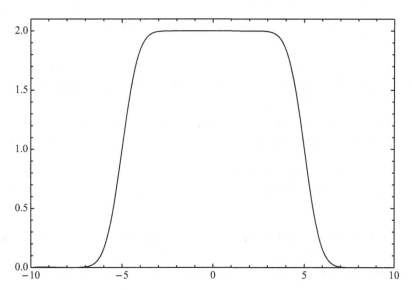

Fig. 4.3   P-Cygni profile for a constant expanding envelope.

This allows us to move beyond Kirchoff's law of $\chi/\eta = 1$. These changes require us to recalculate our normalization factor; thus

```
normalize1[θ0_, xtoη_] := NIntegrate[
    e^-x² 2 π NIntegrate[NIntegrate[If[θ < θ0, -xtoη Sin[θ], If[θ > (π - θ0), 0, xtoη Sin[θ]]],
        {θ, 0, π}], {r, 1, 100}], {x, -10, 10}] / 20
normal = normalize1[0.2, 1]
```

and the energy integral becomes

```
energy1[x_, n_, α_, vr1_, θ0_, xtoη_] := 2 π NIntegrate[NIntegrate[
    If[θ < θ0, -xtoη (e^-(x-Cos[θ] vr1)²)/normal Sin[θ], If[θ > (π - θ0), 0, xtoη (e^-(x-Cos[θ] vr1)²)/normal Sin[θ]]],
    {θ, 0, π}] r^-(n+2) α+2, {r, 1, 100}]
```

as seen in the left figure (4.4).

Case 2: Simple deceleration

In cases where $V(r)$ is not constant our integral is not separable; thus we must adopt a different strategy for computing the integral. Because $V(r) = V_0/\sqrt{r}$, the power function for the velocity gradient is $n = 0.5$.

Because the integrals are not separable we must create a profile function `rprofile3` integrated over the radial dependence.[6]

```
rprofile3[cosθ_, x_, n_, α_, vr1_] := NIntegrate[(e^(-(x-cosθ vr1/√r)^2)/normal) r^(-(n+2) α+2), {r, 1, 100}];
```

It should be noted that the $\cos\theta$ here is a simple variable and not the function call `Cos[θ]`, as it lacks both capitalization and square brackets. This can be a useful trick when computing with complex formulae.

Mathematica complains bitterly if we try to do the second numerical integration directly. We can overcome this by creating a table of the profile at equally spaced values of $\cos\theta$. Our values go from 1 to -1 in steps of 0.1.

```
profilepts = Table[{yy, rprofile3[yy, xx, nx, ax, vx1]}, {yy, 1, -1, -0.1}];
```

We then construct an interpolating polynomial of the function so that we can use `NIntegrate[]` again.

```
kernel = Interpolation[profilepts];
```

The result can then be integrated. Because $\mathrm{d}\cos\theta = -\sin\theta\,\mathrm{d}\theta = \mathrm{d}y$ we must reverse the order of the integration so that the emission terms are positive and the absorption is negative. We can then put this together in one function using the `Module` construct.[7] The final profile function `profile3` becomes

```
profile3[xx_, nx_, ax_, vx1_, θ0_, χtoη_] := Module[. . .
    kernel = Interpolation[profilepts]; NIntegrate[If[θ < θ0,
    -χtoη kernel[Cos[θ]], If[θ > (π - θ0), 0, χtoη kernel[Cos[θ]]]], {θ, 0, π}]]
```

The resulting profile can be seen in the center figure (4.4).

Case 3: Exponential deceleration

In this final case the velocity is not a power function of r. To determine a realistic value n we must fit $V(r)$ to a reasonable power function. This can be achieved by assuming

$$V(r) = V_\infty \left(1 - \Psi e^{(1-r)}\right) \sim Cr^n, \tag{4.47}$$

and calculating a linear fit to $\log V$. In this case, with $V_\infty = 5$ and $\Psi = 0.8$,

```
v∞7 = 5.; Ψ7 = 0.8;
listvel7 = Table[{Log[r], Log[v∞7 (1 - e^(1-Ψ7 r))]}, {r, 1.5, 100, 5}];
lmfv7 = LinearModelFit[listvel7, x, x]
lmfvc7 = lmfv7["BestFitParameters"]
```

---

[6] The function name here is arbitrary. The names used follow those found in the *Mathematica* notebooks.

[7] Here we give the full `Module[]` construct. Subsequent examples will be abbreviated in the text. They can be seen in their full form in the note books.

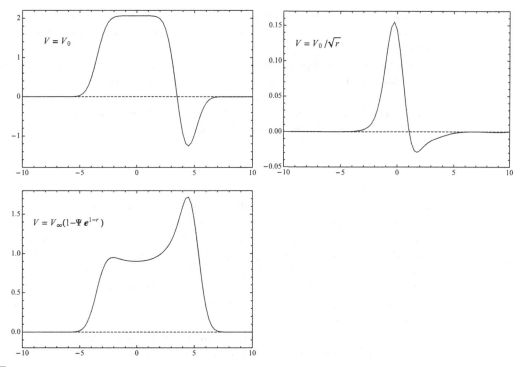

Fig. 4.4   Line shapes of expanding atmospheres.

which yields a slope (and thus the value of n) of 0.261892. The profile can then be calculated in the same manner as case 2. The only change is the reversal of the sign convention for the absorption and emission terms to get the absorption to be on the violet (positive x) side of the plot. This is necessary due to the assumed form of the Doppler formula.

```
profile7[xx_, nx_, ax_, vxl_, Ψ_, θ0_, χtoη_] := Module[...
    If[θ > (π - θ0), 0, NIntegrate[χtoη kernel[Cos[θ]]]], {θ, 0, π}]]
```

The result is again seen in the right figure (4.4).

Although these profiles qualitatively resemble what is observed, they are not quantitatively correct. The reason is that the spectrum lines most likely to be in emission in extended atmospheres are resonance lines of elements and ions. The shapes of these lines are determined by non-LTE effects.

4.2.2 The Sobolev method

Although the profiles we have obtained are qualitatively correct, they fail to produce what is observed for resonance lines that are optically thick. The main problem is that the calculations produce lines that are too strong, predicting higher emission and deeper absorption. Sobolev (1960) recognized that differential motions of moving atmosphere (if present) give a way for photons to escape the envelope without further interaction in the line profile. Because there is a breakdown of the $J = B$ relationship in an expanding atmosphere we can no longer formulate a relationship between optical depth and temperature in the traditional way, and non-LTE methods must be used. In addition to Mihalas (1978), very useful discussions of the details of the Sobolev theory can be found in Shu (1991) and Rybicki (1969).

Even if the stellar envelope is optically thick when static, the internal motions reduce the opacity so that one can "see" farther into a moving atmosphere than into a static one (although not as far as the hypothetical transparent atmosphere we assumed previously). If there is differential motion, the effect is even greater. This is the basis of the Sobolev theory. We therefore need an estimate of the change of velocity along the light of sight. Following Mihalas,

$$Q(r, \mu) = \frac{\partial V}{\partial z} = \mu^2 \frac{\partial V(r)}{\partial r} + \left(1 - \mu^2\right) \frac{V(r)}{r}, \tag{4.48}$$

where $\mu = \cos \theta$ and $V(r)$ is the velocity distribution for the cases considered previously. From this we must calculate the photon escape probability that can then be factored into our line spectra calculations.

For the Sobolev model we cannot simply assume a line shape as we have done thus far. We must specify explicitly the spectroscopic transition used to go along with the continuum models we have developed. In hot stars, the spectrum lines in the optical region are those due to hydrogen, helium, and CNO element ions. Calculation of the transition states are quite complicated even in its simplest form, and elaboration does not add to the understanding of moving atmospheres. We therefore consider only the continuum case for a pure hydrogen atmosphere. Hence we consider only the $H\alpha$ line and treat it as a simple two-level atom, which we will import as a data table.[8]

The Sobolev mechanism needs a velocity gradient within the emitting region in order to work. Because we are in a spherical coordinate system, even the constant velocity case has a gradient, and hence it will show a Sobolev effect. When an atmosphere is static, a photon created in one location is destroyed on average within the thermalization length, but with a velocity gradient the downward transition rate (emission) can exceed the upward transition rate (absorption). The difference between the two is

$$n_i A_i \beta_{ji}, \tag{4.49}$$

where n_i is the population of the upper level, A_i is the Einstein spontaneous emission probability for the transition, and β_{ji} is the photon escape probability. When this probability is nonzero, local thermodynamic equilibrium is violated regardless of the geometry.

The simplest Sobolev model is that of an infinite moving atmosphere with a velocity gradient. In the asymptotic limit the escape probability becomes

$$\beta = |\gamma| \int_0^1 \left(1 - e^{-1/|\gamma|\mu^2}\right) \mu^2 \, d\mu, \tag{4.50}$$

where γ is the velocity gradient in terms of the optical depth. In this case $\beta > \epsilon$, where $\epsilon < 1$ is the ratio of the collisional de-excitation to total de-excitation. The thermalization length is given by

$$\Lambda = \frac{1}{\gamma \sqrt{|\beta + \epsilon(1 - \beta)|}}, \tag{4.51}$$

and it is assumed the bulk velocities of the fluid are much larger than the thermal velocities. The functional form of β in the infinite medium case resembles that for the spherical atmosphere. The spherical case is set up by Rybicki (1969), Mihalas (1978), and Shu (1991), where each focuses on

---

[8] The detailed calculations can be found in **4-3Spheregraybody**.

a particular aspect of the problem. Here we look at how the Sobolev model affects our results of the transparent atmosphere.

The β expression must be calculated on a case by case basis. Given the velocity function $V(r)$, we can determine the derivative $Q(r, \mu)$. The de-excitation ratio ϵ is then

$$\epsilon = \frac{Q(r, \mu)}{\chi L(r)}, \tag{4.52}$$

where $\chi L(r)$ is the atomic line absorption coefficient. The escape probability is then

$$\beta(r) = \frac{1}{\chi L(r)} \int_0^1 \left(1 - e^{-\chi L/Q}\right) Q(r, \mu) \, d\mu. \tag{4.53}$$

One can also account for the reduction of the flux just in front of the central star through the same mechanism following Mihalas,

$$\beta_c(r) = \frac{1}{2\chi L(r)} \int_{\mu_c}^1 \left(1 - e^{-\chi L/Q}\right) Q(r, \mu) \, d\mu. \tag{4.54}$$

The probability function can then be calculated for each case.

4.2.3 Photon escape probability

Because the integration for β involves the $Q(r, \mu)$ functions for each flow model, we must compute separate β functions for each of our three cases. The integrals can only be calculated numerically. In each case we must first import $\chi L(r)$ as a *Mathematica* list of points. These were calculated as a log10 of r, so we must convert the list to be linear in r.

```
logχpoints1 = Import["log10χ1SGB.dat"];
expχHα1 = Interpolation[logχpoints1];
χHα1[r1_] := 10^expχHα1[r1];
```

We also calculate $Q(r, \mu)$ from our velocity function:

```
vcase1 = constant;
que1 = Simplify[(μ² ∂_r vcase1 + (1 - μ²) vcase1 / r)];
quec1[r1_, μ1_, constant1_] := que1 /. {r → r1, μ → μ1, constant → constant1}
```

We can then calculate the escape probability function. Some of these integrals are very time consuming. Rather than recalculating them at each step, we will generate a list which we can interpolate.

```
β1[r1_, constant1_] :=
      1
   ───────
   χHα1[r1]
NIntegrate[(1 - e^-χHα1[r1]/quec1[r1,μ1,constant1]) quec1[r1, μ1, constant1], {μ1, 0, 1}];
```

The escape probability for the other cases can be calculated in the same manner. For the second case

```
logχpoints3 = Import["log10χ3SGB.dat"];
expχHα3 = Interpolation[logχpoints3];
χHα3[r1_] := 10^expχHα3[r1];
vcase3 = v0 / √r ;
que3 = Simplify[(μ² ∂ᵣ vcase3 + (1 - μ²) vcase3 / r)];
quec3[r1_, μ1_, v2_] := que3 /. {r → r1, μ → μ1, v0 → v2}
β3[r1_, v2_] :=
      1
   ─────────
    χHα3[r1]
NIntegrate[(1 - e^(-χHα3[r1]/Abs[quec3[r1,μ1,v2]])) Abs[quec3[r1, μ1, v2]], {μ1, 0, 1}];
```

This last integral has a peculiarity not seen in the first case. The $\beta3$ function does not evaluate all the way out to $r = 100$. By trial and error we find it must be terminated before $r = 45$. The list is then

```
β3list = Table[{r0, β3[r0, 5]}, {r0, 1, 42}];
```

The third case has the same peculiarity and must be terminated before $r = 33$. Therefore

```
logχpoints7 = Import["log10χ7SGB.dat"];
expχHα7 = Interpolation[logχpoints7];
χHα7[r1_] := 10^expχHα7[r1];
vcase7 = vL7 (1 - Φ e^(-r) / e^(-1)) ;
que7 = Simplify[(μ² ∂ᵣ vcase7 + (1 - μ²) vcase7 / r)];
quec7[r1_, μ1_, v2_, Φ1_] := que7 /. {r → r1, μ → μ1, Φ → Φ1, vL7 → v2}
β7[r1_, vL_, Φ2_] :=
      1
   ───────── NIntegrate[(1 - e^(-χHα7[r1]/quec7[r1,μ1,vL,Φ2])) quec7[r1, μ1, vL, Φ2], {μ1, 0, 1}]
    χHα7[r1]
β7list = Table[{r0, β7[r0, 5, 0.8]}, {r0, 1, 32}];
```

A graph of the probability functions can be seen in figure (4.5). It is clear that the integrations fail as the probability of the function reaches 1. The point at which this occurs depends on the model used, so one must adjust the integral limit by trial and error before doing the modeling.

4.2.4 Sobolev correction to spectral line profiles

We are now ready to compute the effect that the Sobolev photon escape mechanism has on line profiles. These calculations can be very resource intensive. The time for doing even the simplest model

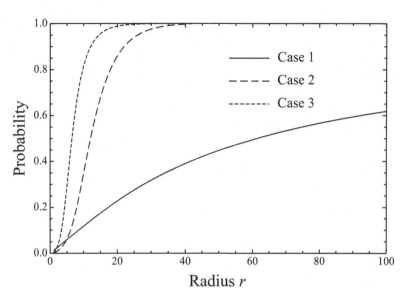

Fig. 4.5 Photon escape probabilities for expanding atmospheres.

is prohibitively long to try on all but the fastest personal computers. We will therefore streamline the procedure as much as possible by replacing direct numerical integrations by a series of discrete evaluations, then integrating the resulting interpolation functions.

If you have studied numerical analysis, you realize that so-called numerical summation procedures are all based on "fitting" the data to a polynomial, operating on the polynomial, and then calculating the results from the newly operated-upon polynomial at the abscissa. Because *Mathematica* has a heritage of symbolic processing, there are many ways to turn a list of numbers into a function. *Mathematica*®, unlike other "numerical" systems such as *Matlab*® and *IDL*®, allows one to specify the functional form. *Mathematica* is also able to take advantage of very sophisticated solution algorithms that include nonequally spaced abscissa methods to ensure a high numerical accuracy. With care we can significantly reduce computation time with a limited loss of numerical accuracy.

The basic approach here is that the local emission intensity as calculated in a particular gray-body temperature profile must be multiplied by a factor $(1 - \beta)$ to compensate for the losses caused by the Sobolev photon escapes. This quantity must be reintegrated along the line of sight, as done in **4-1MovingAtm**.

For each case we will use the Gaussian profiles as before. These profiles are not only easier to calculate, but they will also allow us to compare the Sobolev profiles to our transparent cases. For the first case we first import the calculated emission strength and escape probabilities as interpolated functions:

```
numDEN1 = Import["numDEN1SGB.dat"];
case1R = Interpolation[numDEN1]
β1function = Interpolation[β1list]
βC1function = Interpolation[βC1list]
```

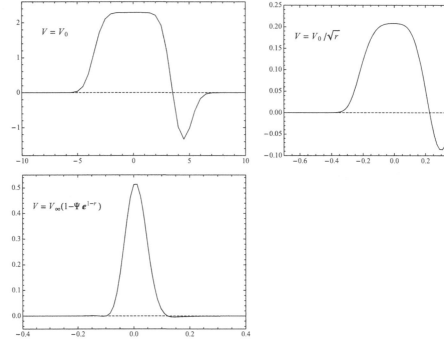

Fig. 4.6 Sobolev line profiles of expanding atmospheres.

We can then calculate the profile integration. The signs have been reversed to produce a profile that is not inverted with respect to the case of no photon escape:

```
profile1β[x_, vr1_, θ0_] := Module[ . . .
    If[θ > (π - θ0), 0, 2 π NIntegrate [termB e^-(x-Cos[θ] vr1)² Sin[θ]]], {θ, 0, π}] ]
points β1 = Table[{x0, profile1β[x0, 5, 0.8]}, {x0, -10, 10, 0.5}];
```

The computation time for this integral is fairly long. *Mathematica* complains about the numerical integration of $\beta 1$ and $\beta C1$, and because a numerical integration must be done on each subroutine call, this contributes to the slowdown.

The resulting profile, seen in figure (4.6), is nearly identical to the transparent case. Because the Sobolev mechanism depends on a velocity gradient, not the velocity, this null result is not unexpected and in fact was predicted by Mihalas.

The second case can be calculated in the same manner, but because our escape probabilities produced positive values only out to $r = 75$ we must take care to have the integrations over only a proper range of the β functions. Another complexity is that the velocity is a function of the radius, so integrations must be done over the angle variables first at a given r and x. From this we produce a list of points from which the interpolation can be done.

```
numDEN3 = Import["numDEN3SGB.dat"];
case3R = Interpolation[numDEN3];
β3function = Interpolation[β31list]
βC3function = Interpolation[βC31list]
```

To account for our limits, we first create the kernel that is the function of r and x.

```
kernel3β[xx_, r2_, vr1_, θ0_] := 2 π NIntegrate[If[θ < θ0,
```

$$- (1 - \beta 3\texttt{function[r2]}) \texttt{ case3R[r2] } e^{-\left(\texttt{xx-Cos[}\theta\texttt{] vr1}/\sqrt{\texttt{r2}}\right)^2} \texttt{Sin[}\theta\texttt{], If}\left[\theta > (\pi - \theta 0),\right.$$

$$0, (1 - \beta \texttt{C3function[r2]}) \texttt{ case3R[r2] } e^{-\left(\texttt{xx-Cos[}\theta\texttt{] vr1}/\sqrt{\texttt{r2}}\right)^2} \texttt{Sin[}\theta\texttt{]]}, \{\theta, 0, \pi\}\right]$$

```
kernel3βlist = Table[{{x2, r0}, kernel3β[x2, r0, 5, 0.8]}, {r0, 1, 40}, {x2, -10, 10, 0.25}];
```

The interpolation list must be generated at a fairly coarse grid so it does not take too much time.

Unfortunately the dynamic range of the data is too large to interpolate in one function with the original variables. *Mathematica* refuses to budge either with `Interpolation[]` or with `ListPlot3D[]`. A log transform of the data would seem worth trying, but some of the data points are negative, so that trick will not work. The first step out of this dilemma is to flatten the form required for multidimensional interpolation so that we can work on both the arguments and the data separately.

```
kernel3βdata = Table[Flatten[kernel3βlist[[ii, jj]]], {ii, 1, 40}, {jj, 1, 81}];
```

Examination of the data shows that the interpolation problem was caused mainly by the function variation in the x-direction rather than in the r direction. Thus as a test we will select a vector of data at constant x, interpolate it, and then integrate it. A reasonable value is $x = 40$, which is near the center of the profile. We can then construct the profile as a single array of points, and normalize the profile.

```
profilex3β[xmin1_, Δx1_, Δr1_, xnmax1_, rnmax1_] := Module[...
      lineprofileβ3[[jj, 2]] = NIntegrate[test[r], {r, 1, rnmax1}], {jj, 1, xnmax1}]]
β3list = Interpolation[lineprofileβ3];
normal = NIntegrate[β3list[x], {x, -10, 10}];
pointsβ3 = lineprofileβ3 / normal; Export["Sobolevβ3.dat", pointsβ3];
```

Comparing the Sobolev profile, figure (4.6), to the Doppler case, figure (4.4), the effect on the profile is profound. The least probability of photon escape is very near the stellar core, and near the core the atmospheric velocity starts high and decreases outward. All line contributions from close to the core are thus near the line center. The P Cygni shape is still present but much contracted in frequency. The apparent half-intensity width of the emission is about 0.1 Doppler width. Thus the Sobolev mechanism narrows the line profile by nearly a factor of 10 more. To make the Sobolev profile resemble the non-Sobolev one would seem to require a vL of 50. That means with a thermal velocity of 20 km/s at vL, of 1000 km/s at the surface of the star would be necessary.

The last case can be calculated following the second case, using a kernel and flattening the interpolation. The result is seen in figure (4.6). The velocity at the core surface for this model with $\Psi = 0.8$ is about $0.2 \times 5 = 1$. As in the second case the Sobolev mechanism narrows the line profile by a factor of 10 more.

From these cases it is clear that, although assuming a constant velocity simplifies the mathematics, it is not a very good gauge of reality, especially in a spherical atmosphere. Any interpretation must be in the context of a Gaussian profile which was chosen for computational simplicity, not reality. In hot stars, Hα being a "resonance" line requires a Voigt profile, but that complicates the modeling and slows down the computations.

In spite of the simplifications, we see that the emission lobes where the material is flowing at right angles to line of sight contribute most to the profiles. The absorption profile is relatively weak and less Doppler shifted in the third case, indicating that the Hα absorbing material in front of the star is fairly near the star core where the Doppler velocity is approximately 20% of the final speed. Because the velocity is also higher there in the absorbing region in front of the star, the third case absorption minimum is deeper and in a much more violet shifted position.

The second case is perhaps the most interesting of all three. The absolute value of the deceleration of this case means that the Sobolev corrections are smaller than for the other two cases and so the profile can be seen to greater line shifts. It also keeps material relatively close to the star core. These results tend to support the opinion of Mihalas that one must be relatively cautious at interpreting Doppler velocities measured from profiles observed in expanding atmospheres.

4.3 Properties of gas degeneracy

4.3.1 Fermions and bosons

One of the early results of quantum mechanics was that microscopic spins of atomic or subatomic particles could be responsible for certain macroscopic properties of particles in bulk. It was noted that there are two types of spinning particles in our universe, bosons and fermions. Bosons are particles that obey Bose–Einstein statistics and have whole integer or zero spins. Fermions are particles that have odd numbers of 1/2 integer spins and obey Fermi-Dirac statistics.[9] At very low temperatures, bosons behave as low temperature extensions of the ideal gas particles, while fermions display a radically different behavior.

A central property of fermions is that only two particles can occupy a given quantum energy state (cell) at a time, a behavior known as the Pauli Exclusion Principle. For bosons there is no such limit, even at absolute zero. To build models of stars composed of degenerate fermions requires a knowledge of the equation of state, just as models of nondegenerate stars usually assume some version of the ideal gas law. To rigorously set up the Fermion equation of state one usually starts with the partition function approach (Kittel and Kroemer, 1980; Reif, 1965); and the Appendix notebook **13FermionsBosons**). For this approach the quantum volume is used as the criterion for degeneracy. Harwit (1988) and Böhm-Vitense (1992) follow the same process more heuristically by comparing the maximum density required by the Pauli exclusion principle and that allowed by the Maxwellian distribution of the ideal gas law. Because that approach is more in line with our computational approach, we follow a similar line of reasoning here.

The ideal gas law is considered a low density equation of state. The sun's central density is some 150 g/cm^3, more than $10\times$ the density of solid osmium (the densest natural element), and the sun's core is still treated as an ideal gas. It might seem odd to consider such a model a "low density" theory, but the calculations bear this out.

---

[9] See the appendix notebook **13FermionsBosons** for more details.

4.3.2 Phase space

Fundamental to the density question is the definition of a six-dimensional construct known as phase space. In the simplest terms, phase space has three spatial coordinates and three momentum coordinates. The volume element in Euclidean space becomes

$$d^6 = \Delta x \, \Delta y \, \Delta z \, \Delta p_x \, \Delta p_y \, \Delta p_z. \tag{4.55}$$

In quantum theory $d^6 = h^3$, where h is the Planck constant. Because of the exclusion principle the quantum volume can have at most two fermions of the same kind, whose spins are directed antiparallel to each other. Assuming the momenta are isotropic, we may add together all momenta of the same $p \pm \Delta p$ to form a spherical shell of volume $4\pi p^2 \Delta p$ in momentum space.

The volume occupies $4\pi p^2 \Delta p / h^3$ quantum cells, and each cell can contain at most two fermions. Thus the fermion density must satisfy

$$n(p, \Delta p) \leq 8\pi \frac{p^2 \Delta p}{h^3}, \tag{4.56}$$

or for nonrelativistic electrons

$$n_e(v, \Delta v) \leq 8\pi \frac{m_e^3 v^2 \Delta v}{h^3}, \tag{4.57}$$

where m_e is the electron mass.

The equivalent expression for an (ideal gas) Maxwell–Boltzmann particle is

$$n_e(v, \Delta v) = n_e \frac{e^{-m_e v^2/2kT}}{\int_0^\infty e^{-m_e v^2/2kT} v^2 \, dv} v^2 \Delta v, \tag{4.58}$$

where the normalization integral in the denominator is

$$\mathtt{nfactor[[1]]} = \int_0^\infty e^{-m_e v^2/2kT} v^2 \, dv = \sqrt{\frac{\pi}{2}} \left(\frac{kT}{m_e}\right)^{3/2}. \tag{4.59}$$

4.3.3 Nonrelativistic degeneracy

The condition for electron degeneracy can be found by taking the ratio n_e/n_{max}. For the nondegenerate region this ratio is less than 1, while the degenerate region lies where the ratio is 1. In *Mathematica* it is more useful to take the log of this ratio. Thus

```
neMB[v_, ne_, temp1_] := Evaluate[ne (e^(- me v^2 / (2 kB temp1)) v^2) / nfactor[[1]]]

nemax[v_] := (8 π me^3 v^2) / h1^3

lg10ratioel[lg10v1_, lg10ne1_, lg10temp2_] :=
  Module[{}, lg = Log[10, neMB[10^lg10v1, 10^lg10ne1, 10^lg10temp2] / nemax[10^lg10v1]];
    lg1 = If[lg > 0, 0, lg]; lg1]
```

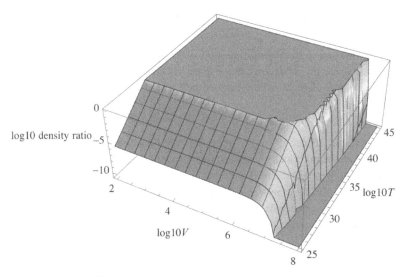

Fig. 4.7   Electron degeneracy at a density of $10^{30}/m^3$.

This function can then be plotted as a log function of velocity lgV, temperature lgT, or density lgne as needed.[10] For example, figure (4.7) shows a plot of velocity and temperature at an electron density of $10^{30}/m^3$. The degenerate region (where $n_e/n_{max} = 0$) is indicated by the light gray region at the top left.

Although the above degeneracy calculation was done for an electron gas, the same can be done for fermionic nuclei. In *Mathematica* this involves simply replacing the electron mass me by the appropriate nuclear mass mN. Because a realistic ionized gas consists of electrons and nuclei, it is useful to compare the degeneracy of the two.

The observable white dwarfs in our galaxy are apparently all by-products of stellar evolution. Their once hydrogen dominated interiors have been burned into helium or heavier elements. Thus to make a realistic comparison we consider a nitrogen white dwarf. We have chosen this because nitrogen-14 has an uneven number of protons and its ion is always a fermion. In figure (4.8) the functions of electrons and nitrogen are compared at a constant temperature of 10^6 K. The smaller degenerate region is that of nitrogen, compared to the larger electron region. The density necessary for nitrogen degeneracy is some five orders of magnitude higher than that of electrons.

It is clear that the degeneracy of heavier nuclei can be ignored in the conditions typically found in most evolved stellar cores and white dwarfs. When deriving the equation of state for pressure, we need to consider only the degeneracy of the electrons because the electron pressure is the dominant one.

4.3.4 Equation of state for a degenerate gas

In elementary physics books one often finds a kinetic theory derivation of the relationship of pressure to the change of momentum when a particle collides elastically with a container wall. We use a similar derivation here, but because we are dealing with fermions there will be a disconnect between the kinetic

---

[10] It is useful to view the resulting 3D graphs from various angles. This can be done within *Mathematica* and can be seen in **4-5Degeneracy**. Only a few basic graphs are presented here.

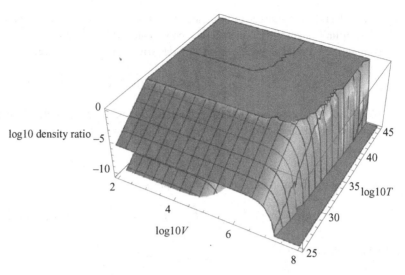

Fig. 4.8 Degeneracy of electrons and nitrogen nuclei compared.

energy of the particles and the temperature of the gas. This disconnect is most pronounced at $T = 0$ K. Because of the Pauli exclusion principle, N particles stack in the energy levels pairwise, so the specific heat is not 0. This heat does not provide motion to the electrons. Adding up the electrons by integration and dividing by the volume gives

$$n_e = \int_0^{p_0} \frac{8\pi p^2}{h^3}\,\mathrm{d}p = \frac{8\pi p_0^3}{3h^3}, \tag{4.60}$$

where p_0 is the momentum corresponding to the Fermi energy.

Because we assume only the electrons in the gas mixture are degenerate, we must obtain an estimate of the electron density given the material density of the nondegenerate ions. If we assume these ions consist of hydrogen, helium, and heavier elements or "metals," then

$$X + Y + Z = 1, \tag{4.61}$$

where X, Y, and Z are the fractions by mass of hydrogen, helium, and the heavier elements respectively. Because each H atom has one electron, the density of H electrons is

$$n_e\,(\mathrm{H}) = \frac{X\rho}{m_p}, \tag{4.62}$$

where ρ is the total density and m_p is the proton mass. For He there are two electrons, and because helium typically consists of four nucleons,

$$n_e\,(\mathrm{He}) = \frac{Y2\rho}{4m_p} = \frac{Y\rho}{2m_p}. \tag{4.63}$$

For the heavier elements we will assume that an element with a mean atomic mass A will contribute $A/2$ electrons; thus the total electron density becomes

$$n_e = \left(X + \frac{Y}{2} + \frac{Z}{2}\right)\frac{\rho}{m_p} = \frac{1}{2}\,(1 + X)\,\frac{\rho}{m_p}. \tag{4.64}$$

The electron pressure can be calculated by determining the number of electrons striking a unit surface area per unit time (N/AT). In spherical coordinates, if there are $n\,(\theta, \phi, v)$ electrons per unit volume coming from a solid angle $d\Omega = \sin\theta\,d\theta\,d\phi$, with velocity v and direction (θ, ϕ), then

$$\frac{N}{AT} = \int v \cos\theta\, n\,(\theta, \phi, v) \sin\theta\, d\theta\, d\phi\, dv. \tag{4.65}$$

If we assume the electrons are scattered elastically, such that the angle of incidence equals the angle of scattered reflection, then the momentum exerted to the surface by a scattered electron is

$$\Delta p = 2p \cos\theta. \tag{4.66}$$

Assuming the electron gas is isotropic, the pressure is then

$$P_e = \int_0^{p_0} \int_0^{2\pi} \int_0^{\pi/2} 2n_e(p) \cos\theta\, v \cos\theta \sin\theta\, d\theta\, d\theta\, d\phi\, dp, \tag{4.67}$$

which simplifies to

$$P_e = \frac{1}{3} \int_0^{p_0} \frac{8\pi p^2}{h^3} pv\, dp. \tag{4.68}$$

In the nonrelativistic case $v = p/m$, and the electron pressure becomes

$$P_e = \frac{8\pi}{15} \frac{p_0^5}{m_e h^3}. \tag{4.69}$$

Substituting equations (4.60) and (4.64),

$$P_e = \frac{h^2}{20 m_e m_p} \left(\frac{3}{\pi m_p}\right)^{2/3} \left(\frac{1+X}{2}\rho\right)^{5/3}. \tag{4.70}$$

Under extremely high pressure for a white dwarf or neutron star the electron velocities are very high, and as such are relativistic. In that case the velocities approach the speed of light c, and the pressure becomes

$$P_e = \frac{2\pi c}{3h^3} p_0^4, \tag{4.71}$$

or with equations (4.60) and (4.64),

$$P_e = \frac{hc}{8m_p} \left(\frac{3}{\pi m_p}\right)^{1/3} \left(\frac{1+X}{2}\rho\right)^{4/3}. \tag{4.72}$$

The relativistic and nonrelativistic have different powers of ρ, intersecting at a density of approximately 10^{10} kg/m^3, as seen in figure 4.9. Near such densities the general form must be used. The fully relativistic form of momentum is $p = mv\gamma\,(v)$; thus the general velocity relation becomes

$$v = \frac{p}{m\sqrt{1 + p^2/m^2 c^2}}. \tag{4.73}$$

The pressure integral thus becomes

$$P_e = \frac{8\pi}{3m_e h^3} \int_0^{p_0} \frac{p}{m\sqrt{1 + p^2/m^2 c^2}}\, dp. \tag{4.74}$$

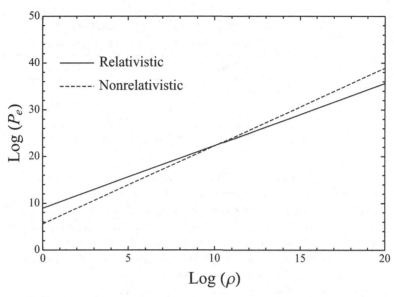

Relativistic and nonrelativistic electron pressure compared.

Substituting $p/m_e c = \sinh u$, the integral becomes

$$P_e = \frac{8\pi m_e^4 c^5}{3h^3} \int_0^{u_0} \sinh^4 u \, du. \tag{4.75}$$

In the transition region this can be written as

$$P_e = \frac{8\pi m_e^4 c^5}{3h^3} f(x), \tag{4.76}$$

where

$$f(x) = \frac{1}{8}\left[x\left(2x^2 - 3\right)\left(x^2 + 1\right)^{1/2} + 3\sinh^{-1} x\right], \tag{4.77}$$

and $x = p_0/m_e c$.

4.4 White dwarfs

4.4.1 Physical properties of white dwarfs

In astrophysics, the main place where fermions become degenerate is in the cores of highly evolved stars that have been subject to lots of thermonuclear processing and then undergo core collapse when they run out of nuclear fuel, commonly known as white dwarfs. Stable degenerate hydrogen cores do not arise because the main sequence stars with sufficient mass to generate hydrogen-produced degenerate electrons have rapid nuclear fusion. As a result they either explode upon collapse (supernovae) or are already too hot to be stable under radiation pressure.

Table 4.1 Properties of Typical White Dwarf Stars

Temperature	$20\,000 \rightarrow 5000$ K
Luminosity	$0.1L_\odot \rightarrow 10^{-4}L_\odot$
Radius	4000 km $\rightarrow 2000$ km
Density	2.4×10^7 g/cm$^3 \rightarrow 1.9 \times 10^6$ g/cm^3
Pressure	4×10^7 N/m$^2 \rightarrow 3 \times 10^6$ N/m^2

White dwarf stars are thought to arise as stars evolved from lower mass main sequence stars after their outer hydrogen-rich envelopes are blown away. Observationally some white dwarfs have been found to be variable stars (ZZ Ceti, pulsating double binary white dwarfs). Some have also been found to have immense magnetic fields, up to 10^9 gauss. These properties reflect origins and evolution under a much richer variety of conditions than might at first be expected.

Although the term "white dwarf" has been around since the initial discovery of Sirius B and 40 Er B, the term is actually a misnomer. One might expect white dwarfs to be white ($10\,000$ K or hotter) but most are actually cooler. The preferred term is degenerate star, which is adopted in their spectral classification (DA, DB, DC, etc.).[11]

4.4.2 The Chandrasekhar limit

In Section 3.1 we noted that stars were polytropic of index n if they satisfied the condition[12]

$$P = c\rho^\gamma, \tag{4.78}$$

where $n = 1/(\gamma - 1)$. From equations (4.70) and (4.72) it is clear that a degenerate star is polytropic with index $n = 3/2$ in the nonrelativistic case and $n = 3$ in the relativistic case. In the relativistic case this places an upper limit on the mass a white dwarf can have, known as Chandrasekhar's limit.

To find this limit let us assume a star is completely degenerate. Evaluating equation (4.72) numerically the pressure is then

$$P = K\left(\frac{1+X}{2}\rho\right)^{4/3} = K\mu^{4/3}\rho^{4/3}, \tag{4.79}$$

where $K = 1.23154 \times 10^{10}$ Pa and $\mu = (1+X)/2$.

The mass of the star is found by

$$M = 4\pi \int_0^R \rho(r)r^2 \, dr, \tag{4.80}$$

but because the star is polytropic it is more useful to substitute the polytropic variables from equations (3.20)–(3.22), specifically $z = Ar$ and $\rho = \rho_c\omega^n$, which yields

$$M = \frac{4\pi\rho_c}{A^3} \int_0^{z_0} \omega^n z^2 \, dz, \tag{4.81}$$

---

[11] This is not the only common misnomer in astrophysics. For example, asteroid means "little star" but the correct term is minor planet. The term "asteroid" was coined by William Herschel (discoverer of Uranus) for those objects discovered by others that moved like planets but were not large enough to show disks.

[12] Here again c is a free parameter.

where $\omega(z_0) = 0$. Substituting the differential form of the Lane–Emden equation (3.45),

$$\omega^n z^2 = -\frac{d}{dz}\left(z^2\frac{d\omega}{dz}\right),$$
(4.82)

then

$$M = -\frac{4\pi\rho_c}{A^3}\left(z^2\frac{d\omega}{dz}\right)_{z=z_0}.$$
(4.83)

From equation (3.20),

$$A = \sqrt{\frac{4\pi G}{K(n+1)}}\,\frac{\rho_c^{(n-1)/2n}}{\mu^{2/3}},$$
(4.84)

where $\mu = (1+X)/2$; thus we find

$$M = -4\pi\mu^2\left[\frac{(n+1)K}{4\pi G}\right]^{3/2}\rho_c^{(3-n)/2n}\left(z^2\frac{d\omega_n}{dz}\right)_{z=z_0}.$$
(4.85)

In the extreme relativistic case $n = 3$ and the mass is constant regardless of central density. At this critical value

$$M = -4\pi\mu^2\left[\frac{K}{\pi G}\right]^{3/2}\left(z^2\frac{d\omega_n}{dz}\right)_{z=z_0}.$$
(4.86)

The value of the derivative at $z = z_0$ can be determined numerically, and it is found that

$$-\left(z^2\frac{d\omega_n}{dz}\right)_{z=z_0} = 2.01824.$$
(4.87)

If we assume that a star fuses all of its hydrogen into higher elements before becoming a white dwarf, then $X = 0$ and $\mu = 0.5$. The mass for a relativistic degenerate star becomes $M = 1.435M_\odot$.

As seen in Section 3.3 the mass-radius relation for a polytropic star is

$$\frac{R_1}{R_2} = \left(\frac{M_1}{M_2}\right)^{(n-1)/(n-3)}.$$
(4.88)

For $1 < n < 3$ the radius of the star decreases with increasing mass. Because a nonrelativistic degenerate star is a polytrope of index $n = 3/2$, it is reasonable to presume that the radius of a white dwarf is a decreasing function of its mass up to its fully relativistic limit of $M \sim 1.4M_\odot$.

We can demonstrate this relationship in *Mathematica*.[13] The simple white dwarf model consists of purely degenerate matter, so there is no temperature dependence to the equation of state. Only the hydrostatic equations matter; thus in spherical coordinates

$$\frac{dP}{dr} = -\frac{\rho GM(r)}{r^2},$$
(4.89)

$$\frac{dM}{dr} = 4\pi r^2\rho,$$
(4.90)

where P, M, ρ, and r have their usual meanings. Because there is a singularity at $r = 0$, we will start the solution at $r = 1$ m with some central density cenden. Because the degenerate matter will vary

[13] See **4-6ModelWD**.

from nonrelativistic to relativistic with increasing mass, we must use the interpolated function

$$\texttt{fint[x\_]} := \frac{1}{8}\left(x\left(2x^2 - 3\right)\sqrt{\left(x^2 + 1\right)}\ + 3\,\texttt{ArcSinh[x]}\right)$$

With $\texttt{constant1} = 8\pi m_e^4 c^5/3h^3$ and $\texttt{constant2} = 8\pi m_e^3 m_p c^3/3h^3$, then

```
pressure[x2_] := constant1 fint[x2]
```

$$\texttt{mix[ex1\_]} := \frac{2}{1 + \texttt{ex1}}$$

```
massdensity[x2_, ex2_] := mix[ex2] constant2 x2^3
```

From these we can determine the density as a function of pressure:

```
eosinv[ex2_] := Table[{pressure[x3], massdensity[x3, ex2]}, {x3, 0, 10, 0.01}]
eosinvf = Interpolation[eosinv[exvalue]]
```

We can also generate the equation of state with P as the dependent variable, from which we can determine the central pressure:

```
eos[ex2_] := Table[{massdensity[x3, ex2], pressure[x3]}, {x3, 0, 10, 0.01}]
eosinterp = Interpolation[eos[exvalue]]
```

Because we assume a nitrogen white dwarf, the hydrogen mass ratio $\texttt{exvalue} = 0$. The solution to the hydrostatic equations is then found via $\texttt{NDSolve[]}$; thus

$$\texttt{sol2 = NDSolve}\Big[$$
$$\Big\{\texttt{pres'[r]} == -\frac{\texttt{eosinvf[pres[r]] gG mass[r]}}{r^2}, \texttt{mass'[r]} == 4\pi r^2\,\texttt{eosinvf[pres[r]]},$$
$$\texttt{pres[1]} == \texttt{prescenter1}, \texttt{mass[1]} == \frac{4\pi}{3}\,\texttt{cenden}\Big\}, \{\texttt{pres, mass}\}, \{r, 1, 30\,000\,000\}\Big]$$
```
pressure = sol2[[1, 1, 2]]; masseq = sol2[[1, 2, 2]];
```

This generates interpolated function solutions for pressure and mass as a function of radius. The density function is found by $\texttt{eosinvf[pressure[r]]}$. These can be plotted to see results for various central densities.

Through various trials one finds a reasonable range of cenden values is 5×10^6 kg/m^3 to 5×10^{12} kg/m^3. At the smallest central density the mass is about 0.03 solar masses and the radius is about 5.4 Earth radii. At the largest central density it is about $1.37M_\odot$ and $0.25R_\oplus$. We can now create a Do[]

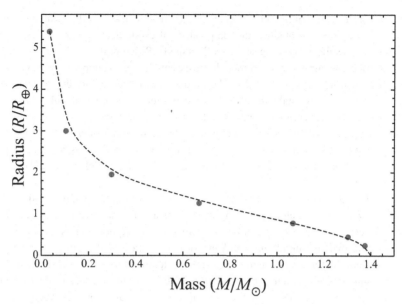

Fig. 4.10   Mass vs. radius for a degenerate star.

loop to generate a list of masses and radii

```
ρcenter = Table[5. × 10^(5+jj), {jj, 1, 7}];
prescenter = eosinterp[ρcenter];
massradius = Table[{0, 0}, {jj, 1, 7}];
Do[Clear[pres, mass, maxrad, radius, pressure, sol1];

    sol1 = NDSolve[{pres'[r] == - (eosinvf[pres[r]] gG mass[r]) / r^2, ...},

    {pres, mass}, {r, 1, 30 000 000}]; ...

    massradius[[ii]] = {maxmass, earthradii};, {ii, 1, 7}];
```

The result is seen in figure (4.10). It is clear that as the mass approaches the critical limit of $1.4M_\odot$ the radius of the star approaches 0. Beyond this mass the pressure of the degenerate electrons is not strong enough to oppose gravitational collapse.

4.4.3 Crystallization in white dwarfs

According to Van Horn in a review (1968a) and paper (1986b), the notion of crystalline structure in white dwarfs goes back to Kirzhnits (1960) and independently to Salpeter (1961). Later Mestel and and Ruderman (1967) showed that crystallization will begin when the central temperature falls below

$$T \sim 3 \times 10^5 \left(\frac{\rho}{10^6}\right) Z^{5/3} \text{ K}, \tag{4.91}$$

where T is the temperature, ρ is the density in g/cc, and Z is the mean nuclear charge. The pronounced Z dependence means that the temperature of crystallization depends very much on the composition, which results in "cooling sequences" in the H–R diagram.

Because our white dwarf models are essentially "zero temperature" nonevolving models, crystallization will not appear naturally. The production of "hot" white dwarfs such as the nuclei of planetary nebulae is well established; therefore evolutionary cooling models (which we do not consider here owing to their computational complexity) must consider crystallization as well as strong convection through the surface layers as pointed out by Lamb and Van Horn (1975) in their seminal study of C^{12} white dwarfs. In that paper the crystallization was described as being appropriate for a body-centered cubic (bcc) lattice ionic crystal. To include this in their model they calculated the Madelung and phonon energies.

Because the progenitor of a white dwarf has a high temperature and pressure, the constituents are ionized, with the electrons being degenerate and the positive ions nondegenerate. In a gaseous state, the electron degeneracy supplies the main pressure support against gravity through spin repulsion. That means that although Lamb and Van Horn model carbon white dwarfs, the crystal structure is ionic and not covalent as are diamonds at terrestrial temperatures and pressures. In fact, Lamb and Van Horn assume a body-centered (bcc) structure. The main uncertainty in their calculations was the value of the parameter Γ, which is the ratio of the Coulombic repulsion energy to the kinetic energy. Early estimations put $\Gamma \sim 60$, which gives the onset temperature equation above. By 1975 the favored value was $\Gamma \sim 160$, with consequent modifications to the onset temperature. These modifications established that the phase change is discontinuous and of first order. Hence the model of an ionic crystal is ensured and neither positive or negative ions are free to migrate at random. They can, however, oscillate so that phonons will be present. The source of the support against gravity then becomes electrostatic energy (so-called Madelung energy) of both ions and electrons. The crystals therefore have a specific heat that slows the contraction and cooling of the white dwarfs so their visible lifetimes are on the order of 10^{10} years instead of the much shorter times predicted assuming a gaseous state. Thus white dwarfs are among the oldest stars in the galaxy.

The ionic crystal structure most studied is that of a body-centered crystal characterized by alternating sign charges within the lattice. The total lattice energy is defined as the energy required to separate the crystal into individual ions to infinite separation. Kittel (1985) in a model of NaCl considers a central field repulsion plus an alternating Coulombic term of the form (in cgs units):

$$U_{tot} = N \left(z\lambda \, e^{-R/\rho} - \frac{\alpha q^2}{R} \right), \qquad (4.92)$$

where z is the number of "nearest" neighbors, R is the spacing, q is the charge, N the number of particles, and α is the so-called Madelung constant. At equilibrium $dU/dR = 0$, and the equilibrium separation R_0 can be determined in terms of λ and ρ. That leaves

$$U_{tot} = -\frac{N\alpha q^2}{R_0} \left(1 - \frac{\rho}{R_0} \right). \qquad (4.93)$$

Although Kittel proceeds to show the relatively easy solution to the one-dimensional case, he does not solve the 3D case. Hassani (2003) provides a discussion of three dimensional simple cubic cases that Kittel ignores. Historically in the calculation of the Madelung energy it was noticed that there is a common factor called the Madelung constant involved in the series representing the energy. Each Madelung case (still in cubic form) having a different dimension has a different Madelung constant.

Like the number π, the underlying numerical nature of each constant has become of interest purely for the mathematics it represents.

With x, y, and z as integers, the original series ignoring the central potential as postulated by Kittel and Crandall (1996) is

$$M = \sum_{(x,y,z)\neq 0} \frac{(-1)^{x+y+z}}{\sqrt{x^2 + y^2 + z^2}}. \tag{4.94}$$

Crandal gives the definition with $r^2 = x^2 + y^2 + z^2$ such that the subscript indicates the dimension and the exponent is s,

$$M_3(s) = \sum \frac{(-1)^{x+y+z}}{r^{2s}}. \tag{4.95}$$

The standard Madelung constant would be $M_3(1/2)$. Weisstein (2008) in *MathWorld* uses a different notation, where

$$b_n(2s) = \sum_{x,y,z=-\infty}^{\infty} \frac{(-1)^{x+y+z}}{\left(x^2 + y^2 + z^2\right)^s}, \tag{4.96}$$

so that the absolute value of $b_3(1)$ is equal to $M_3(1/2)$.

Before calculating the 3-dimensional case, it is useful to look at the constants of lower dimensions. The 1-dimensional case, as found in Kittel (1995), is

$$M_1(1/2) = 2 \sum_{j=-1}^{\infty} \frac{(-1)^j}{j}. \tag{4.97}$$

Mathematica finds the solution analytically:

```
2 Sum[(-1)^j / j, {j, ∞}]
```

```
-2 Log[2]
```

which yields $M_1(1/2) = -2\ln(2)$, in agreement with Kittel. Hassani (2003) shows the same result numerically:

```
oneD[n_] := N[-2 Sum[ (-1)^(j+1) / j , {j, 1, n}], 20]
```

This series converges very slowly, differing about 2% from the analytical solution after 50 terms, and 0.02% after 5000 terms.

For the 3-dimensional case there is no closed solution for $b_3(1) = M_3(1/2)$, so the constant must be calculated numerically. Weisstein (2008) gives the constant as

$$b_3(1) = -1.74756459463318\cdots \tag{4.98}$$

but gives no source or method. Hassani gives a partitioned sum using the 1D and 2D sums, plus one more for 3D. Thus

```
oneD[n_] := N[ Sum[ (-1)^j / j, {j, 1, n}], 30]
```

$$\text{twoD[n\_]} := N\Big[\sum_{j=1}^{n} \sum_{k=1}^{n} \frac{(-1)^{j+k}}{\sqrt{j^2 + k^2}}, 30\Big]$$

$$\text{threeD[n\_]} := N\Big[\sum_{m=1}^{n} \sum_{j=1}^{n} \sum_{k=1}^{n} \frac{(-1)^{m+j+k}}{\sqrt{m^2 + j^2 + k^2}}, 30\Big]$$

Thus, the constant becomes

```
αα3[n0_] := (6 oneD[n0] + 12 twoD[n0] + 8 threeD[n0]);
```

This yields a value within 1% of Weisstein's value after 50 terms, but the summation converges slowly. There are more efficient solutions, some of which can be found in **4-7WDCrystal**.

The main interest in crystallization is that it provides a means of storing a large amount of the thermal energy produced in the late stages of stellar evolution prior to the actual white dwarf stage. It is this heat that gets released slowly to thwart the rapid cooling that would occur if the star were a fluid and not a solid. The specific heat in the crystal is provided by phonons, the quantized lattice vibrations whose properties are analogous to photons. There are two main simple models of phonon specific heat. Following Kittel (1986) we use the phonon model of Debye.

Kittel uses the Planck distribution to derive the total energy U in phonon oscillations in a crystal, and from that he finds the specific heat $\partial U/\partial T$ to be

$$C_V = 9Nk \left(\frac{T}{\theta}\right)^3 \int_0^{\theta/T} \frac{x^4 e^x}{(e^x - 1)^2} dx, \qquad (4.99)$$

where N is the number of particles in the crystal, T is the temperature in Kelvin, k is the Boltzmann constant, and θ is the Debye temperature:

$$\theta = \frac{\hbar v}{k} \left(\frac{6\pi^2 N}{V}\right)^{1/3}, \qquad (4.100)$$

where V is the volume and v is the speed of sound in the crystal. This temperature θ is density dependent, and the speed of sound is determined by the elastic constants of the material. The specific heat has a T^3 dependence as modified by the integral. This means that as the star cools the specific heat decreases, which releases more heat.

The Debye temperature is very high in a carbon white dwarf because the density is so high. If we scale the Debye temperature from diamond we can estimate θ for the C^{12} white dwarf of 1×10^8 g/cc:

$$\theta\text{wd} = N\Big[2300 \left(\frac{1 \times 10^8}{3}\right)^{1/3}\Big]$$

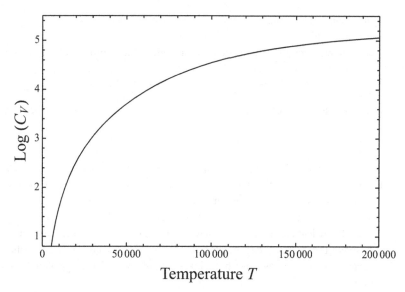

Fig. 4.11 Specific heat vs. temperature for a white dwarf.

which yields a Debye temperature of 740 209 K. The specific heat becomes

$$\mathbf{cvgm} = 9 \times 1.4 \times 10^{-19} \; \frac{1 \times 10^{8}}{12 \times 1.66 \times 10^{-24}}$$

or 6.3253×10^{12} J/g. Solving for the integral in *Mathematica*, one finds an analytical solution so long as $x < 0$. Because this is not the case we need, we must solve it numerically:

$$\mathbf{debyeCvwd[temp\_] := NIntegrate}\!\left[\frac{\mathbf{x^4\,e^x}}{\mathbf{(e^x - 1)^2}},\; \mathbf{\{x,\,0,\,\Theta wd\,/\,temp\}}\right]\!\left(\frac{\mathbf{temp}}{\mathbf{\Theta wd}}\right)^{\mathbf{3}}\mathbf{cvgm}$$

A plot of the specific heat C_V as a function of temperature, seen in figure (4.11), shows how it decreases with decreasing temperature.

4.5 Neutron stars

4.5.1 History of neutron stars

Soon after the discovery of the neutron by Chadwick in 1932, it was realized that one could construct large self-cohesive structures of those particles via the strong nuclear force. As pointed out by Harwit (1988) and Böhm-Vitense (1992), as the density of an object increases beyond that of a white dwarf the Fermi energy μ_e rises. At Fermi energies of 13 to 24 MeV and densities of 10^{14} to 10^{17} kg/m^3 it reaches the energy necessary to drive the inverse beta decay, which is capable of converting nuclei into neutron ensembles as long as energy is quickly supplied to the process by something as energetic as a supernova explosion.

Traditional beta decay is the decay of a neutron into a proton, electron, and anti-neutrino:

$$n \rightarrow p + e^- + \bar{\nu}_e, \tag{4.101}$$

while inverse beta decay is absorption of an electron by a proton, producing a neutron and neutrino:

$$p + e^- \rightarrow n + \nu_e. \tag{4.102}$$

In this latter case, $\mu_n = \mu_p + \mu_e$ and as more neutrons form, μ_p decreases while μ_n and μ_e increase. Eventually the chemical potential of the neutrons $\mu_n \rightarrow$ neutron rest mass and neutrons "drip" out of nuclei, a process known appropriately enough as neutron drip.

Until 1967, neutron stars could only be characterized without actually observing one. For a simple model all one had to do was to take a white dwarf model where the electrons are degenerate and change it to make the neutrons degenerate instead. In this way both white dwarfs and neutron stars can be described as degenerates. Together with black holes they form a general class known as compact stellar objects.

In 1967 Jocelyn Bell, then a graduate student of Anthony Hewish, detected radio pulses coming from the sky. These were traced to rapid rotation magnetic fields from objects called pulsars, some of which were associated with supernovae. As a flurry of observational and theoretical activity soon showed, the new objects fit the theoretical properties of neutron stars so well that the convergence of scientific conclusion in that direction was truly astounding (Manchester and Taylor, 1977). Hewish was awarded the Nobel prize in 1974 for the work that led up to the pulsar discovery.

Perhaps the most dramatic subsequent discovery was that of the binary pulsar PSR 1913+16 by Hulse and Taylor in 1974. The binary pulsar is a perfect laboratory for testing theories of relativistic gravity (Manchester and Taylor, 1977; Will, 1993; Green, 1985). Hulse and Taylor were also awarded the Nobel prize for their discovery in 1993.

4.5.2 A simple model

A neutron star model can be constructed following the approach used for a white dwarf. Again only the hydrostatic equations (4.89) and (4.90) apply; thus

$$\frac{dP}{dr} = -\frac{\rho GM(r)}{r^2}, \tag{4.103}$$

$$\frac{dM}{dr} = 4\pi r^2 \rho. \tag{4.104}$$

The density is again a degenerate gas, but one of neutrons rather than electrons. Following the same process as before in *Mathematica*[14], we define constant1 $= 8\pi m_n^4 c^5 / 3h^3$ and constant2 $= 8\pi m_n^3 m_p c^3 / 3h^3$ in terms of the neutron mass m_n rather than the electron mass. Because we are dealing with neutrons there is no dependence on the hydrogen mass ratio and mix $= 1$. The pressure and density

---

[14] See **4-7ModelNS**.

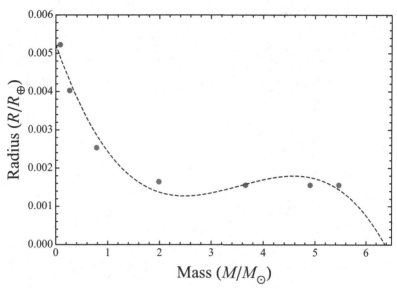

Fig. 4.12   Mass vs. radius for a neutron star.

functions are then

```
pressure[x2_] := constant1 fint[x2]
```

$$\texttt{mix[ex1\_]} := \frac{2}{1 + \texttt{ex1}}$$

```
massdensity[x2_] := constant2 x2³
```

from which we determine density as a function of pressure. For a given central density we can again calculate a table of masses and radii via a Do[] loop.

For neutron stars a good range for central density is 5×10^{15} to 5×10^{21} kg/m^3. Thus

```
ρcenter = Table[6. × 10¹⁴⁺ʲʲ, {jj, 1, 7}];
prescenter = eosinterp[ρcenter];
massradius = Table[{0, 0}, {jj, 1, 7}];
Do[Clear[pres, mass, maxrad, radius, pressure, sol1]; sol1 = NDSolve[...];
massradius[[ii]] = {maxmass, earthradii};, {ii, 1, 7}];
```

The resulting mass–radius plot can be seen in figure (4.12). It is clear that this relationship is strongly nonlinear.

To find the mass limit for a neutron star our calculations can be fit to a cubic polynomial, which can then be solved to find the zero intercept:

```
lmf1 = LinearModelFit[massradius, {1, x, x², x³}, x]
fiteq = Normal[lmf1]
Solve[fiteq == 0, x]
```

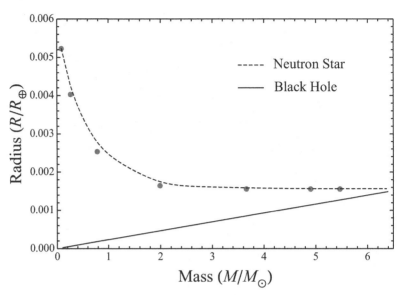

Fig. 4.13 Neutron star and black hole compared.

The result is a critical limit at $6.37359 M_\odot$. Although figure (4.12) shows a reasonable agreement between our calculated radii and the cubic fit, it can be argued that seven data points is hardly sufficient in calculating an upper mass limit.

Another approach to determining the mass limit is to compare the radius of a neutron star to the "radius" of a black hole of equal mass. In simple terms the size of a black hole can be determined by the radius of its event horizon, which is given by

$$R_{bh} = \frac{2GM}{c^2},$$ (4.105)

where M is the mass and c is the speed of light.[15] Comparison of the simple black hole radius to that of a neutron star, as seen in figure (4.13), shows the two intersect around $6.4 M_\odot$. At higher masses the neutron star becomes a black hole. It must be emphasized that our simple neutron star is a zero-temperature one. As long as the neutron star is hot, it can avoid the black hole developing at its center. Most young neutron stars are still hot from having been in a supernova and still emit x-rays either continually or in pulses. It is also clear from the graph that the radius of a neutron star is relatively close to that of a black hole. A realistic neutron star model must account for general relativity, which has not been done here. Calculations taking general relativity into account derive an upper mass limit closer to $3 M_\odot$.

4.5.3 Neutron star cooling

Young neutron stars cool by neutrino emission. Thus when both of these processes are active,

$$n \rightarrow p + e^- + \bar{\nu}_e$$ (4.106)

$$p + e^- \rightarrow n + \nu_e$$ (4.107)

[15] Astute readers might notice that a mass of this radius has an escape velocity of c under Newtonian gravity. The general relativity calculation follows a completely different approach, but the result is the same.

it is called the direct Urca process. At typical Fermi temperatures ($\sim 10^{12}$ K), the fermions must have momenta close to the Fermi momenta. The neutrino and anti-neutrino moments are $\sim k_b T/c$ and are small compared with the Fermi momenta. Momentum conservation requires

$$p_{F_e} + p_{F_p} \geq p_{F_n}. \tag{4.108}$$

This leads to

$$n_p \geq \left(\frac{1}{2}\right)^3 n_H \tag{4.109}$$

with

$$X_p = \text{proton fraction} = \frac{n_p}{n_p + n_n} \tag{4.110}$$

that

$$X_p > \frac{1}{9} \simeq 11\%. \tag{4.111}$$

If $\mu_e > m_\mu c^2$ (muon rest energy), $P_{F_e} \neq p_{F_p}$ and

$$X_p > 1/9 \text{ and } X_p < \frac{1}{\left[\left(1 + 2^{-1/3}\right)^3 + 1\right]} \simeq 15\%. \tag{4.112}$$

The neutrino luminosity for direct Urca is

$$E_{Urca} = 4 \times 10^{27} \left(Ye\frac{n}{n_s}\right)^{1/3} T_9^6 \text{ erg cm}^{-3}\text{s}^{-1}, \tag{4.113}$$

where T_9 is the temperature in units of $10^9 K$. The neutrino luminosity for a modified Urca process where an additional nucleon is involved in the interaction is

$$E_{Urca} = 9 \times 10^{21} \left(Ye\frac{n}{n_s}\right)^{1/3} T_9^8 \text{ erg cm}^{-3}\text{s}^{-1}. \tag{4.114}$$

Thus protons can exist in neutron stars of high temperature. Once the neutron star cools, however, there are no protons.

Exercises

4.1 Determine the line profiles for an expanding atmosphere with the following velocity dependence:
1. $V(r) = V_\infty \left(1 - \Psi \frac{r_0}{r}\right)$
2. $V(r) = r$
3. $V(r) = r^2$
4. $V(r) = r^3$

4.2 Determine the photon escape probability for the above cases.

4.3 Using **4-3Spheregraybody** run cases 4, 5, and 6.

4.4 Determine numerically the value of

$$\left(z^2 \frac{d\omega_n}{dx} \right)_{z=z_0}$$

for $n = 3$ and $n = 3/2$.

4.5 Show that in the classical limit ($n = 3/2$) a degenerate star does not have an upper mass limit. (Hint: We have seen that the mass of a polytropic star depends on its central density ρ_c. Because $r = z/A$ also depends on the central density, it is possible to derive a relationship between the mass and radius of a polytrope.)

4.6 From the simple model of a neutron star as a nonrelativistic mass of degenerate neutrons (and therefore polytropic):

1. Derive the mass–radius relationship this model.
2. Given the upper mass limit discussed in the text, estimate the minimum radius of a neutron star.
3. Is this a realistic estimate?

4.7 The relativistic equation for the Fermi energy does not include the mass of the particle directly:

$$\varepsilon_F = \frac{5hc}{16\pi} \left(\frac{N}{V} \right)^{1/3} = kT_F,$$

where N/V is the number density of the neutrons and h, c, and k have their usual meanings. We can use the mass of the neutron to calculate the number density for neutron stars with the following central densities:

$$\rho = 10^{14}, 10^{15}, 10^{16}, 10^{17}, 10^{18} \text{ g/cc}$$

and then calculate the relativistic Fermi energies.

1. One assumes that neutrons cease being neutrons when the Fermi energy exceeds the neutron rest energy. Does this happen for any of these cases?
2. What kinetic temperatures would be necessary for the neutrons to degenerate in each case?

4.8 Given two white dwarfs, one of $0.3M_\odot$ and one of $1.3M_\odot$, assume both are made of Fe^{56} and are of uniform density. Both start with $T = 10^9$ K and end at $T = 10^3$ K. Calculate the specific heat and thermal conductivity for these cases.

4.9 Using **4-7WDCrystal** model a white dwarf with the Einstein model of specific heat.

4.10 With **4-8ModelNS**, compare the properties such as radius, etc. between a $1.1M_\odot$ white dwarf and $1.1M_\odot$ neutron star, and comment on the likelihood of each case.

4.11 Carry out the $H\alpha$ analysis using the Voigt profile and a temperature and velocity field appropriate to a Wolf–Rayet star of nitrogen or carbon type.

5

General relativity and applications

In the previous chapter we had seen that neutron stars can approach densities similar to the limit for black holes. It is therefore necessary to understand the relativistic approach to gravity, known as general relativity. As we will see, general relativity describes gravity not as a force, but as a warping of space and time itself. In this chapter we present an overview of general relativity by extending what we know of special relativity. We will then derive the solution for a central mass, from which we will present the classical tests of general relativity.

5.1 From special relativity to general relativity

5.1.1 Rotations and boosts

In Chapter 1 it was shown that an inertial coordinate frame $S'(x', t')$ moving at a speed v a relative to a coordinate frame $S(x, t)$ is related by the Lorentz transformation:

$$x' = \frac{x - vt}{\sqrt{1 - v^2/c^2}} \qquad t' = \frac{t - vx/c^2}{\sqrt{1 - v^2/c^2}}. \tag{5.1}$$

This transformation ensures that the speed of light c is the same in all inertial frames, which is the basis for special relativity. It can be written more succinctly as

$$x' = \gamma x - \beta \gamma ct \qquad ct' = -\beta \gamma x + \gamma ct, \tag{5.2}$$

where $\beta = v/c$ and $\gamma = 1/\sqrt{1 - \beta^2}$.

Because ct and x are each dimensionally lengths, we can consider them as two components of a space–time vector. The Lorentz transformation can then be expressed in matrix form as the linear transformation

$$\begin{pmatrix} x' \\ ct' \end{pmatrix} = \begin{pmatrix} \gamma & -\beta\gamma \\ -\beta\gamma & \gamma \end{pmatrix} \begin{pmatrix} x \\ ct \end{pmatrix}. \tag{5.3}$$

Such a linear transformation is similar to the rotation of a two-dimensional vector. If we consider two frames R and R' related by a rotational angle θ, as seen in figure (5.1), then

$$\begin{pmatrix} x' \\ y' \end{pmatrix} = \begin{pmatrix} \cos\theta & \sin\theta \\ -\sin\theta & \cos\theta \end{pmatrix} \begin{pmatrix} x \\ y \end{pmatrix}. \tag{5.4}$$

Because the range of $\gamma = 1 \rightarrow \infty$ and the range of $\beta\gamma = -\infty \rightarrow \infty$, we can let

$$\gamma = \cosh\phi \qquad \beta\gamma = \sinh\phi. \tag{5.5}$$

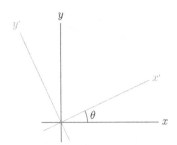

Fig. 5.1 Rotated coordinate frames.

Thus

$$\begin{pmatrix} x' \\ ct' \end{pmatrix} = \begin{pmatrix} \cosh\phi & -\sinh\phi \\ -\sinh\phi & \cosh\phi \end{pmatrix} \begin{pmatrix} x \\ ct \end{pmatrix}. \tag{5.6}$$

The Lorentz transformation is thus a hyperbolic rotation, also known as a Lorentz boost. The rotational nature of such a boost can be seen setting $\phi = \mathrm{i}\theta$, where $\mathrm{i} = \sqrt{-1}$. Given the trigonometric identities $\cosh(\mathrm{i}\theta) = \cos\theta$ and $\sinh(\mathrm{i}\theta) = \mathrm{i}\sin\theta$ we find

$$\begin{pmatrix} \mathrm{i}ct' \\ x' \end{pmatrix} = \begin{pmatrix} \cos\theta & \sin\theta \\ -\sin\theta & \cos\theta \end{pmatrix} \begin{pmatrix} \mathrm{i}ct \\ x \end{pmatrix}. \tag{5.7}$$

The Lorentz transformation is a rotational transformation not of a Cartesian coordinate frame, but of a Minkowski frame where time can be treated as an imaginary coordinate.[1]

The four-dimensional line element in Minkowski space is given by

$$\mathrm{d}s^2 = (\mathrm{i}c\mathrm{d}t)^2 + \mathrm{d}x^2 + \mathrm{d}y^2 + \mathrm{d}z^2 = -c^2\mathrm{d}t^2 + \mathrm{d}x^2 + \mathrm{d}y^2 + \mathrm{d}z^2. \tag{5.8}$$

The hyperbolic or imaginary character of a Lorentz boost means the time component of $\mathrm{d}s^2$ must be opposite in sign to the spatial components. However, just as the magnitude of a vector is unchanged by the rotation of its coordinate frame, the magnitude of the space–time 4-vector $\mathrm{d}s$ is unchanged under a Lorentz "rotation." In general any 4-vector is Lorentz invariant if its magnitude is unchanged under a rotation or boost.

From this invariant we can define the invariant proper time. For an object that is stationary,

$$\mathrm{d}s^2 = -c^2\mathrm{d}t^2. \tag{5.9}$$

Thus we can define the proper time

$$\mathrm{d}\tau^2 = -\frac{\mathrm{d}s^2}{c^2}. \tag{5.10}$$

5.1.2 Matrices, summation notation, and metrics

In general the Lorentz transformation can be expressed as a 4×4 matrix. Because matrix notation can be cumbersome, it is usually written in a compact form known as summation notation. For a general

[1] This is not to say that time is intrinsically imaginary. What is important is that space–time is Minkowskian, not Cartesian. Treating time as imaginary is just one way to represent this behavior.

matrix A, each element of the matrix can be denoted by a pair of indices:

$$
A = \begin{pmatrix}
a_{11} & a_{12} & a_{13} & \cdots & a_{1n} \\
a_{21} & a_{22} & a_{23} & \cdots & a_{2n} \\
a_{31} & a_{32} & a_{33} & \cdots & a_{3n} \\
\vdots & \vdots & \vdots & \ddots & \vdots \\
a_{m1} & a_{m2} & a_{m3} & \cdots & a_{mn}
\end{pmatrix},
\tag{5.11}
$$

where the first index represents the row of the element and the second the column. A linear transformation

$$
\begin{pmatrix} x'_1 \\ x'_2 \end{pmatrix} = \begin{pmatrix} m_{11} & m_{12} \\ m_{21} & m_{22} \end{pmatrix} \begin{pmatrix} x_1 \\ x_2 \end{pmatrix}
\tag{5.12}
$$

is then written as

$$
x'_i = \sum_{j=1}^{2} m_{ij} x_j = m_{ij} x_j.
\tag{5.13}
$$

In summation notation, because x_j and m_{ij} have j in common, it is assumed that j is summed over all values. The i is not summed; therefore this equation represents the linear transformation for both x_1 and x_2.

For our space–time coordinates, we assign the index notation[2]

$$
x^{\mu} = \left(x^0, x^1, x^2, x^3\right) = (ct, x, y, z).
\tag{5.14}
$$

If we then denote the Lorentz transformation matrix as $\Lambda^{\mu}{}_{\nu}$, then

$$
(x^{\mu})' = \Lambda^{\mu}{}_{\nu} x^{\nu}.
\tag{5.15}
$$

The line element ds^2 can also be written in summation notation as

$$
ds^2 = \eta_{\mu\nu}\, dx^{\mu}\, dx^{\nu},
\tag{5.16}
$$

where

$$
\eta_{\mu\nu} = \begin{pmatrix}
-1 & 0 & 0 & 0 \\
0 & 1 & 0 & 0 \\
0 & 0 & 1 & 0 \\
0 & 0 & 0 & 1
\end{pmatrix},
\tag{5.17}
$$

and is known as the metric of our Minkowski space.[3] As we will see, the metric is perhaps the most important matrix in general relativity. It tells us how space and time are connected. For example, the dot product[4] in Minkowski space is

$$
\mathbf{A} \cdot \mathbf{B} = \eta_{\mu\nu} A^{\mu} B^{\nu} = -A_t B_t + A_x B_x + A_y B_y + A_z B_z.
\tag{5.18}
$$

---

[2] It is important to note that these indices are *superscripts*, not exponents. Although this might seem an awkward notation, the difference between superscripts and exponents is generally clear within context. It is also possible to define the components as x_{μ}. For orthogonal coordinates in Minkowski space it makes no difference; however, in general relativity there is a distinct difference between the two.

[3] There are many variations to the Minkowski metric. Here we use the sign convention $(-+++)$. Other popular sign conventions are $(+---)$ and $(+++-)$. Physically this makes no difference; however, when comparing different texts it is important to know the sign convention used.

[4] In tensor formalism the dot product is sometimes referred to as an inner product, which is the *contraction* of the outer product. Hence given a tensor $C^{\mu}{}_{\nu} = A^{\mu} B_{\nu}$, the contraction $C^{\nu}{}_{\nu}$ is the dot product of $\mathbf{A}$ and $\mathbf{B}$.

Unlike Cartesian space, Minkowski space allows $A^2 = \mathbf{A} \cdot \mathbf{A}$ to be positive (spacelike), negative (timelike), or null (lightlike).

The metric can also be used to raise or lower the indices of our 4-vectors, such that $x_\mu = \eta_{\mu\nu} x^\nu$ and $x^\mu = \eta^{\mu\nu} x_\nu$. From this it is clear that

$$x^\mu = \eta^{\mu\tau} x_\tau = \eta^{\mu\tau} \eta_{\tau\nu} x^\nu. \tag{5.19}$$

Thus

$$\eta^{\mu\tau} \eta_{\tau\nu} = \delta^\mu{}_\nu, \tag{5.20}$$

where $\delta^\mu{}_\nu = 1$ if $\mu = \nu$ and 0 otherwise.[5] This is a general property of all metrics.

All of the usual vector equations can be written in this notation. For instance if we define a 4-vector derivative

$$\partial_\mu = \left(\frac{1}{c} \frac{\partial}{\partial t}, \frac{\partial}{\partial x}, \frac{\partial}{\partial y}, \frac{\partial}{\partial z} \right), \tag{5.21}$$

then the wave equation for light becomes

$$\nabla^2 \phi - \frac{1}{c^2} \frac{\partial^2 \phi}{\partial t^2} = \eta^{\mu\nu} \partial_\mu \partial_\nu \phi = \partial^\nu \partial_\nu \phi. \tag{5.22}$$

5.1.3 The Lorentz group

An important property of the Lorentz transformation is that it can be applied more than once. Suppose we have two inertial coordinate frames A and B, where B moves with a velocity v_1 relative to A, just as before. Now suppose there is a third inertial frame C that moves with v_2 relative to B. The frames A and B are connected by a Lorentz transformation, as are B and C. It must also be true that A and C are connected by a Lorentz transformation. That is

$$B = \Lambda_1 A, \qquad C = \Lambda_2 B. \tag{5.23}$$

Thus

$$C = \Lambda_2 \Lambda_1 A = \Lambda_3 A. \tag{5.24}$$

In other words, the product of two Lorentz transformations must also be a Lorentz transformation, a property known as closure.

It is also true that Lorentz transformations are associative. That is

$$\Lambda_1 (\Lambda_2 \Lambda_3) = (\Lambda_1 \Lambda_2) \Lambda_3. \tag{5.25}$$

They are also invertible, in that you can transform from $B \to A$ as well as $A \to B$, and of course you can trivially transform a coordinate frame to itself, known as an identity transformation.

Any mathematical structure that has the properties of closure, associativity, invertibility, and identity is known as a group. Although not often discussed in the language of group theory, many mathematical structures such as vectors under addition, real numbers under multiplication, and rotations in space are groups.

---

[5] Also known as the Kronecker delta.

The group of Lorentz transformations $\Lambda^\mu{}_\nu$ is known as the Lorentz group, and consists of the set of boosts along any velocity plus the set of all rotations in 3-dimensional space, as well as any combination of rotation and boost. Because the set of spatial rotations is a group in its own right (actually a subgroup of the Lorentz group), it is sometimes noted by $\Lambda^i{}_j$, where the Latin indices i and j are limited to the spatial components.[6] The Lorentz group is itself a subgroup of Lorentz transformations plus translations in space–time, known as the Poincaré group. The Poincaré transformation is written as

$$(x^\mu)' = \Lambda^\mu{}_\nu x^\nu + a^\mu. \tag{5.26}$$

The structure of a group provides a concise description of the underlying symmetry of the physics. By expressing quantities in the language of groups one can more readily uncover their broad properties. Consider the concept of velocity. From Chapter 1 we have seen that $v = dx/d\tau$, where τ is the proper time. The 4-velocity u is then

$$u^\mu = \frac{dx^\mu}{d\tau}. \tag{5.27}$$

Multiplying this by the proper mass m_o we find the 4-momentum is

$$p^\mu = m_o \frac{dx^\mu}{d\tau} = (E, \mathbf{p}). \tag{5.28}$$

In the 4-vector formulation energy and momentum are combined.

5.1.4 The equivalence principle and curved space

Legend has it that Galileo demonstrated that two objects of different mass, when dropped simultaneously from the tower of Pisa, hit the ground at the same time. Galileo's experiment is perhaps the most famous demonstration of the principle of equivalence. It is often stated that near the surface of Earth $\mathbf{F} = m\mathbf{g}$, but this is more correctly written as

$$\mathbf{F} = m_i \mathbf{a} = m_g \mathbf{g}, \tag{5.29}$$

where m_i is an object's inertial mass and m_g is its gravitational mass. The principle of equivalence asserts that $m_i = m_g$. To the limits of measurement the two masses appear to be equal. More recent experiments place an upper limit of difference as

$$\frac{|m_g - m_i|}{m_i} < 5 \times 10^{-13}. \tag{5.30}$$

It is therefore reasonable to presume the equivalence principle is valid.

A consequence of the principle is that if you happen to be in free fall you experience weightlessness. From your own perspective you cannot distinguish between falling in a gravitational field and moving at a constant velocity. Likewise, if you were to be accelerated at a constant 9.8 m/s^2 it would feel like gravity. The equivalence principle implies a new kind of symmetry between inertial motion and motion under gravity.

To explore this symmetry, let us consider a coordinate frame accelerating at a constant $\mathbf{g}$. An object "at rest" in such a frame would experience a constant gravitational "force" downward, just as if it were resting on Earth. This accelerating frame is related to an inertial coordinate frame by a Poincaré

---

[6] It is common practice to use Greek indices when considering space–time and Latin indices when considering only spatial terms. We follow that practice here.

transformation; thus we can ask how the metric of the accelerating frame compares to an inertial Minkowski metric.

We can begin by determining the motion of a constantly accelerating object in an inertial frame. For acceleration to be constant in the object's frame,

$$\mathbf{a} = \frac{d\mathbf{u}}{d\tau} = \text{constant}, \tag{5.31}$$

where $\mathbf{u}$ is the 4-velocity of the function in the inertial frame. The magnitude of the 4-velocity is constant because

$$u^{\mu}u_{\mu} = \frac{ds^2}{d\tau^2} = -c^2, \tag{5.32}$$

and the acceleration is orthogonal to the velocity because

$$\frac{d}{d\tau}\left(u^{\mu}u_{\mu}\right) = 2u^{\mu}a_{\mu} = 0. \tag{5.33}$$

If we take the acceleration to be $\mathbf{a} = \mathbf{g}$ along the x axis, then we have a set of equations

$$u^{\mu}u_{\mu} = -c^2, \qquad u^{\mu}a_{\mu} = -u^0 a_0 + u^1 a_1 = 0, \qquad a^{\mu}a_{\mu} = g^2. \tag{5.34}$$

Solving these equations we find

$$t = \frac{c}{g}\sinh\left(\frac{g\tau}{c}\right), \qquad x = \frac{c}{g}\cosh\left(\frac{g\tau}{c}\right). \tag{5.35}$$

From these coordinates, if we transform from the inertial frame to the co-accelerating frame, we find the metric becomes

$$ds^2 = g_{\mu\nu}dx^{\mu}dx^{\nu} = -c^2\left(1 + gx/c^2\right)dt^2 + dx^2 + dy^2 + dz^2. \tag{5.36}$$

The metric $g_{\mu\nu}$ of the accelerating frame differs from that of the inertial frame $\eta_{\mu\nu}$. By the equivalence principle, the object in free fall does not perceive gravity, but the metric of its reference frame is not Minkowskian. Because the metric tells us about the behavior of space and time, the effects of gravity must be equivalent to a deformation of the metric. Gravity is therefore not a force in the traditional sense, but rather a byproduct of the shape of space and time. To describe gravity in terms of the metric, we must be able to write the laws of physics in the language of metrics. This requires a new set of mathematical tools.

5.2 Mathematical toolbox

5.2.1 Tensors

Consider a displacement dr between two points. Such a displacement is a vector in our 4-dimensional space. Typically vectors are defined simplistically as a quantity with magnitude and direction, but vectors are defined formally by the way in which they transform. A vector must be independent of the coordinate system used to describe it. Thus given two coordinates A and A', one can in principle find $x'^{\mu} = f(x^{\mu})$. By the chain rule for differentiation the vector in A' is found to be

$$(dx^{\mu})' = \frac{\partial (x^{\mu})'}{\partial x^{\nu}}dx^{\nu}. \tag{5.37}$$

This transformation is the defining property of a vector. Thus we may say that anything that transforms as

$$V'^\mu = \frac{\partial x'^\mu}{\partial x^\nu} V^\nu \tag{5.38}$$

is a vector. A vector can also be written as

$$V'_\mu = \frac{\partial x^\nu}{\partial x'^\mu} V_\nu. \tag{5.39}$$

The two forms are called contravariant and covariant vectors respectively, but we will simply refer to both forms as vectors.

Vectors are a special case of a more general form called tensors.[7] Vectors are known as rank-1 tensors, because their components have 1 super- or subscript. Scalars such as ds^2 are known as rank-0 tensors. The behavior of higher rank tensors can be seen by taking the direct or outer product of two vectors. If we define the outer product $T^{\mu\nu} = V^\mu W^\nu$, then

$$T'^{\mu\nu} = V'^\mu W'^\nu = \frac{\partial x'^\mu}{\partial x^\alpha} \frac{\partial x'^\nu}{\partial x^\beta} V^\alpha W^\beta = \frac{\partial x'^\mu}{\partial x^\alpha} \frac{\partial x'^\nu}{\partial x^\beta} T^{\alpha\beta}. \tag{5.40}$$

Any quantity that transforms in this way is a rank-2 covariant tensor. Tensors of rank 2 or higher can be covariant, contravariant, or mixed tensors. In general tensors can be of any rank and transform in a similar way. A mixed rank-3 tensor, for example, transforms as

$$T'^\lambda{}_{\mu\nu} = \frac{\partial x'^\lambda}{\partial x^\alpha} \frac{\partial x^\beta}{\partial x'^\mu} \frac{\partial x^\gamma}{\partial x'^\nu} T^\alpha{}_{\beta\gamma}. \tag{5.41}$$

Lorentz transformations for tensors vary in form depending on their rank. Scalars (rank-0) are Lorentz invariant. Vectors (rank-1) transform as we have seen:

$$V'^\mu = \Lambda^\mu{}_\nu V^\nu. \tag{5.42}$$

Tensors of rank-2 transform as

$$T'^{\mu\nu} = \Lambda^\mu{}_\alpha \Lambda^\nu{}_\beta T^{\alpha\beta}, \tag{5.43}$$

and rank-n tensors follow the same pattern.

Tensors can be described by the symmetry of their components. A symmetric tensor has the property

$$S_{\alpha\beta} = S_{\beta\alpha}, \tag{5.44}$$

while for an antisymmetric tensor

$$A_{\alpha\beta} = -A_{\beta\alpha}. \tag{5.45}$$

Any tensor can be written as a sum of symmetric (denoted by () around indices) and antisymmetric (denoted by [] around indices) parts. Given a general tensor $T_{\alpha\beta}$, we can define

$$S_{\alpha\beta} = T_{(\alpha\beta)} = \frac{1}{2}\left(T_{\alpha\beta} + T_{\beta\alpha}\right), \tag{5.46}$$

and

$$A_{\alpha\beta} = T_{[\alpha\beta]} = \frac{1}{2}\left(T_{\alpha\beta} - T_{\beta\alpha}\right). \tag{5.47}$$

---

[7] An introduction to the simplest form of tensors is given in **5-0Tensors**, where it is shown how *Mathematica* can handle Cartesian tensors, i.e., those for which the covariant/contravariant distinction does not have to be made.

The general tensor is then

$$T_{\alpha\beta} = T_{(\alpha\beta)} + T_{[\alpha\beta]}. \tag{5.48}$$

The same can be done for tensors of other ranks. For a rank-3 tensor

$$T_{(\alpha\beta)} = \frac{1}{3!}\left(T_{\alpha\beta\gamma} + T_{\beta\gamma\alpha} + T_{\gamma\alpha\beta} + T_{\beta\alpha\gamma} + T_{\gamma\gamma\beta} + T_{\gamma\beta\alpha}\right), \tag{5.49}$$

$$T_{[\alpha\beta]} = \frac{1}{3!}\left(T_{\alpha\beta\gamma} + T_{\beta\gamma\alpha} + T_{\gamma\alpha\beta} - T_{\beta\alpha\gamma} - T_{\gamma\gamma\beta} - T_{\gamma\beta\alpha}\right). \tag{5.50}$$

As we shall see, symmetry of tensors is important when discussing differential forms.

Perhaps the most useful property of tensors is that tensor equations are independent of their frame of reference. If an equation between two tensors is true in one coordinate frame, then it is true in all coordinate frames. It is therefore useful to express the laws of physics in their tensor forms.[8]

5.2.2 Maxwell's equations in tensor form

From Maxwell's equations[9]

$$\nabla \cdot \mathbf{B} = 0, \quad \nabla \times \mathbf{E} + \frac{1}{c}\frac{\partial \mathbf{B}}{\partial t}, \tag{5.51}$$

we can define a vector potential $\mathbf{A}$ by

$$\mathbf{B} = \nabla \times \mathbf{A}, \quad \text{because} \quad \nabla \times (\nabla \times \mathbf{A}) \equiv 0, \tag{5.52}$$

and a scalar potential ϕ by

$$-\nabla\phi = \mathbf{E} + \frac{1}{c}\frac{\partial \mathbf{A}}{\partial t}, \quad \text{because} \quad \nabla \times (\nabla\phi) \equiv 0. \tag{5.53}$$

The scalar and vector potentials can be represented as a 4-vector where

$$A_\mu = \eta_{\mu\nu}A^\nu = (-\phi, \mathbf{A}), \tag{5.54}$$

from which one can define a 4×4 matrix with elements

$$F_{\mu\nu} = \frac{\partial A_\nu}{\partial x^\mu} - \frac{\partial A_\mu}{\partial x^\nu}. \tag{5.55}$$

In matrix form this becomes

$$F_{\mu\nu} = \begin{pmatrix} 0 & -E_x & -E_y & -E_z \\ E_x & 0 & B_z & -B_y \\ E_y & -B_z & 0 & B_x \\ E_z & B_y & -B_x & 0 \end{pmatrix}, \tag{5.56}$$

and is a rank-2 tensor known as the field strength tensor or Faraday tensor.[10] The ways the Faraday tensor can be written takes slightly different forms depending on whether the indices are raised or lowered, but can easily be converted between forms, such as

$$F^{\mu\nu} = \eta^{\mu\alpha}\eta^{\nu\beta}F_{\alpha\beta}. \tag{5.57}$$

---

[8] When an equation is expressed in tensor form it is sometimes referred to as the covariant form of the equation. This has nothing to do with covariant tensors.

[9] Recall Section 1.4.1.

[10] For a more detailed discussion of the tensor representation of Maxwell's equations, see Goldstein, Poole, and Safko (2002).

It is also often written in its dual form $^\star F^{\mu\nu}$, where

$$^\star F^{\mu\nu} = \frac{1}{2}\epsilon^{\mu\nu\lambda\sigma} F_{\lambda\sigma},\qquad (5.58)$$

where ϵ is known as the Levi–Civita symbol, and

$$\epsilon^{\mu\nu\lambda\sigma} = \begin{cases} 1 & \text{for an even permutation of } 0,1,2,3 \\ -1 & \text{for an odd permutation of } 0,1,2,3 \\ 0 & \text{otherwise} \end{cases} \qquad (5.59)$$

In matrix form the dual Faraday tensor is

$$^\star F^{\mu\nu} = \begin{pmatrix} 0 & B_x & B_y & B_z \\ -B_x & 0 & E_z & -E_y \\ -B_y & -E_z & 0 & E_x \\ -B_z & E_y & -E_x & 0 \end{pmatrix}. \qquad (5.60)$$

If the charge and current density are written in their 4-vector form, where

$$J^\mu = (c\rho, \mathbf{J}), \qquad (5.61)$$

then Maxwell's equations can be written in their tensor form as

$$\partial_\mu F^{\mu\nu} = \frac{4\pi}{c} J^\nu, \qquad \partial_\mu {}^\star F^{\mu\nu} = 0, \qquad (5.62)$$

where $\partial_\mu = \partial/\partial x^\mu$. The Lorentz force equation for electric charges can also be written in tensor form as

$$\frac{\mathrm{d}p^\mu}{\mathrm{d}\tau} = eF^\mu{}_\nu u^\nu, \qquad (5.63)$$

where p^μ is the 4-momentum, u^μ is the 4-velocity, and e is the electric charge.

Although the tensor forms of equations are always valid, they are used almost exclusively in the context of special and general relativity. Because the vacuum speed of light c is invariant within relativity, tensor forms are often expressed in units where $c = 1$. We will often follow that tradition here.

5.2.3 Differential forms

Given a scalar function $f(x, y, z)$, the differential of f is given by

$$\mathrm{d}f = \frac{\partial f}{\partial x}\mathrm{d}x + \frac{\partial f}{\partial y}\mathrm{d}y + \frac{\partial f}{\partial z}\mathrm{d}z. \qquad (5.64)$$

Thus we can say that $\mathrm{d}f$ is a differential form of f. This particular differential form is also a rank-1 tensor, called a 1-form. In general, p-forms are covariant tensors of rank p that are completely anti-symmetric. Because our tensor equations can be put in antisymmetric covariant form (or already have that form), the tensor equations of special and general relativity can be expressed as differential forms. Scalars become 0-forms, 4-vectors become 1-forms, and so on.

Differential forms are not simply a mathematical shorthand similar to summation notation, but rather a rich mathematical language with a power all its own. Put simply, differential forms can be differentiated and integrated independent of any geometric structure. They can be applied in flat and curved space equally. For our purposes we do not require differential forms, but we provide a brief overview as they are widely used in general relativity.

In differential forms one uses the exterior derivative rather than the usual partial derivative. We have seen that the two are related because for a scalar ϕ the exterior derivative is $\mathrm{d}\phi$ and

$$(\mathrm{d}\phi)_\mu = \partial_\mu \phi. \tag{5.65}$$

Here the exterior derivative of a 0-form (scalar) is a 1-form (vector). It is in fact the gradient of ϕ. This "raising" behavior applies in general, in that the exterior derivative of a p-form yields a $(p + 1)$-form,

$$(\mathrm{d}A)_{\mu_1 \cdots \mu_{p+1}} = (p + 1)\, \partial_{[\mu_1} A_{\mu_2 \cdots \mu_{p+1}]}. \tag{5.66}$$

For a 1-form $\mathbf{A}$,

$$(\mathrm{d}\mathbf{A})_{\mu\nu} = \left(\partial_\mu A_\nu - \partial_\nu A_\mu \right). \tag{5.67}$$

For any p-form $\mathbf{A}$,

$$\mathrm{d}\left(\mathrm{d}\mathbf{A}\right) = 0. \tag{5.68}$$

This identity (often written as $\mathrm{d}^2 = 0$) arises from the antisymmetric nature of the exterior derivative and the fact that partial derivatives commute, $\partial_\mu \partial_\nu = \partial_\nu \partial_\mu$, when acting on a function.

Higher p-forms can also be generated from lower ones by an antisymmetric product known as the wedge product. The wedge product of two 1-forms $\mathbf{A}$ and $\mathbf{B}$ is given by[11]

$$\mathbf{A} \wedge \mathbf{B} = 2A_{[\mu} B_{\nu]} = A_\mu B_\nu - A_\nu B_\nu, \tag{5.69}$$

while the wedge of a p-form $\mathbf{A}$ and a q-form $\mathbf{B}$ is

$$\mathbf{A} \wedge \mathbf{B} = \frac{(p+q)!}{p!q!} A_{[\mu_1 \cdots \mu_p} B_{\mu_{p+1} \cdots \mu_{p+q}]}. \tag{5.70}$$

The wedge product follows the usual rules of addition and distribution, but either commutes or anti-commutes depending on the values of p and q:

$$\mathbf{A} \wedge \mathbf{B} = (-1)^{pq}\, \mathbf{B} \wedge \mathbf{A}. \tag{5.71}$$

The wedge product of 1-forms is similar to the cross product of vectors. They operate in similar ways and their magnitudes are the same. The difference is that a wedge of 1-forms (vectors) is a 2-form, not another vector. In 3-dimensional space the wedge and cross products are related by the Hodge duality operator. The Hodge dual $\star\mathbf{A}$ of a p-form $\mathbf{A}$ in an n-dimensional space[12] is given by

$$(\star\mathbf{A})_{\mu_1 \cdots \mu_{n-p}} = \frac{1}{p!} \epsilon^{\nu_1 \cdots \nu_p}{}_{\mu_1 \cdots \mu_{n-p}} A_{\nu_1 \cdots \nu_p}. \tag{5.72}$$

Unlike the other operations, the Hodge dual does depend on the metric, and a necessary condition is that $p < n$. Applying the dual operation twice returns the original p-form within a sign,

$$\star\star\mathbf{A} = (-1)^{s+p(n-p)}\, \mathbf{A}, \tag{5.73}$$

where s is the number of minus signs in the eigenvalues of the metric. For Minkowski space $s = 1$. In 3-dimensional space the Hodge dual of the 1-form wedge product is the cross product; thus

$$\star (\mathbf{A} \wedge \mathbf{B})_i = \epsilon_i{}^{jk} A_j B_k = (\mathbf{A} \times \mathbf{B})_i. \tag{5.74}$$

This connection between wedge product and cross product exists only for 3-dimensional space.

---

[11] Recall the [] notation for antisymmetric tensors here.
[12] To be precise, an n-dimensional manifold.

This property of dualness is quite clear in electromagnetism. Maxwell's equations can be written as

$$d\mathbf{F} = 0, \qquad d\,(\star\mathbf{F}) = 4\pi\,(\star\mathbf{J})\,, \tag{5.75}$$

where $\mathbf{F} = F_{\mu\nu}$ is the 2-form of the Faraday tensor and $\mathbf{J}$ is the 1-form of the charge-current density. The Hodge dual of the Faraday tensor is given by $\mathbf{F} \to \star\mathbf{F}$ and $\star\mathbf{F} \to -\mathbf{F}$; therefore in the absence of charge the field equations reduce to

$$d\mathbf{F} = d\,(\star\mathbf{F}) = 0. \tag{5.76}$$

In other words, the electromagnetic field is invariant under Hodge duality. Physically this means the electric and magnetic fields are interchangeable. This underlying symmetry is broken by the presence of charge.

5.2.4 Covariant differentiation

To express equations in tensor form we must be careful to ensure that operations on equations maintain a tensor form. This does not mean that the tensor rank must remain the same, only that operations on tensors must consistently yield tensors. Consider, for example, the partial derivative of a scalar function ϕ. Under a change of coordinates,

$$\frac{\partial\phi}{\partial x^\mu} \to \frac{\partial\phi}{\partial x'^\mu} = \frac{\partial x^\nu}{\partial x'^\mu}\frac{\partial\phi}{\partial x^\nu}. \tag{5.77}$$

As a result, $V_\mu = \partial\phi/\partial x^\mu$ transforms as a rank-1 covariant tensor. This pattern does not hold in general. If we take the derivative of V_μ, we find

$$\frac{\partial V^\nu}{\partial x^\mu} \to \frac{\partial V'^\nu}{\partial x'^\mu} = \left(\frac{\partial x^\lambda}{\partial x'^\mu}\frac{\partial}{\partial x^\lambda}\right)\left(\frac{\partial x'^\nu}{\partial x^\sigma}V^\sigma\right). \tag{5.78}$$

Thus

$$\frac{\partial V'^\nu}{\partial x'^\mu} = \frac{\partial x^\lambda}{\partial x'^\mu}\frac{\partial x'^\nu}{\partial x^\sigma}\frac{\partial V^\sigma}{\partial x^\lambda} + \frac{\partial x^\lambda}{\partial x'^\mu}\frac{\partial^2 x'^\nu}{\partial x^\sigma\partial x^\lambda}V^\sigma. \tag{5.79}$$

In Cartesian Minkowski space the second term of the right-hand side vanishes. It does not, however, vanish for a general metric. As a result, $\partial_\mu V^\nu$ does not consistently transform as a tensor. We must therefore define a new "tensor friendly" derivative. This is done by defining a covariant derivative, given by

$$\nabla_\mu V^\nu = \partial_\mu V^\nu + \Gamma^\nu_{\mu\lambda}V^\lambda, \tag{5.80}$$

where the elements of $\Gamma^\nu_{\mu\lambda}$ are known as connection coefficients.

The addition of the connection term ensures that nontensor term of equation (5.79) is canceled out. For the covariant derivative

$$\nabla'_\mu V'^\nu = \frac{\partial x'^\nu}{\partial x^\alpha}\frac{\partial x^\beta}{\partial x'^\mu}\nabla_\beta V^\alpha, \tag{5.81}$$

and thus transforms as a rank-2 tensor. For lowered indices, the covariant derivative takes the form

$$\nabla_\mu V_\nu = \partial_\mu V_\nu - \Gamma^\lambda_{\mu\nu}V_\lambda, \tag{5.82}$$

and

$$\nabla'_\mu V'_\nu = \frac{\partial x^\alpha}{\partial x'^\mu}\frac{\partial x^\beta}{\partial x'^\nu}\nabla_\alpha V_\beta. \tag{5.83}$$

The same process applies for tensors of higher rank, where a $+\Gamma$ term is applied for each raised index and a $-\Gamma$ for each lowered one. Similar to the exterior derivative of forms, the covariant derivative raises the rank of a tensor by one.

Because the metric $g_{\mu\nu}$ of a general space is a rank-2 tensor, one can take its covariant derivative to find

$$\nabla_\sigma g_{\mu\nu} = 0, \qquad \nabla_\sigma g^{\mu\nu} = 0. \tag{5.84}$$

The vanishing of metric derivatives is known as metric compatibility and is a central feature of covariant differentiation.

The specific form of the connection coefficients depend on the metric of the space, and is given by

$$\Gamma^\sigma_{\mu\nu} = \frac{1}{2} g^{\sigma\rho} \left(\partial_\mu g_{\nu\rho} + \partial_\nu g_{\rho\mu} - \partial_\rho g_{\mu\nu} \right). \tag{5.85}$$

Given any metric $g_{\mu\nu}$ we can calculate $\Gamma^\sigma_{\mu\nu}$ and thus the covariant derivatives. As a result, most equations expressed in Minkowski space can be applied to all metric spaces by substituting $\partial_\mu \to \nabla_\mu$. For example, the equations for electromagnetic fields in general relativity can be found from the special relativity equations (5.62) to yield

$$\nabla_\mu F^{\mu\nu} = 4\pi J^\nu, \qquad \nabla_\mu \star F^{\mu\nu} = 0. \tag{5.86}$$

It is important to note, however, that $\Gamma^\sigma_{\mu\nu}$ is not a measure of a metric's curvature. Although the connections vanish for Cartesian Minkowski space, they do not vanish for general Minkowski space, such as when the spatial terms are expressed in spherical coordinates.

5.3 Einstein's field equations

5.3.1 The geodesic equation

It is commonly said that a line is the shortest distance between two points. A more precise way of stating this is that a geodesic is the shortest distance between two points. The two are the same in the case of the surface of a desk, but are different on the surface of a globe, where the shortest distance between two points is an arc of a great circle.[13]

For a space of metric $g_{\mu\nu}$, the line element is given by

$$ds^2 = g_{\mu\nu} \, dx^\mu \, dx^\nu, \tag{5.87}$$

and the distance between two points a and b along a particular path is then

$$L = \int_a^b ds. \tag{5.88}$$

The value of the distance will be larger or smaller depending on the particular path, but the geodesic is the path for which L has the extremal value.[14] To find the geodesic we must find the extrema of this functional. This is similar to the Lagrangian method of classical mechanics. For a system in which

---

[13] This is why, for example, a flight from New York to Tokyo travels into the arctic circle rather than due west.

[14] The length L is a minima only when $g_{\mu\nu}$ is positive definite. It can also be a maxima or saddle point of the functional.

energy is conserved, one may define a Lagrangian $L = T - V$, where $T(\dot{x})$ is the kinetic energy of the system and $V(x)$ is the potential energy. The motion of the system is that for which L is an extrema.

Consider the simple case of a single particle starting at time t_1 and ending at t_2, this means

$$\delta \int_{t_1}^{t_2} L \, dt = \int_{t_1}^{t_2} \delta L \, dt = 0. \tag{5.89}$$

Now because T depends on the velocity of an object, and V depends on the position of the object, we can say

$$L \equiv L(\dot{x}, x). \tag{5.90}$$

Thus, because

$$\delta L(\dot{x}, x) = \frac{\partial L}{\partial x} \delta x + \frac{\partial L}{\partial \dot{x}} \delta \dot{x} \tag{5.91}$$

we have

$$\int_{t_1}^{t_2} \left[\frac{\partial L}{\partial x} \delta x + \frac{\partial L}{\partial \dot{x}} \delta \dot{x} \right] dt. \tag{5.92}$$

Now then,

$$\delta \dot{x} = \frac{d}{dt} (\delta x). \tag{5.93}$$

Thus, we can integrate the second term of our equation by parts

$$\int_{t_1}^{t_2} \left[\frac{\partial L}{\partial \dot{x}} \delta \dot{x} \right] dt = \int_{t_1}^{t_2} \left[\frac{\partial L}{\partial \dot{x}} \frac{d}{dt} (\delta x) \right] dt \tag{5.94}$$

$$= \frac{\partial L}{\partial \dot{x}} \delta x \Big|_{t_1}^{t_2} - \int_{t_1}^{t_2} \left[\frac{d}{dt} \left(\frac{\partial L}{\partial \dot{x}} \right) \delta x \right] dt. \tag{5.95}$$

Because any varied path must have the same starting and end point as our actual path, $\delta x = 0$ at t_1 and t_2; therefore the first term of our integral vanishes. This means that our total integral becomes

$$\delta \int_{t_1}^{t_2} L \, dt = \int_{t_1}^{t_2} \left[\frac{\partial L}{\partial x} - \frac{d}{dt} \left(\frac{\partial L}{\partial \dot{x}} \right) \right] \delta x \, dt = 0. \tag{5.96}$$

For this to be true in general it must be the case that

$$\frac{d}{dt} \left(\frac{\partial L}{\partial \dot{x}} \right) - \frac{\partial L}{\partial x} = 0. \tag{5.97}$$

This is the Lagrange equation for one dimension. This yields for a single particle of mass m

$$m \frac{d^2 x}{dt^2} + \frac{dV}{dx} = 0. \tag{5.98}$$

For a geodesic in a metric space the derivation is significantly more tedious, but results in the geodesic equation

$$\frac{d^2 x^\sigma}{ds^2} + \Gamma^\sigma_{\mu\nu} \frac{dx^\mu}{ds} \frac{dx^\nu}{ds} = 0. \tag{5.99}$$

The solution to the geodesic equation finds the shortest distance in space–time between two points in an arbitrary geometry.

5.3.2 Newtonian gravity as a metric equation

Newton's gravitational equations can be expressed as

$$\nabla^2\phi = 4\pi G\rho, \tag{5.100}$$

$$\mathbf{F} = -m\nabla\phi, \tag{5.101}$$

where ϕ is the gravitational potential, ρ is the mass density, and G is the gravitational constant. By the equivalence principle the mass m of the test particle cancels out, and its equation of motion can be written as

$$\frac{\mathrm{d}^2 x^i}{\mathrm{d}t^2} + \frac{\partial\phi}{\partial x^i} = 0, \tag{5.102}$$

where $i = 1, 2, 3$.[15]

As we have seen in Section 5.1, gravitational acceleration can be expressed as an effect of the metric rather than a force. Expressed in this way our test particle must be in a locally inertial frame of reference. Its free-fall motion must therefore follow a geodesic, and the equation of motion under gravity must be expressible as a metric equation.

The geodesic equation is typically expressed in terms of the line element $\mathrm{d}s$,

$$\frac{\mathrm{d}^2 x^\sigma}{\mathrm{d}s^2} + \Gamma^\sigma_{\mu\nu}\frac{\mathrm{d}x^\mu}{\mathrm{d}s}\frac{\mathrm{d}x^\nu}{\mathrm{d}s} = 0, \tag{5.103}$$

but if we change the parameter from s to $x^0 = ct$, then

$$\frac{\mathrm{d}x^\mu}{\mathrm{d}s} = \frac{\mathrm{d}x^\mu}{\mathrm{d}x^0}\frac{\mathrm{d}x^0}{\mathrm{d}s}, \tag{5.104}$$

and the geodesic equation becomes

$$\frac{\mathrm{d}^2 x^i}{\mathrm{d}t^2} + \left(\Gamma^i_{\mu\nu} - \Gamma^0_{\mu\nu}\frac{1}{c}\frac{\mathrm{d}x^i}{\mathrm{d}t}\right)\frac{\mathrm{d}x^\mu}{\mathrm{d}t}\frac{\mathrm{d}x^\nu}{\mathrm{d}t}. \tag{5.105}$$

For nonrelativistic motion,

$$\frac{\mathrm{d}x^i}{\mathrm{d}x^0} = \frac{1}{c}\frac{\mathrm{d}x^i}{\mathrm{d}t} \ll 1. \tag{5.106}$$

Therefore in the nonrelativistic limit the geodesic equation to first order becomes

$$\frac{\mathrm{d}^2 x^i}{\mathrm{d}t^2} + c^2\Gamma^i_{00} = 0. \tag{5.107}$$

From equation (5.85) we find

$$\Gamma^i_{00} = \frac{1}{2}g^{i\sigma}\left(2\frac{\partial g_{00}}{\partial x^0} - \frac{\partial g_{00}}{\partial x^\sigma}\right). \tag{5.108}$$

We can then assume that the metric deviates only weakly from the inertial Minkowski metric, and to first order

$$g_{\mu\nu} = \eta_{\mu\nu} + h_{\mu\nu}, \tag{5.109}$$

---

[15] As before, we shall use Latin indices to represent the three spatial coordinates, and Greek indices to represent all four (space and time) coordinates.

where $\eta_{\mu\nu}$ is the Minkowski metric and $|h_{\mu\nu}| \ll 1$. If we further assume the metric is time independent, the connection term reduces to

$$\Gamma^i_{00} = \eta^{i\mu} \frac{1}{2} \frac{\partial h_{00}}{\partial x^\mu}. \tag{5.110}$$

Comparing to equation (5.102), we find

$$h_{00} = -\frac{2}{c^2}\phi, \tag{5.111}$$

and the metric becomes

$$g_{00} = -\left(1 + \frac{2}{c^2}\phi\right). \tag{5.112}$$

As before, gravity is expressed not as a force, but as a property of the metric.

We can take this concept further by expressing the gravitational field equation in terms of the metric. That is,

$$\nabla^2 g_{00} = -\frac{8\pi G}{c^2}\rho. \tag{5.113}$$

The mass density can be expressed as an energy term, $E = T_{00} = -c^2\rho$, which yields[16]

$$\nabla^2 g_{00} = \frac{8\pi G}{c^4} T_{00}. \tag{5.114}$$

This represents the gravitational field equation in the Newtonian limit.

It is important to note that this formulation of Newton's equations represents a fundamental shift in the way we view gravity. The Laplacian derivative of the metric, which measures its deviation from the "flat" Minkowski metric, is proportional to the mass-energy within our space. Put simply, the presence of mass-energy changes the shape of space and time, and gravity is an effect of that change. This is famously summarized as "Geometry tells matter how to move, and matter tells geometry how to curve."[17]

This does not, however, represent the complete and final form for general relativity. This represents only the nonrelativistic limit of the complete equations. To finish the job, we will need a covariant form of the derivative term as well as a complete description of an energy tensor.

5.3.3 The stress–energy tensor

In classical mechanics motion is determined by forces acting upon masses. However, for fluids we must define a stress vector $\mathbf{T}$ acting at a given point. With a force $\mathbf{F}$ acting upon a surface $\Delta \mathbf{A}$, where the unit vector $\mathbf{n}$ is normal to $\Delta \mathbf{A}$, the stress vector becomes

$$\mathbf{T} = \lim_{\Delta \mathbf{A} \to 0} \frac{\Delta \mathbf{F}}{\Delta \mathbf{A}} = \frac{d\mathbf{F}}{d\mathbf{A}}. \tag{5.115}$$

The stress vector is related to the Cauchy stress tensor T^{ij}, where

$$T^j = T^{ij} n_i. \tag{5.116}$$

---

[16] We introduce the minus sign here to agree with the formal definition of the stress–energy tensor.
[17] Misner, Thorne, and Wheeler (1973).

Physically the diagonal terms T^{ii} represent the pressure components, while the off-diagonal terms represent the sheer forces.

This can be generalized by noting that the force across an area boundary is equivalent to the momentum flux density across that boundary. Thus the T^{ij} elements represent momentum flux across spatial coordinates (pressure and sheer). Extending this idea to $T^{\mu\nu}$, we define the stress–energy tensor. The components $T^{i0} = T^{0i}$ are therefore the momentum density of our fluid, and T^{00} is the mass–energy density. The stress–energy tensor fully describes the matter–energy fields within a region.

Perhaps the simplest stress-energy tensor is the so-called "dust" model of matter, given by

$$T^{\mu\nu} = \rho u^\mu u^\nu, \qquad (5.117)$$

where ρ is the mass density, and u^μ is the 4-velocity of the dust particles. If the dust is at rest, the only nonvanishing term is the mass–energy density $T^{00} = \rho c^2$. Here there is no pressure term, as the dust is considered to be noninteracting. Pressure p can be included by assuming matter to be an ideal fluid. In this case

$$T^{\mu\nu} = \left(\rho + \frac{p}{c^2}\right) u^\mu u^\nu + p g^{\mu\nu}, \qquad (5.118)$$

where p and ρ are governed by an equation of state.

More complex is the stress–energy tensor for electromagnetic fields. This is not the Faraday tensor $F^{\mu\nu}$, but rather

$$T_{\mu\nu} = \frac{1}{4\pi}\left(\frac{1}{4}g_{\rho\sigma}F_{\alpha\beta}F^{\alpha\beta} - F_{\rho\alpha}F_\sigma^\alpha\right), \qquad (5.119)$$

where T_{00} is the energy density and T_{0i} is the Poynting vector.

Regardless of its form, the stress–energy tensor is symmetric ($T^{\mu\nu} = T^{\nu\mu}$) and its divergence[18] vanishes. That is

$$\nabla_\nu T^{\mu\nu} = 0. \qquad (5.120)$$

Equation (5.120) is a tensor form of the more familiar mass conservation law from classical mechanics,

$$\frac{\partial \rho}{\partial t} + \nabla \cdot (\rho \mathbf{v}) = 0. \qquad (5.121)$$

Not all symmetric tensors have a vanishing divergence. This fact plays an important role in Einstein's equations, as we shall see.

5.3.4 The Einstein equations

We have seen how $g_{\mu\nu}$ and $\Gamma^\rho_{\mu\nu}$ are affected by the curvature of space–time, but neither of these quantify curvature in a direct way. Metrics do not need to be Minkowski in order to be "flat," and $\Gamma^\rho_{\mu\nu} \neq 0$ does not necessarily mean that space–time is curved. What is needed is a way to quantify local curvature. This can be done by looking at the commutation of covariant derivatives.

Suppose we wanted to move from point A to point B, but were limited only to motion east–west and north–south. On a perfectly flat field, the order in which we travel does not matter. If we travel 300 meters east, then 200 meters north to get from A to B, we could equally have traveled 200 meters north then 300 meters east. Traveling east + north is the same as traveling north + east because our field is

---

[18] Note that this is the divergence of a tensor, not a vector. The standard divergence $\nabla \cdot \mathbf{A} = \nabla_\nu A^\nu$, and this follows a similar form, hence the term divergence.

flat. However, on a larger scale this is not true. If one travels $1,000$ km due east, then 2000 km due north, the point at which you would arrive is different from the one reached by traveling 2000 km north and then 1000 km east. Because the surface of the earth is curved, the order matters.

Mathematically the curvature can be determined by looking at the difference between traveling north+east versus east+north. This is done through the covariant derivatives. The derivative of a vector $\nabla_\beta V_\alpha$ is a measure of the rate of change along the β coordinate. Thus our north+east analogy is equivalent to taking a derivative along north then east, versus east then north. If space–time is curved, then these are not the same; hence

$$\nabla_\gamma \left(\nabla_\beta V_\alpha \right) \neq \nabla_\beta \left(\nabla_\gamma V_\alpha \right). \tag{5.122}$$

This difference allows us to define a curvature tensor known as the Ricci tensor,[19]

$$R_{\mu\nu} = \partial_\epsilon \Gamma^\epsilon_{\mu\nu} - \partial_\nu \Gamma^\epsilon_{\mu\epsilon} + \Gamma^\delta_{\mu\nu} \Gamma^\epsilon_{\epsilon\delta} - \Gamma^\delta_{\mu\epsilon} \Gamma^\epsilon_{\delta\nu}. \tag{5.123}$$

From this we can also define a Ricci curvature scalar,

$$R = R^\mu_\mu = g^{\mu\nu} R_{\mu\nu}, \tag{5.124}$$

which is a measure of local curvature.

In the nonrelativistic limit,

$$R_{00} \to \nabla^2 g_{00}, \tag{5.125}$$

and given that $R_{\mu\nu}$ is symmetric like the stress energy tensor $T_{\mu\nu}$, it is tempting to suggest[20]

$$R_{\mu\nu} \propto T_{\mu\nu} \tag{5.126}$$

as the gravitational field equation. However, for the stress tensor $\nabla_\nu T^{\mu\nu} = 0$, although this is not true in general for the Ricci tensor. We must therefore find a similar tensor with the same classical limit. This can be done by combining the Ricci tensor and scalar to define a new tensor

$$G^{\mu\nu} = R^{\mu\nu} - \frac{1}{2} g^{\mu\nu} R, \tag{5.127}$$

where

$$\nabla_\nu G^{\mu\nu} = 0. \tag{5.128}$$

This tensor is known as the Einstein tensor. It is symmetrical, has the same classical limit as $R^{\mu\nu}$, and its divergence vanishes. It therefore satisfies all the conditions necessary for a general theory of gravity.

The gravitational field equations can therefore be expressed as

$$G_{\mu\nu} = \frac{8\pi G}{c^2} T_{\mu\nu}. \tag{5.129}$$

This equation holds for raised and mixed indices as well. If we multiply the equation by $g^{\lambda\mu}$ we find

$$R^\lambda_\nu - \frac{1}{2} \delta^\lambda_\nu R = \frac{8\pi G}{c^2} T^\lambda_\nu. \tag{5.130}$$

---

[19] The covariant derivatives actually define a more general tensor known as the Riemann tensor. Because of the metric and stress–energy tensor are symmetric, we need only the Ricci tensor, which is a contracted symmetric form of the Riemann tensor.

[20] In fact, Einstein first proposed this before getting it right.

Contracting this we find

$$R = -\frac{8\pi G}{c^2} T. \tag{5.131}$$

The Einstein field equations can be expressed as

$$R_{\mu\nu} = \frac{8\pi G}{c^2} \left(T_{\mu\nu} - \frac{1}{2} g_{\mu\nu} T \right). \tag{5.132}$$

For regions absent of matter (or electromagnetic fields) both $T_{\mu\nu}$ and T vanish. Thus

$$R_{\mu\nu} = 0, \tag{5.133}$$

which are the vacuum field equations for gravity.

Because $G_{\mu\nu}$ and $T_{\mu\nu}$ are symmetric, there are only 10 degrees of freedom rather than 16. In general the Einstein field equations form a set of 10 coupled, second-order nonlinear differential equations. Fortunately there are ways to simplify the equations in many cases.

5.4 Solving Einstein's equations

5.4.1 Calculating the Ricci tensor

In general there are 40 independent connection terms $\Gamma^\sigma_{\mu\nu}$. It is therefore useful to determine which terms if any are nonvanishing in order to simplify calculations. This can be done by comparing two forms of the geodesic equation. From equation (5.99) the geodesic equation can be written as

$$\ddot{x}^\sigma + \Gamma^\sigma_{\mu\nu} \dot{x}^\mu \dot{x}^\nu = 0, \tag{5.134}$$

where $\ddot{x} = d^2x/ds^2$ and $\dot{x} = dx/ds$. The geodesic equation can also be expressed as the solution to the Lagrange equation:

$$\frac{d}{ds} \left(\frac{\partial L}{\partial \dot{x}^\mu} \right) - \frac{\partial L}{\partial x^\mu} = 0, \tag{5.135}$$

where

$$L = \frac{1}{2} g_{\mu\nu} \dot{x}^\mu \dot{x}^\nu. \tag{5.136}$$

This is an extremal equation and equivalent to the geodesic equation. Comparing these two forms one can determine the nonvanishing connection terms as differential equations.

As an example, consider the case of a static spherical metric. Because the metric is time independent, the metric terms $g_{\mu\nu}$ must be functions of the spatial terms alone. By spherical symmetry the off-diagonal terms of the metric must vanish, and the metric must be spatially isotropic. Thus we can write

$$ds^2 = -f(\mathbf{x})dt^2 + g(\mathbf{x}) \left(dx^2 + dy^2 + dz^2 \right). \tag{5.137}$$

Because we have spherical symmetry, it is useful to use spherical coordinates; thus

$$dx^2 + dy^2 + dz^2 = dr^2 + r^2\, d\theta^2 + r^2 \sin^2 \theta\, d\phi^2. \tag{5.138}$$

The Lagrangian can then be written as[21]

$$2L = -e^{N(r)} \left(\dot{x}^0\right)^2 + e^{P(r)}\dot{r}^2 + r^2\dot{\theta}^2 + r^2 \sin^2\theta\dot{\phi}^2. \tag{5.139}$$

For the x^0 term, equation (5.135) yields

$$\frac{\partial L}{\partial \dot{x}^0} = -2e^N\dot{x}^0, \qquad \frac{\partial L}{\partial x^0} = 0, \tag{5.140}$$

and

$$\ddot{x}^0 e^N + \dot{x}^0 \dot{N}e^\alpha = 0. \tag{5.141}$$

Because

$$\dot{N} = \frac{\mathrm{d}N}{\mathrm{d}s} = \frac{\mathrm{d}N}{\mathrm{d}r}\frac{\mathrm{d}r}{\mathrm{d}s} = N'\dot{r}, \tag{5.142}$$

equation (5.141) becomes

$$\ddot{x}^0 + N'\dot{x}^0\dot{x}^1 = 0. \tag{5.143}$$

From equation (5.134), the geodesic equation is

$$\ddot{x}^0 + \Gamma^0_{\mu\nu}\dot{x}^\mu\dot{x}^\nu = 0. \tag{5.144}$$

Comparison with equation (5.143) shows that all connection terms vanish except $\Gamma^0_{01} = \Gamma^0_{10}$; thus

$$\ddot{x}^0 + 2\Gamma^0_{01}\dot{x}^0\dot{x}^1 = 0, \tag{5.145}$$

and

$$\Gamma^0_{01} = \frac{N'}{2}. \tag{5.146}$$

Following the same procedure for the $x^1 = r$ variable, the nonvanishing connections are

$$\Gamma^1_{00} = \frac{N'}{2}e^{N-P}, \qquad\qquad \Gamma^1_{11} = \frac{P'}{2}, \tag{5.147}$$

$$\Gamma^1_{22} = -r\,e^{-P}, \qquad\qquad \Gamma^1_{33} = -r\sin^2\theta\,e^{-P}. \tag{5.148}$$

For the $x^2 = \theta$ components,

$$\Gamma^2_{12} = \Gamma^2_{21} = \frac{1}{r} \qquad\qquad \Gamma^2_{33} = -\frac{1}{2}\sin 2\theta, \tag{5.149}$$

and for $x^3 = \phi$,

$$\Gamma^3_{13} = \Gamma^3_{31} = \frac{1}{r}, \qquad\qquad \Gamma^3_{23} = \Gamma^3_{32} = \cot\theta. \tag{5.150}$$

For static spherical symmetry, only 9 out the the original 40 connection terms are nonvanishing.

---

[21] The metric functions N and P are written as exponentials so that the typical boundary condition that $g_{\mu\nu} \to \eta_{\mu\nu}$ for $r \to \infty$ means that the functions vanish at the boundary.

5.4.2 Static vacuum solution with spherical symmetry

We are now able to calculate the metric of a central mass M. We will assume the mass to be spherical and at the origin of our coordinate system. In the case of Newtonian gravity,

$$\nabla^2 \phi = 4\pi G\rho, \tag{5.151}$$

where ϕ is the gravitational potential. For a spherical mass M at the origin, the mass density at a point outside the mass is $\rho = 0$, and the potential is

$$\phi(r) = -\frac{GM}{r}. \tag{5.152}$$

As seen in Section 5.3, the classical limit of the metric term g_{00} becomes

$$g_{00} = -\left(1 + \frac{2}{c^2}\phi\right) = -\left(1 + \frac{2GM}{c^2 r}\right); \tag{5.153}$$

therefore a general solution must reduce to this in the classical limit.

For the general solution we will also assume points outside the central mass. For these points the stress–energy tensor $T_{\mu\nu} = 0$, and Einstein's equations reduce to their vacuum form

$$R_{\mu\nu} = 0. \tag{5.154}$$

Only three of these equations are nonvanishing:

$$-\frac{e^{N-P}}{r}\left(P_{,} - \frac{1}{r}\right) - \frac{e^N}{r^2} = 0, \tag{5.155}$$

$$-\frac{N'}{r} - \frac{1}{r^2}\left(1 - e^P\right) = 0, \tag{5.156}$$

$$-\frac{1}{2}r^2 e^{-P}\left[N'' - \frac{1}{2}P'N' + \frac{1}{2}(N')^2 + \frac{N' - P'}{r}\right] = 0. \tag{5.157}$$

Combining these equations, this reduces to

$$\frac{dP}{dr} = -\frac{dN}{dr} = \frac{1}{r}\left(1 - e^P\right). \tag{5.158}$$

Solving for $P(r)$, we find

$$e^P = g_{11} = \frac{Ar}{1 + Ar}, \tag{5.159}$$

where A is the constant of integration. Solving for $N(r)$ we find $N = -P + B$; thus

$$e^N = -g_{00} = e^B e^{-P}. \tag{5.160}$$

There are two boundary conditions to be applied to these solutions. The first is that for large distances from the central mass its gravitational attraction should be vanishingly small. Thus it should be that $g_{\mu\nu} \to \eta_{\mu\nu}$ as $r \to \infty$. That is,

$$\lim_{r\to\infty} g_{00} = -1 \qquad\qquad \lim_{r\to\infty} g_{11} = 1; \tag{5.161}$$

therefore the constant $B = 0$ and $N = -P$.

The second condition is that metric should reduce to the Newtonian approximation. Comparing the general g_{00} to the Newtonian form above, we find

$$A = -\frac{c^2}{2GM}.$$ (5.162)

The general metric is then

$$ds^2 = -\left(1 - \frac{2GM}{c^2 r}\right) c^2 dt^2 + \left(1 - \frac{2GM}{c^2 r}\right)^{-1} dr^2 + r^2 d\theta^2 + r^2 \sin^2 \theta\, d\phi^2.$$ (5.163)

This metric is the unique solution[22] of a spherically symmetric space–time for points where there is no mass–energy, and is known as the Schwarzschild metric. The Schwarzschild metric is valid for a range of spherical objects from planets to black holes, as we shall see.

The quantity GM/c^2 has dimensions of length. As such one may define the geometric mass

$$m = \frac{GM}{c^2}$$ (5.164)

for any mass, which equates it to a length. This equivalence between mass and geometry is a direct consequence of general relativity. Metrics are often expressed in terms of geometric mass (and setting $c = 1$); thus the Schwarzschild metric is often written as

$$ds^2 = -\left(1 - \frac{2m}{r}\right) dt^2 + \left(1 - \frac{2m}{r}\right)^{-1} dr^2 + r^2 d\theta^2 + r^2 \sin^2 \theta\, d\phi^2.$$ (5.165)

5.4.3 Metrics and the tetrad formalism

For our purposes it is generally useful to express our solutions in the line element form

$$ds^2 = g_{\mu\nu}\, dx^\mu\, dx^\nu,$$ (5.166)

where $g_{\mu\nu}$ is the metric. However, deriving solutions via the metric form can be computationally inefficient. For this reason, computation packages such as CARTAN calculate in the *vielbein* form commonly referred to as the tetrad. The vielbein is a transformation rule between the coordinate 1-form and a general basis; thus

$$\omega^\sigma = e^\sigma_\mu\, dx^\mu.$$ (5.167)

The tetrad is related to the metric by

$$g_{\mu\nu} = \eta_{\sigma\lambda} e^\sigma_\mu e^\lambda_\nu,$$ (5.168)

where $\eta_{\sigma\lambda}$ is the flat (Minkowski) metric.

For diagonal metrics (those where $g_{\mu\nu} = 0$ if $\mu \neq \nu$), then the tetrad elements are simply the square root of the metric elements. For example, the tetrad for preceding Schwarzschild metric is

$$e^1_r = \left(1 - \frac{2m}{r}\right)^{-1/2}, \quad e^2_\theta = r, \quad e^3_\phi = r \sin\theta, \quad e^4_t = \left(1 - \frac{2m}{r}\right)^{1/2}.$$ (5.169)

---

[22] Although the metric is a unique physical solution, the metric can be mapped to a variety of different forms.

For this reason the tetrad of an orthonormal frame can be thought of as the square root of the metric. This is not true for general metrics with off-diagonal elements, but because the metrics we will consider will be diagonal (or can be expressed in diagonal form), we do not go into the intricacies of the tetrad formalism here.

5.4.4 The Schwarzschild solution as derived by CARTAN/*Mathematica*

Mathematica is capable of doing very advanced tensor calculations through an add-on package known as CARTAN.[23] Here we show how CARTAN derives the Schwarzschild solution. As we shall see, it is much easier than calculating it by hand.

Once CARTAN is installed, it can by loaded by the command

```
<< Cartan.m
```

You may be prompted for a working directory and asked whether you want a transcript of your session. CARTAN will then ask a series of questions to determine the form of our geometry:

```
Geometry given by : 1) interactive input 2) input from file
```

Here we wish to input things directly, so we choose 1.

```
Number of dimensions[4] :
```

The default dimension is 4, so we may either enter "4" or simply hit enter.

```
Give an ordered list, e.g., {x, y, z, t}, of coordinates :
```

Because we want to use spherical coordinates, we enter {r,theta,phi,t}.

```
Type of tetrad frame :
        1) orthonormal, 2) Newman - Penrose, 3) general
Selection[1] :
```

Here we want the default orthonormal frame.

```
Metric signature[{1, 1, 1, -1}] :
```

Our metric is positive for spatial coordinates and negative for time, so again we choose the default. We also choose the default for the coordinate orientation,

```
Orientation[-1] :
```

---

[23] The CARTAN package is part of the *Tensors in Physics* add-on system written by Harald H. Soleng. It can be purchased through the Wolfram website. It is highly recommended for performing general relativity calculations in *Mathematica*.

CARTAN then asks for the tetrad type to be interpreted in matrix notation, for which we choose diagonal,

Type of tetrad :
           1) diagonal, 2) ' lower half ', 3) ' upper half ', 4) general
Selection : 1

Next one must enter the tetrad itself. For a spherically symmetric geometry our tetrad takes the form

$$\left\{ e^{\lambda(r)} , r , r\sin\theta , e^{\mu(r)} \right\},$$

so we must enter

(1 r) : E^(lambda[r])
(2 theta) : r
(3 phi) : r Sin[theta]
(4 t) : E^(mu[r])

Here the uppercase "E" must be used to denote the special *Mathematica* symbol for the exponential "e." Finally, CARTAN asks if the manifold has torsion. The default is no.

At this point the program pauses, waiting for your commands. The basic syntax of CARTAN is to use C before a quantity we wish to calculate and S before a quantity we wish to show. For example, to have CARTAN calculate the Einstein tensor $G_{\mu\nu}$, one enters

CEinstein

followed by the usual *Mathematica* shift-return. Although it looks as if this does nothing, the calculations of the Einstein equations for metric are performed. To see the results one enters SEinstein, which yields

$$G_{11} = \frac{-e^{2\,lambda[r]} + 2\,r\,mu'[r] + 1}{e^{2\,lambda[r]}\,r^2}$$

$$G_{22} = -\frac{e^{-2\,lambda[r]}\left(r\,lambda'[r]\,mu'[r] + lambda'[r] - r\,mu''[r] - r\,mu'[r]^2 - mu'[r]\right)}{r}$$

$$G_{33} = -\frac{e^{-2\,lambda[r]}\left(r\,lambda'[r]\,mu'[r] + lambda'[r] - r\,mu''[r] - r\,mu'[r]^2 - mu'[r]\right)}{r}$$

$$G_{44} = \frac{e^{-2\,lambda[r]}\left(-1 + e^{2\,lambda[r]} + 2\,r\,lambda'[r]\right)}{r^2}$$

One can also show the metric by entering SMetric, which yields

$$g_{rr} = e^{2\,lambda[r]}$$

$$g_{thetatheta} = r^2$$

$$g_{phiphi} = r^2\,Sin[theta]^2$$

$$g_{tt} = -e^{2\,mu[r]}$$

The boundary condition for our metric is that it must be "flat" at infinity; thus $\lambda(r) \to 0$ and $\mu(r) \to 0$ as $r \to \infty$. This condition must be imposed within the Einstein tensor above. Because we seek a vacuum solution,

$$G_{\mu\nu} = 0, \tag{5.170}$$

and thus

$$G_{11} + G_{44} = \frac{2e^{-2\lambda}}{r^2}\left(\mu' + \lambda'\right) = 0 \tag{5.171}$$

Thus $\mu(r) + \lambda(r) = $ constant, which applying our boundary condition means $\lambda(r) = -\mu(r)$. This can be imposed within CARTAN by the usual *Mathematica* replacement construct:

```
Einstein = Einstein /. DifferentialRules[lambda[r] → -mu[r]];
```

This simplifies the Einstein tensor quite a bit, which can be seen by again invoking `SEinstein`,

$$G_{11} = \frac{e^{2\,mu[r]}\left(1 - e^{-2\,mu[r]} + 2\,r\,mu'[r]\right)}{r^2}$$

$$G_{22} = -\frac{e^{2\,mu[r]}\left(-2\,mu'[r] - 2\,r\,mu'[r]^2 - r\,mu''[r]\right)}{r}$$

$$G_{33} = -\frac{e^{2\,mu[r]}\left(-2\,mu'[r] - 2\,r\,mu'[r]^2 - r\,mu''[r]\right)}{r}$$

$$G_{44} = \frac{e^{2\,mu[r]}\left(-1 + e^{-2\,mu[r]} - 2\,r\,mu'[r]\right)}{r^2}$$

Setting any of these equal to zero provides the differential equation necessary to determine the metric. From $G_{11} = 0$, for example, we find

$$1 - e^{2\mu(r)} + 2r\mu'(r) = 0. \tag{5.172}$$

The solution can be found by `DSolve`,

```
sol1 = DSolve[1 - e^{-2 mu[r]} + 2 r mu'[r] == 0, mu[r], r];
mU = sol1[[1, 1, 2]]
```

which gives

$$\mu(r) = \frac{1}{2}\ln\left(1 - \frac{e^{2C_1}}{r}\right). \tag{5.173}$$

Comparing our solution to the Newtonian limit, we find

$$e^{2C_1} = \frac{2GM}{c^2}, \tag{5.174}$$

and our derivation of the Schwarzschild metric is complete.

CARTAN remains active once it is called. A new session can be initiated by entering `NewSession[]`. If you want to quit CARTAN, use the command `KillAll`, and then exit *Mathematica*.[24]

5.5 Tests of general relativity

5.5.1 Perihelion advance of planetary orbits

A central prediction of Newtonian gravity is that planets move in elliptical orbits. Excluding the gravitational attraction from other orbiting bodies, a planet's elliptical orbit should remain constant, and its orientation unchanged. In actuality a planet's orbit undergoes a small amount of precession due not only to the gravitation of other planets, but also to the oblateness of the sun. This precession is typically measured in terms of the advance of a planet's perihelion. In addition, there is a rate of perihelion advance that is not accounted for by Newtonian gravity. This was first observed by Urbain Le Verrier, who calculated in 1859 that the advance of Mercury's perihelion disagreed with Newtonian gravity by about 43 arc seconds per century. This additional advance is due to the deviation of general relativity from classical gravity.

Consider the case of a single planet orbiting a perfectly spherical sun. If we assume $M_p \ll M_\odot$, the metric can be taken as the Schwarzschild metric and the planet will follow a geodesic path. For the Schwarzschild metric the Lagrangian is

$$L = -\frac{1}{2}\left(1 - \frac{2m}{r}\right)\dot{t}^2 + \frac{1}{2}\left(1 - \frac{2m}{r}\right)^{-1}\dot{r}^2 + \frac{r^2}{2}\left(\dot{\theta}^2 + \sin^2\theta\,\dot{\phi}^2\right) = 1. \tag{5.175}$$

From the Lagrangian equation (5.135), we find

$$\ddot{\theta} + \frac{2}{r}\dot{r}\dot{\theta} - \sin\theta\cos\theta\,\dot{\phi}^2 = 0, \tag{5.176}$$

$$\ddot{\phi} + \frac{2}{r}\dot{r}\dot{\phi} + 2\cot\theta\,\dot{\theta}\dot{\phi} = 0, \tag{5.177}$$

$$\frac{d}{ds}\left[\left(1 - \frac{2m}{r}\right)\dot{t}\right] = 0. \tag{5.178}$$

From the last equation we define a energy per unit mass term:

$$\left(1 - \frac{2m}{r}\right)\dot{t} = E = \text{constant}. \tag{5.179}$$

We can also choose the orbital plane by setting

$$\theta = \frac{\pi}{2}, \qquad \dot{\theta} = 0, \tag{5.180}$$

which yields for ϕ

$$\ddot{\phi} + \frac{2}{r}\dot{r}\dot{\phi} = \frac{d}{ds}\left(r^2\dot{\phi}\right) = 0, \tag{5.181}$$

---

[24] The use of CARTAN can is demonstrated in more detail in **5-0Tensors**, **5-1SsCartan1** and **5-2ScCartan2**. Additional details can be found in the PDF ebook included in the *Tensors in Physics* package.

from which we define the classical angular momentum per unit mass,

$$h = \left(r^2 \dot{\phi}\right) = \text{constant}. \tag{5.182}$$

Because

$$\dot{r} = \frac{dr}{ds} = \frac{dr}{d\phi} \frac{d\phi}{ds} = \frac{h}{r^2} \frac{dr}{d\phi}, \tag{5.183}$$

equation (5.175) can be written as

$$\frac{h^2}{r^4} \left(\frac{dr}{d\phi}\right)^2 = E^2 - \left(1 - \frac{2m}{r}\right) \left(1 + \frac{h^2}{r^2}\right). \tag{5.184}$$

Letting $u = 1/r$ and differentiating by ϕ, one finds

$$\frac{d^2u}{d\phi^2} + u = \frac{m}{h^2} + 3mu^2, \tag{5.185}$$

which is the relativistic equation of motion for a central force.

Under Newtonian gravity the equation of motion is

$$\frac{d^2u}{d\phi^2} + u = \frac{m}{h^2}, \tag{5.186}$$

which has the solution

$$u = u_0 \left(1 + e \cos \phi\right), \tag{5.187}$$

where $u_0 = m/h^2$ and e is the eccentricity of the elliptical orbit. This can be used to find an approximate relativistic solution. Substituting the classical solution into equation (5.185), and discarding higher orders of e we find

$$\frac{d^2u}{d\phi^2} + u - \frac{m}{h^2} - 6e\frac{m^3}{h^4} \cos \phi = 0. \tag{5.188}$$

This has a solution

$$u = u_0 \left(1 + e \cos \phi\right) + \frac{3em^3}{h^4} \phi \sin \phi. \tag{5.189}$$

The additional term is relatively small, so we can use the small angle identity

$$\cos \phi + \alpha\phi \sin \phi \cong \cos \left(\phi - \alpha\phi\right). \tag{5.190}$$

The relativistic solution is then

$$u \cong \frac{m}{h^2} \left(1 + e \cos \omega\phi\right), \tag{5.191}$$

where

$$\omega = 1 - \frac{3m^2}{h^2}. \tag{5.192}$$

Perihelion for the planet occurs at r_{min}, or when u is a maximum; therefore when $\omega\phi = 2\pi n$. Given perihelion at $\omega\phi = 0$, the next perihelion occurs at $\omega\phi = 2\pi$, or

$$\phi = \frac{2\pi}{1 - 3m^2/h^2} \cong 2\pi \left(1 + \frac{3m^2}{h^2}\right); \tag{5.193}$$

hence the angle through which the perihelion precesses with each revolution is

$$\Delta\phi = \phi - 2\pi = 6\pi \frac{m^2}{h^2}. \tag{5.194}$$

Although this derivation is useful in demonstrating the existence of perihelion advance in general relativity, to determine an actual advance we must eschew the common theorist practice of setting $G = c = 1$ and $GM/c^2 = m$. Rather, these values should be expressed explicitly. In this way, equation (5.185) is written as

$$\frac{\mathrm{d}^2 u}{\mathrm{d}\phi^2} + u = \frac{GMc^2}{c^2 h_r^2} + \frac{3GMu^2}{c^2}, \tag{5.195}$$

where

$$h_r = r^2 \frac{\mathrm{d}\phi}{\mathrm{d}s} \tag{5.196}$$

is the relativistic angular momentum.[25] This compares to the Newtonian form of the equation,

$$\frac{\mathrm{d}^2 u}{\mathrm{d}\phi^2} + u = \frac{GM}{h_N^2}, \tag{5.197}$$

where

$$h_N = r^2 \frac{\mathrm{d}\phi}{\mathrm{d}t} \tag{5.198}$$

is the classical angular momentum.

We can now determine the perihelion advance in *Mathematica*.[26] To compare with astronomical measurements, we must first express our constants in units of AU and years. This is easily done from known constants to find that for the sun

$$\texttt{gMc2} = \frac{GM}{c^2} = 9.87154 \times 10^{-9} \text{ AU}. \tag{5.199}$$

As shown by Grossman (1996), the angular momentum in classical theory is

$$h_N = \pm \frac{2\pi}{P_{years}} a_{AU}^2 \sqrt{\left| 1 - e^2 \right|}, \tag{5.200}$$

where a is the semi major axis, e is the orbital eccentricity, and P is the period. Kepler's third law gives

$$GM = \left(\frac{2\pi}{P_{years}} \right)^2 a_{AU}^3. \tag{5.201}$$

This means

$$\frac{GM}{h_N^2} = \frac{1}{a_{AU} \left(1 - e^2 \right)}. \tag{5.202}$$

Because we are concerned only with closed orbits, $e < 1$, we need not worry about the absolute value constraint.

From equation (5.195), the perihelion advance per revolution is

$$\Delta\phi = 6\pi \frac{\left(GM/c^2 \right)^2}{\left(h_r/c \right)^2} \approx 6\pi \frac{GM/c^2}{a \left(1 - e^2 \right)}. \tag{5.203}$$

---

[25] Note that the derivative here is $\mathrm{d}s = c\mathrm{d}\tau$, hence the necessity of c^2/h_r^2 rather than merely $1/h_r^2$ in equation 5.195.
[26] See **5-3GRtests** for more details.

For Mercury's orbit, Seidelmann (1992) gives values based on the I.A.U. tables (Epoch J2000.0), which are

$$\text{aM} = a_{AU} = 0.3870983098 \text{ AU} \qquad \text{eccM} = e = 0.2056317524914 \qquad (5.204)$$

The advance per orbit is then given by

```
constantN1[semi_, ecc_] := ─────────────
                            semi (1 - ecc²)
                                  1

Δφ = 6 π gMc2 constantN1[aM, eccM]
```

which gives $\Delta\phi = 5.01913 \times 10^{-7}$. This is of course in radians per orbit. The canonical value is expressed in seconds of arc per century,

```
               Δφ              100
ΔφsecCentury = ────── 3600 ─────────
               Degree         mercPyr
```

where `mercPyr` $= 0.24084445$ yrs is again taken from Seidelmann (1992). This gives

$$\Delta\phi = 42.985 \text{ arcseconds/century.} \qquad (5.205)$$

This canonical value can then be compared to value obtained by solving equation (5.195) directly in *Mathematica*. The Newtonian case, equation (5.197) can be solved via DSolve[],

```
sol1 = DSolve[{u''[φ] + u[φ] == gMh2, u[0.] == ─────────────,
                                                aM (1 - eccM)
                                                      1

    u'[0.] == 0.}, u[φ], φ]
```

Here the `==` operation sets the boundary conditions for our solution, given as Mercury's observed parameters. The result is

$$u = \frac{1}{r} = 2.69738 + 0.554667 \cos\phi \qquad (5.206)$$

This can be plotted by

```
orbN[φ1_] := (1 / sol1[[1, 1, 2]]) /. φ -> φ1
PolarPlot[orbN[θ1], {θ1, 0, 14}]
```

with the result seen in figure (5.2).

If we try solving the relativistic equation via DSolve, we find *Mathematica* fails to find a solution. We must therefore use NDSolve. Here we solve it over 500 Newtonian periods

```
sol4 = NDSolve[{u3''[φ] == gMh2 - u3[φ] + 3 gMc2 u3[φ]², u3[0] == ─────────────,
                                                                   aM (1 - eccM)
                                                                         1

    u3'[0] == 0.}, u3[φ], {φ, 0, 500 × 2 π}, MaxSteps → 10⁷, StartingStepSize → 0.1]
```

This produces an interpolation function, which when plotted looks almost identical to figure (5.2).

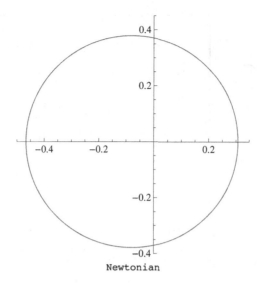

Newtonian

Fig. 5.2   Newtonian orbit of Mercury.

It is difficult from a raw orbit calculation over short time scales to determine what deviations the GR formulation produces. There are actually two types of perturbations that occur. The first ones are basically periodic terms while the second ones are secular terms describing changes that can be fit to low order powers of time. Although it is true that extremely long periodicities can produce nonlinear secular terms, it is usually agreed that the periodic terms are periodicities compared to multiples of the main orbit period. The derivation of the perihelion advance of Mercury as given above is basically a derivation of one secular perturbation. The numerical solution we have obtained contains both secular and periodic terms. This can be seen by plotting the radial difference between the relativistic and Newtonian cases over 40 Newtonian periods, seen in figure (5.3). The sinusoidal features are due to the periodic terms in the solution, while the increase in amplitude is due to the secular terms as the relativistic solution drifts away from the Newtonian one.

Here we are concerned only with the secular changes. We therefore evaluate the radial differences at intervals of $\pi/10$, then perform a moving average over 20 intervals (one period).[27]

```
uR[φz_] = sol4[[1, 1, 2]] /. φ → φz;

uN[φ1_] = sol3[[1, 1, 2]] /. φ -> φ1

diffu = Table[Abs[(uR[φz] - uN[N[Mod[φz, 2 π]]])], {φz, 0., 500 × 2 π, π / 10}];

mAv1 = MovingAverage[diffu, 20];
```

As seen in figure (5.4), the linear progression of the secular term becomes apparent.

[27] Here sol3 is the NDSolve solution for the Newtonian orbital equation, following sol1.

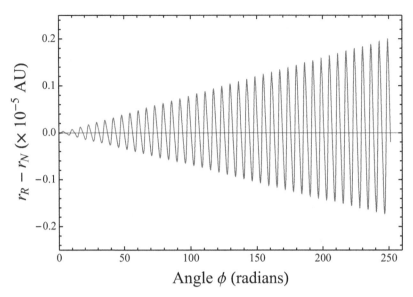

Fig. 5.3 Radial deviation between the relativistic and Newtonian orbits of Mercury.

In Newtonian coordinates:

$$u = \frac{1 - e \cos(\phi - \phi_0)}{a(1 - e^2)} = \frac{1}{a(1 - e^2)} - \frac{e \cos \phi_0 \cos \phi}{a(1 - e^2)} - \frac{e \sin \phi_0 \sin \phi}{a(1 - e^2)}. \tag{5.207}$$

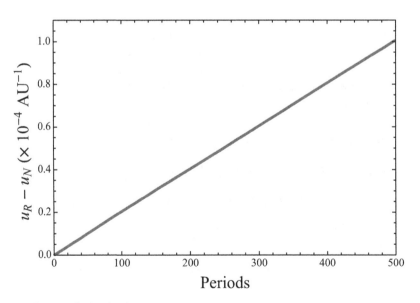

Fig. 5.4 Average deviation of $u = 1/r$ for the orbit of Mercury.

Written in this form, the proper regression is clear, but ϕ_0 is a function of time. We therefore must do multiple least squares fits to obtain the average ϕ_0 as a function of ϕ. We begin with an NDSolve solution for u,

```
sol5 = NDSolve[{u5''[φ] == gMh2 - u5[φ] + 3 gMc2 u5[φ]², u5[0] == 1/(aM (1 - eccM)),

    u5'[0] == 0.}, u5[φ], {φ, 0, 1000 π}, MaxSteps → 10⁷, StartingStepSize → 0.1]
uu5[φz_] := sol5[[1, 1, 2]] /. φ → φz;
```

We then calculate a fit for a selected region, for example near $\phi = 5$ periods,

```
ulist1 = Table[{φ5, uu5[φ5]}, {φ5, 0, 10}];
fituφ1 = LinearModelFit[ulist1, {1, Cos[φL], Sin[φL]}, φL]
qφ1 = fituφ1["BestFitParameters"]
errqφ1 = fituφ1["ParameterErrors"]
```

Once such a fit is calculated for several regions, we can do a linear regression to determine the anomaly parameter. Thus

```
total = {{5, φ01}, {35, φ02}, {85, φ03}, {155, φ04}, {250, φ05}}
φu = LinearModelFit[total, y, y, IncludeConstantBasis → False]
φu1 = φu["BestFitParameters"][[1]]
errqφ5 = φu["ParameterErrors"][[1]]
```

Converting our result to its canonical form, this yields

$$\Delta\phi = 41.1966 \pm 0.8241 \text{arcseconds/century}. \tag{5.208}$$

5.5.2 Light deflection

In Minkowski space the line element for light is given by

$$ds^2 = -c^2 dt^2 + dx^2 + dy^2 + dz^2. \tag{5.209}$$

Dividing this by dt^2,

$$\frac{ds^2}{dt^2} = -c^2 + \frac{dx^2}{dt^2} + \frac{dy^2}{dt^2} + \frac{dz^2}{dt^2} = -c^2 + v^2, \tag{5.210}$$

and since $v = c$ for light, the line element becomes $ds = 0$. In other words, light follows a null geodesic. This holds true for all metrics, therefore the path of light near our (spherical) sun is described by a null geodesic of the Minkowski metric.

In the previous section we derived the geodesic motion for a planet. Light is governed by the same geodesic equations, but since

$$h = r^2 \dot{\phi} = r^2 \frac{d\phi}{ds}, \tag{5.211}$$

it must be that $h = \infty$ for light. The null geodesic equation hence reduces to

$$\frac{d^2 u}{d\phi^2} + u = 3mu^2. \tag{5.212}$$

It is possible for light to orbit sufficiently dense objects[28] but we will here consider the scattering problem, where starlight grazes the region of a central mass.

In the absence of mass, equation (5.212) reduces to

$$\frac{d^2 u}{d\phi^2} + u = 0, \tag{5.213}$$

which has a solution

$$u = A \cos \phi, \tag{5.214}$$

which is the equation for a straight line. If we let $\phi = 0$ when the trajectory is closest to the origin (just grazing the sun) at a distance R_0, then $A = 1/R_0$, and our equation becomes

$$u = \frac{\cos \phi}{R_0}. \tag{5.215}$$

We can then find an approximate solution for the general case by substituting the massless solution into the right-hand side of equation (5.212), which yields

$$\frac{d^2 u}{d\phi^2} + u = \frac{3m \cos^2 \phi}{R_0^2}. \tag{5.216}$$

The solution to this equation is

$$u = \frac{1}{r} = \frac{\cos \phi}{R_0} + \frac{m}{R_0^2} \left(\cos^2 \phi + 2 \sin^2 \phi \right), \tag{5.217}$$

which can be written as

$$R_0 = r \cos \phi + \frac{m}{R_0 r} \left(r^2 \cos^2 \phi + 2r^2 \sin^2 \phi \right). \tag{5.218}$$

Expressing this solution in Cartesian coordinates, $x = r \cos \phi$ and $y = r \sin \phi$, we find

$$R_0 = x + \frac{m}{R_0} \frac{x^2 + 2y^2}{\sqrt{x^2 + y^2}}. \tag{5.219}$$

In the massless case, the light trajectory is a line parallel to the y-axis. When mass is present, the trajectory follows an asymptotically straight line when initially far from the mass, is deflected through an angle δ near the mass, and then again approaches a straight trajectory far from the mass. The angle between these two asymptotes is the angle of deflection. For large values of y, the trajectory approaches

$$x = R_0 - \frac{2m}{R_0} y; \tag{5.220}$$

---

[28] Seen by the solution $u = 3m$, which we will explore further in the next chapter.

therefore the angle of deflection is

$$\delta = \frac{4m}{R_0}. \tag{5.221}$$

For light that grazes the solar surface, the angle deflection is $\delta = 1.76582$ arcseconds. Newtonian gravity also predicts the gravitational deflection of light, but it is smaller than the relativistic prediction by a factor of 2. Confirmation of the relativistic deflection was first achieved by Eddington in 1919.

5.5.3 Gravitational redshift

As light leaves the surface of a star it must overcome the star's gravitational attraction. By conservation of energy the light must lose energy to the gravitational potential of the star. Because the energy of a photon is $E = h\nu = hc/\lambda$, a loss of energy means the wavelength of the light must increase as it moves farther from the star. In other words, the light is gravitationally redshifted.

Although this simplistic approach gives a qualitative understanding of the effect, the actual amount of gravitational redshift depends specifically on the metric. To calculate this effect within the Schwarzschild metric, consider two stationary points P_1 and P_2. For a stationary point,

$$dr = d\theta = d\phi = 0; \tag{5.222}$$

hence the for each point

$$ds_i^2 = -\left(1 - \frac{2m}{r_i}\right)c^2 dt^2. \tag{5.223}$$

Because the proper time is given by $d\tau = -ds/c$,

$$d\tau_i = \sqrt{1 - \frac{2m}{r_i}} dt. \tag{5.224}$$

If a light pulse travels from P_1 to P_2, each point will observe the period between successive wave crests to be $\Delta\tau_i$, whereas in the coordinate frame of the light the period is Δt. These are related by

$$\Delta t = \left(1 - \frac{2m}{r_1}\right)^{-1/2} \Delta\tau_1 = \left(1 - \frac{2m}{r_2}\right)^{-1/2} \Delta\tau_2. \tag{5.225}$$

From this we find

$$\frac{\Delta\tau_1}{\Delta\tau_2} = \sqrt{\frac{1 - 2m/r_1}{1 - 2m/r_2}}. \tag{5.226}$$

Because the frequency $\nu = 1/\Delta\tau$, the observed frequencies are related by

$$\nu_2 = \sqrt{\frac{1 - 2m/r_1}{1 - 2m/r_2}}\,\nu_1. \tag{5.227}$$

If we consider ν_1 to be the surface of a star and ν_2 to be a distant observer this reduces to

$$\nu = \sqrt{1 - \frac{2GM}{c^2 R}}\,\nu_0. \tag{5.228}$$

Often this is expressed in terms of a redshift factor

$$z = \frac{\nu - \nu_0}{\nu_0} \simeq \frac{GM}{c^2 R} \tag{5.229}$$

Gravitational redshift was verified using the Mössbauer effect by Pound and Rebka in 1959. In principle it is possible to measure the gravitational redshift of sunlight. Substituting the solar mass and radius into our equation, we get a redshift factor for the sun[29] of 2.14023×10^{-6}. Often this is expressed in velocity units, multiplying by the speed of light. Thus

$$z = 0.641626 \text{ km/s}. \tag{5.230}$$

In actual practice, such a measurement of the solar redshift from Earth is very difficult. Due to the eccentricity of its orbit, Earth has a small time-dependent radial velocity relative to the sun. Additionally there are various solar atmospheric motions. The radial velocity of Earth is given (Green, p149) as

$$v_r = eV_o \sin \phi, \tag{5.231}$$

where $e \sim 0.0167$ is the orbital eccentricity, $V_o \sim 30$ km/s is the circular velocity, and ϕ is the true anomaly. If we assume $\sin \phi$ has an average $\pi/2$, the result has an uncertainty of 0.5 km/s. The solar thermal velocity is about 0.9 km/s, and the pulsational velocity is about 10^{-4} km/s. Incorporating these motions into our calculations does not significantly change the redshift value; however, the average deviation becomes

$$\Delta z = \sqrt{(0.9)^2 + (0.5)^2 + (0.0004)^2} = 1.0247 \text{ km/s}, \tag{5.232}$$

which is twice the redshift factor itself. The adopted solar value for the redhift is 0.635 km/s, which is an observationally derived value scaled from white dwarf observations.

Because of their compact size and relatively high mass, attempts to observe the gravitational redshift in white dwarfs have been made since the 1930s with some success on many of the brighter white dwarfs (Lang, 1992). In those white dwarfs that are members of astrometric binaries, the masses derived from the gravitational redshifts provide an experimental test of general relativity (Will, 1993).

Green (1985) has derived the general relativity effects on traditional astronomical measurements with particular clarity. Green's final derivation (p. 275) yields a redshift z as a function of three terms: 1) a systematic coordinate Doppler effect, 2) a gravitational redshift term, and 3) a second-order Doppler effect caused by a difference in the velocity of the emitting region and the observing point. In the quoted observational redshifts for white dwarfs, it is assumed that the coordinate and observer Doppler effects have been removed so that only the gravitational redshift term remains.

In Chapter 4 we derived the radius of a degenerate star as a function of its mass, as seen in figure (4.10). Our fitted mass–radius function can be substituted into equation (5.229) to calculate a theoretical relationship between mass and redshift[30]. This can be compared with those white dwarfs whose mass has been determined from an astrometric binary orbit (Lang, pp. 541–543). The result is seen in figure (5.5). Observationally if the mass is not determined, it is the radius that is directly determined from the luminosity. Comparison between measured and theoretical redshifts as a function of radius can also be seen in figure (5.5).

The correlation between theoretical and observed redshifts is not very high; thus the white dwarf redshift as a "proof" of GR is not given high marks. One important factor not taken into account in either the observations or the theory was that of chemical composition. Whether that is capable of bringing the observations into agreement cannot be determined at this time.

---

[29] See **5-3GRTests** for details.
[30] See **4-6ModelWD**.

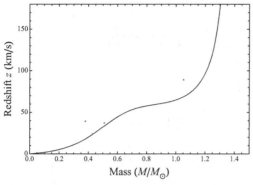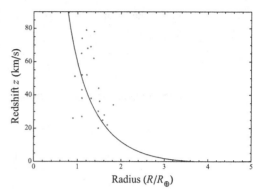

Fig. 5.5   Theoretical and observed gravitational redshifts compared.

5.5.4 Time delay

This test was first proposed by Shapiro (1964) and looks at the travel time of light in a gravitational field. Because space–time is curved by a gravitational field, the travel time near a large mass such as the Sun is longer than it would be in flat space. To demonstrate this we again consider the Schwarzchild metric:

$$\mathrm{d}s^2 = -\left(1 - \frac{2m}{r}\right)\mathrm{d}t^2 + \left(1 - \frac{2m}{r}\right)^{-1}\mathrm{d}r^2 + r^2\mathrm{d}\theta^2 + r^2\sin^2\theta\,\mathrm{d}\phi^2. \tag{5.233}$$

For simplicity we can consider the path to be within the plane $\theta = \pi/2$, and because $\mathrm{d}s = 0$ for light our metric reduces to

$$-\left(1 - \frac{2m}{r}\right)\mathrm{d}t^2 + \left(1 - \frac{2m}{r}\right)^{-1}\mathrm{d}r^2 + r^2\mathrm{d}\phi^2 = 0. \tag{5.234}$$

The null geodesic path of light in this metric is given by equation (5.218). Because the deviation of light is small, we can consider the path to be a straight line; thus

$$R_0 = r\cos\phi, \tag{5.235}$$

where R_0 is the point of closest approach to the central mass for the light path. Differentiating this we find

$$r\sin\phi\,\mathrm{d}\phi + \mathrm{d}r\cos\phi = 0. \tag{5.236}$$

Thus

$$r^2\mathrm{d}\phi^2 = \cot^2\phi\mathrm{d}r^2 = \frac{R_0^2}{r^2 - R_0^2}\mathrm{d}r^2 \tag{5.237}$$

The geodesic equation thus becomes

$$\mathrm{d}t^2 = \left[\left(1 - \frac{2m}{r}\right)^{-2} + \left(1 - \frac{2m}{r}\right)^{-1}\frac{R_0^2}{r^2 - R_0^2}\right]\mathrm{d}r^2. \tag{5.238}$$

Taking the square root to first order,

$$dt \simeq \frac{\pm r}{\sqrt{r^2 - R_0^2}} \left(1 + \frac{2m}{r} - \frac{mR_0^2}{r^3} \right) dr. \tag{5.239}$$

Integrating this from two points R_1 and R_2, we find the travel time is

$$
\begin{aligned}
T = & \left(\sqrt{R_1^2 - R_0^2} + \sqrt{R_2^2 - R_0^2} \right) \\
& + 2m \ln \left[\frac{1}{R_0^2} \left(\sqrt{R_1^2 - R_0^2} + R_1 \right) \left(\sqrt{R_2^2 - R_0^2} + R_2 \right) \right] \\
& - m \left(\frac{1}{R_1} \sqrt{R_1^2 - R_0^2} + \frac{1}{R_2} \sqrt{R_2^2 - R_0^2} \right)
\end{aligned}
\tag{5.240}
$$

The first term of this equation is the time expected for flat space, that is, the classical travel time. The latter two terms are deviations from that time. Within our solar system time delay is on the order of 100 microseconds.[31]

5.6 Gravitational waves

Unlike Newtonian gravity, general relativity allows for gravitational waves which propagate at the speed of light. These waves are analogous to electromagnetic waves, but have a somewhat different form. To demonstrate these wave solutions, consider a weak-field approximation to the metric,

$$g_{\mu\nu} = \eta_{\mu\nu} + h_{\mu\nu}, \tag{5.241}$$

where $\eta_{\mu\nu}$ is the Minkowski metric for flat space, and $h_{\mu\nu}$ is a small perturbative term. If we ignore higher order terms of $h_{\mu\nu}$, the connection coefficients for covariant differentiation become

$$\Gamma^{\alpha}_{\mu\nu} = \frac{1}{2}\eta^{\alpha\beta} \left(\partial_\mu h_{\nu\beta} + \partial_\nu h_{\beta\mu} - \partial_\beta h_{\mu\nu} \right). \tag{5.242}$$

If we express our metric in cartesian coordinates, substitution into the Einstein field equation yields

$$G_{\mu\nu} = \left(-\frac{1}{c^2}\frac{\partial^2}{\partial t^2} + \nabla^2 \right) \gamma_{\mu\nu} = 0, \tag{5.243}$$

where

$$\gamma_{\mu\nu} = h_{\mu\nu} - \frac{1}{2}\eta_{\mu\nu} h. \tag{5.244}$$

For a simple plane wave traveling along the x-axis, the solution becomes

$$\gamma_{\mu\nu} = \epsilon_{\mu\nu} \cos\left(kx - \omega t \right), \tag{5.245}$$

---

[31] See **5-3GRtests** for a numerical example of this calculation.

Fig. 5.6 Time evolution of a plus-polarized gravitational wave.

where

$$\epsilon_{\mu\nu} = \begin{pmatrix} 0 & 0 & 0 & 0 \\ 0 & 0 & 0 & 0 \\ 0 & 0 & \epsilon_{22} & \epsilon_{23} \\ 0 & 0 & \epsilon_{32} & \epsilon_{33} \end{pmatrix}, \tag{5.246}$$

with $\epsilon_{33} = -\epsilon_{22}$ and $\epsilon_{23} = \epsilon_{32}$.

If we set $\epsilon_{23} = \epsilon_{32} = 0$, then $\gamma_{\mu\nu} = h_{\mu\nu}$ and our line element becomes

$$ds^2 = -c^2 dt^2 + dx^2 + (1 + h_{22})\, dy^2 + (1 - h_{22})\, dz^2, \tag{5.247}$$

which is known as a plus-polarized gravitational wave. If we set $\epsilon_{22} = \epsilon_{33} = 0$, then again $\gamma_{\mu\nu} = h_{\mu\nu}$ and the line element becomes

$$ds^2 = -c^2 dt^2 + dx^2 + dy^2 + dz^2 + 2h_{23}\, dy\, dz, \tag{5.248}$$

known as a cross-polarized wave. The cross-polarized wave is simply a plus-polarized wave rotated by 45° in the yz-plane. This can be seen by making the subsitution

$$dy' = \frac{1}{\sqrt{2}}\,(dy + dz) \qquad dz' = \frac{1}{\sqrt{2}}\,(-dy + dz), \tag{5.249}$$

which yields the plus-polarized form seen in figure 5.6.[32]

Since the wave equation is linear for the weak-field approximation, a general gravitational plane wave can be written as a superposition of plus and cross polarized waves. Thus the general plane wave metric becomes

$$ds^2 = -c^2 dt^2 + dx^2 + (1 + h_{22})\, dy^2 + (1 - h_{22})\, dz^2 + 2h_{23}\, dy\, dz. \tag{5.250}$$

In its general form, equation (5.243) becomes

$$\left(-\frac{1}{c^2}\frac{\partial^2}{\partial t^2} + \nabla^2\right)\gamma_{\mu\nu} = -\frac{16\pi G}{c^4} T_{\mu\nu}. \tag{5.251}$$

If the source of a gravity wave is slowly moving, $T^{00} \simeq \rho c^2$, and equation (5.251) has the solution

$$\gamma^{ij} = -\frac{2G}{c^4 r}\frac{d^2}{dt^2} I^{ij}_{ret}, \tag{5.252}$$

where

$$I^{ij} = \int \rho x^i x^j d^3 x \tag{5.253}$$

---

[32] To see 3D plots of a gravitational wave, see **5-7GRWaves**.

is the second moment of the mass distribution, and is evaluated at the retarded time. For a simple example, consider two stars of equal masses m co-orbiting with a separation distance of $2a$. The position of these stars can be expressed as

$$x^i = \pm (0, a \cos \omega t, a \sin \omega t), \qquad (5.254)$$

and our metric solution becomes

$$ds^2 = -c^2 dt^2 + dx^2 + \left(1 + k \cos 2\omega t'\right) dy^2 + \left(1 - k \cos 2\omega t'\right) dz^2 + 2k \sin 2\omega t' dy\, dz, \qquad (5.255)$$

where $t' = t - r/c$ is the retarded time, and

$$k = \frac{8Gma^2\omega^2}{c^4 r}. \qquad (5.256)$$

The angular frequency of the gravitational wave is 2ω, which is twice that of the orbiting stars.

Efforts to observe gravitational waves directly are ongoing; however, gravitational waves have been indirectly observed through the energy loss of gravitational radiation. For two masses m_1 and m_2 in circular orbits, the power emitted by gravitational radiation is

$$\frac{dE}{dt} = \frac{32G^4}{5c^5} \frac{(m_1 m_2)^2 (m_1 + m_2)}{a^5}. \qquad (5.257)$$

This loss of energy from the system means that the stars gradually inspiral, reducing their distance of separation and shortening their orbital period. Peters and Matthews (1963) have shown the rate of period decrease is given by

$$\frac{dP}{dt} = -\frac{192\pi}{5c^5} \left(\frac{2\pi G}{P}\right)^{5/3} \left(1 + \frac{73}{24}e^2 + \frac{37}{96}e^4\right) \frac{m_1 m_2}{(m_1 + m_2)^{1/3} \left(1 - e^2\right)^{7/2}}, \qquad (5.258)$$

where e is the orbital eccentricity. In 1974 Russell Alan Hulse and Joseph Hooton Taylor discovered the pulsar PSR 1913+16. The pulsar had a (rotational) pulse period of about 59 milliseconds, yet systematic variation of the pulses indicated the pulsar was a binary companion. Given the regularity of pulsars rotation, this allowed Hulse and Taylor to measure the orbital parameters of the binary system with sufficient precision to verify orbital decay due to the emission of gravitational waves. For their work Hulse and Taylor were awarded the 1993 Nobel prize in Physics.

Manchester and Taylor (1977, p92) show a beautiful radial velocity curve for PSR 1913+16. We have digitized their fitted curve to obtain the "center of mass motion compensated data" used below.[33] A plot of this data is seen in figure (5.7).

In the radial velocity curve only the pulsar is "visible." That makes this case easier to analyze than the two-line binary considered in Chapter 1, but leaves information about the companion unknown. Additionally, the radial velocities in a system of two stars does not allow us to determine their masses because the inclination is not derivable unless the system is an eclipsing binary. However, if the stars are close enough that the periapses are advancing via GR or tidal forces, one can determine the masses. Such is the case for PSR 1913+16.

In Chapter 1 we presented a spectroscopic orbit solution for HD108613 using the improved Lehmann–Filhes method detailed by Smart (1960). In the notebook **1-2SpectBin** not only is there an implementation of the Smart method but a complete error analysis is given that cannot be found elsewhere. In **5-6binarypulsar** we carry out a complete binary pulsar solution with the necessary GR corrections

---

[33] See **5-6binarypulsar** where data are imported as binarypulsar.csv.

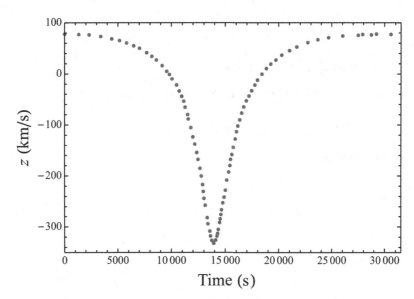

Fig. 5.7 Radial velocity of PSR 1913+16.

brought about by the use of the Schwarzschild isotropic metric. In lieu of an error analysis in this notebook, we find that our results compare favorably with the independent solutions of Will (1993) and Green (1985). In our analysis we work with both the periapsis and apoapsis parts of the velocity curve to make them equal whereas most analyses chose to analyze that part of the velocity curve that seems to have the smaller errors. That may in turn lead to systematic effects which our approach should minimize.

Spectroscopic methods can obtain the orbital period, the eccentricity, and the argument of the periapsis ω uniquely. The equation for orbital radial velocity (where z is along the line of sight) is

$$\frac{dz}{dt} = \frac{na\sin i}{\sqrt{1 - e^2}} \left(\cos(\nu + \omega) + e\cos\omega \right). \tag{5.259}$$

Here n is the daily (not yearly) motion, a is the semi-major axis, i is the inclination, ν is the true anomaly, e the eccentricity, and ω is the argument of periapsis.

The first task is to determine the period. The time scale starts when the radial velocity is maximum positive and ends a little over a period later. So we use a search to determine the period first in seconds and then in days. The indexing of our data complicates the determination of the period as ct must start as 1, not 0.

```
dataSS = Import["binarypulsar.csv"]; dblnum = Length[dataSS];
maxRV = dataSS[[1, 2]]

ct = 1
Do[If[dataSS[[ii, 2]] ≠ maxRV, ct = ct + 1], {ii, 1, dblnum}]
periodS = dataSS[[ct - 1, 1]]; periodD = periodS / (24 × 3600)
```

This yields a period of $27{,}907 \pm 332$ seconds, or $0.000884319 \pm 0.0000105276$ years.

To find $e \cos \omega$ we must convert dataSS to a table in order to determine the extrema of our velocity range

```
vR = Table[0, {ii, 1, dblnum}];

Do[vR[[ii]] = dataSS[[ii, 2]] , {ii, 1, dblnum}]
```

The maximum and minimum can then be found as

```
α1 = Max[vR]
```
```
β1 = Abs[Min[vR]]
```

The amplitude of orbital motion is then

$$\text{kay1} = \frac{\alpha 1 + \beta 1}{2} = 205.12, \tag{5.260}$$

and

$$e \cos \omega = \frac{\alpha 1 - \beta 1}{\alpha 1 + \beta 1} = -0.617021. \tag{5.261}$$

To determine $e \sin \omega$ we need to determine the area of our curve. To facilitate the use of Kepler's third law, we first convert our velocity curve to real time in years

```
vRT = Table[{0, 0}, {ii, 1, dblnum}];

Do[vRT[[ii]] = {dataSS[[ii, 1]] / (24 × 3600 × 365.25), vR[[ii]]},
  {ii, 1, dblnum}];
```

We then construct a smooth curve interpolation and determine its zeroes and minima.

```
curveRV = Interpolation[vRT]

root1 = FindRoot[curveRV[tt] == 0, {tt, .00032}]

root2 = FindRoot[curveRV[tt] == 0, {tt, .00052}]

root3 = FindRoot[curveRV[tt] == -β1, {tt, .00043}]

zero1 = root1[[1, 2]]; zero2 = root2[[1, 2]]; zero3 = root3[[1, 2]];
```

The areas can then be calculated as

$$\text{areaA = Abs}\left[\int_0^{\text{zero1}} \text{curveRV[tt] dtt}\right]; \quad \text{areaB = Abs}\left[\int_{\text{zero2}}^{\text{periodY}} \text{curveRV[tt] dtt}\right];$$

$$\text{areaC = Abs}\left[\int_{\text{zero3}}^{\text{zero2}} \text{curveRV[tt] dtt}\right]; \quad \text{areaD = Abs}\left[\int_{\text{zero1}}^{\text{zero3}} \text{curveRV[tt] dtt}\right];$$

Following Green (p476), $e \sin \omega$ can be expressed in our notation as

$$e \sin \omega = \left(\frac{\text{areaB} - \text{areaA}}{\text{areaB} + \text{areaA}}\right) \frac{2\sqrt{\alpha 1 \, \beta 1}}{\alpha 1 + \beta 1} = -0.0302459. \tag{5.262}$$

From equations (5.261) and (5.262) we can solve for e and ω. This yields[34]

$$\omega = 178.825°, \quad e = 0.617145. \tag{5.263}$$

The semi-major axis a and the stellar masses cannot be determined by classical means. This is because the inclination is unknown and cannot be determined by radial velocity. However, one can determine $a \sin i$ and the mass function as

$$a \sin i = \frac{P}{4\pi} (\alpha 1 + \beta 1) \sqrt{1 - e^2} = 716, 856 \text{ km}, \tag{5.264}$$

and

$$\frac{m_2^3 \sin^3 i}{(m_1 + m_2)^2} = \frac{3.993 \times 10^{-20} (a \sin i)^3}{P^2} = 0.140993 M_\odot, \tag{5.265}$$

where m_1 is the visible star in solar masses, m_2 is the "unseen" star, a is the semi-major axis in kilometers, and P is the period in days.

The semi-major axis and stellar masses can be determined through general relativity, specifically the effects of periapsis advance and time delay. Because the orbits of the binary system are not highly eccentric, and their separation distance is large enough that tidal interactions can be ignored, we can use the approximate equations for periapsis advance, just as we have done for the perihelion of Mercury. Thus

$$a^3 = P^2 (m_1 + m_2), \qquad \Delta\phi = \frac{6\pi \, GM_\odot c^2 (m_1 + m_2)}{a (1 - e^2)}. \tag{5.266}$$

Will (1993) quotes PSR 1913+16 with period $P = 0.323$ days, eccentricity $e = 0.617$, and periapsis advance $\Delta\phi = 4.226$ arc seconds per year. Solving these,

$$\textbf{Solve}\left[\left\{\textbf{ai} = \sqrt[3]{\textbf{per}^2 \; \textbf{mM}} \;, \; \textbf{adv} == \frac{\textbf{6} \; \pi \; \textbf{gMc2} \; \textbf{mM}}{\textbf{ai} \left(\textbf{1 - ecc}^2\right)}\right\}, \; \{\textbf{ai, mM}\}\right],$$

which yields $a = 0.01303$ AU and $m_1 + m_2 = 2.8286 M_\odot$.

For the time delay, Green (pp. 275, 487) notes that there will be a coordinate time difference between when an observer "sees" an emission event and when it happens at the coordinate origin. If we take the coordinate origin is the center of mass for the system, then

$$t_0 = t_e + \frac{\rho}{c} + \frac{2m}{c} \ln\left(\frac{r_e + r_0 + \rho}{r_e + r_0 - \rho}\right), \tag{5.267}$$

where r_0 is the distance of the center of mass, ρ is the distance of the pulsar at the time of emission, and m is the Schwarzschild semi-radius for the companion star. Using the fact that the pulsar is moving with respect to the center of mass, the ratio of $dt/d\tau$ at the time of emission can be expressed as

$$\frac{dt_e}{d\tau_e} = 1 - \frac{m_2}{2 (m_1 + m_2)} \frac{m}{a} + \frac{Gm_2}{c^2} \left(\frac{m_1 + 2m_2}{m_1 + m_2}\right) \frac{1}{r}. \tag{5.268}$$

This equation has a constant term and a variable term. Ignoring the constant term, we can integrate this with the formula for r in terms of the eccentric anomaly and the eccentricity of the orbit over one orbital

---

[34] Here we use Smart's convention. See **5-6binarypulsar** for details.

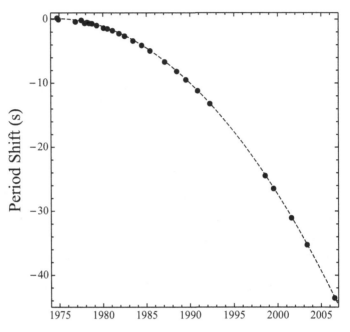

Fig. 5.8   Theoretical vs. observed period decay of PSR 1913+16.

period to get the accumulated time delay as the pulsar orbits. Green (p. 488) gives this expression as

$$\Delta T = \frac{Gm_2}{ac^2}(m_1 + m_2)\frac{m}{a} + \frac{Gm_2}{c^2}\left(\frac{m_1 + 2m_2}{m_1 + m_2}\right)\frac{eP}{2\pi} \tag{5.269}$$

This delay is determined from observation as $\Delta T = 4.4$ milliseconds; thus with the eccentricity and semi-major axis known we can solve for the individual masses

$$\texttt{Δtime} = \frac{\mathbf{4.4 \times 10^{-3}}}{\mathbf{365.25 \times 24 \times 3600}};$$

$$\texttt{Solve}\left[\left\{(\texttt{mm}_1 + \texttt{mm}_2) == \texttt{mM1, Δtime} == \frac{\texttt{gMc2 mm}_2}{\texttt{ai1}}\left(\frac{\texttt{mm}_1 + 2\ \texttt{mm}_2}{\texttt{mm}_1 + \texttt{mm}_2}\right)\frac{\texttt{ecc1 per}}{2\ \pi}\right\}, \{\texttt{mm}_1, \texttt{mm}_2\}\right]$$

This gives two sets of solutions, but only for one set are both masses positive. The result is $m_1 = 1.4154M_\odot, m_2 = 1.41319M_\odot$. These masses are almost equal in mass, and both are at the Chandrasekhar limits for white dwarfs. It thus appears in this particular case that due to angular momentum a single collapsing object could not form. However, eventually the pair will merge and form a Type I supernova.

With the masses and eccentricity known, it is now possible to determine the rate of period decay predicted by equation (5.258). Our calculated value is -75.8 microseconds/year, compared to the observed value of -76.5 microseconds/year. A comparison of the theoretical and observed period decay can be seen in figure 5.8.

Exercises

5.1   Show that $F_{\mu\nu}$ transforms as a rank-2 tensor under a Lorentz transformation.

5.2   Calculate the Ricci scalar for the surface of a sphere.

5.3   Use CARTAN to verify the structure of the isotropic Schwarzschild metric, and derive the auxiliary equations to go with it.

5.4   Calculate the observed solar redshift by calculating the redshift of light leaving the sun plus the blue shift of light reaching Earth.

5.5   Explore computationally (using NDSolve) the behavior of $h \neq 0$ particle orbits encountering a Schwarzschild singularity, using the equation from **5-4Orbsin2Times**:

$$\left(\frac{dr}{d\tau}\right)^2 = 2\epsilon + \frac{2GM}{r} - \frac{h^2}{r^2} + \frac{2h^2m}{r^3}$$

5.6   Starting with the equation for the photon paths from **5-4Orbsin2Times**

$$\left(\frac{du}{d\tau}\right)^2 + u^2 = \frac{2\epsilon}{h^2} + \frac{2GM}{h^2}u + \frac{2GM}{c^2}u^3$$

use NDSolve to work out the topological properties of u as a function of h, and compare with the classical equivalent.

5.7   Given the masses and eccentricity of the binary pulsar PSR 1913+16, and the equation for period decay, determine the time until the pulsar and its companion merges. Compare this to the time for Jupiter and the Sun.

5.8   For two masses in a circular orbit, the power emitted by their gravitational waves is

$$\frac{dE}{dt} = -\frac{32}{5}\frac{G^4}{c^5}\frac{(m_1 m_2)^2 (m_1 + m_2)}{r^5},$$

where r is their distance of separation. Calculate the approximate power radiated by PSR 1913+16. Compare this to the power radiated by the sun and Jupiter.

5.9   In **5-6binarypulsar** it is claimed that there are no apparent tidal influences on the binary pulsar orbit.

1. Calculate the distance between the star centers at closest approach.
2. Compare the radii of the components assuming neutron stars given by the theory in **4-8ModelNS**.
3. Comment on whether you think ignoring the tidal forces is justified or not.
4. Based on mass alone, either component could be a white dwarf. Use your separation results to argue that this is not likely.

5.10   Look up Roche limit on *Wikipedia*. Use that theory to comment on tidal disruption of the binary pulsar as the gravitational radiation makes the two components draw closer.

5.11   In Danby (1988) there is a discussion (p. 117) about the tidal potential of a homogeneous ellipsoid. Find this expression (or the equivalent using *Wikipedia* under "Roche sphere") and use Robertson's GR weak field approximation (**5-5twobody GR**) to develop a zero-order tidal potential for the binary pulsar (**5-6binarypulsar**). Use this formulation to comment on whether tidal forces need to be included in the orbital analysis.

6

Binaries and clusters

Although the two-body problem in Newtonian dynamics is easily solved, the motion of three or more gravitating objects has no analytical solution. To explore the behavior of clusters (the N-body problem) we must approach them computationally. In this chapter we focus on the Hamiltonian approach, as *Mathematica* is especially suited for obtaining practical solutions of Hamiltonian equations. We will start with a simple Newtonian problem and show how it is transformed into Hamiltonian form. We will then move to a description of the many ways *Mathematica* and its available tools support the study of differential equations in general and Hamiltonian equations in particular. A good foundation to the Hamiltonian approach can be found in Abraham and Marsden (1980), which covers analytical dynamics, Hamiltonian dynamics, and its application to celestial mechanics.

As we will see, the N-body problem often leads to chaotic motion, specifically deterministic chaos. Deterministic chaos has its roots in the celestial mechanics of the solar system, which is one example we will explore. The first text to hint at chaos (as we now understand it) was in Poincaré's book *Les méthodes nouvelles de la Méchanique celeste*. We will here follow its English translation (Poincaré, 1993). A modern perspective of the subject can be found in Diacu and Holmes (1996) and Peterson (1993), although references on the subject are vast and a comprehensive list is not covered here.

6.1 Variational mechanics

The name variational mechanics comes from the fact that both the Lagrangian and Hamiltonian equations of motion originate in the minimization of some appropriate quantity using calculus of variations. In modern times, the wide availability of numerical integration procedures means the study of dynamical systems in Newtonian form is often taken for granted. But in the early part of the twentieth century, long before real galactic or relativistic mechanics existed, astronomers were very much concerned with solving exceedingly complex dynamical theories such as the motion of Earth's moon under the perturbation of the planets. Lacking electronic computers, the most advanced analytical theories were relied upon.

Most undergraduate physics and astrophysics students have an acquaintance with advanced Newtonian mechanics and its relationship to Lagrange's equations. Some students have seen discussions of Hamiltonians in nonrelativistic classical mechanics situations. The class in which students most frequently encounter Hamiltonians is quantum mechanics, but the quantum formulation is not always helpful for formulating dynamical problems in other fields. Here we trace the evolution of a simple dynamical situation from Newtonian mechanics through Lagrangians and Hamiltonians to a general theory of canonical transformations. This will lead us to the Poisson or Poincaré bracket formulation necessary for understanding advanced gravitational situations such as lunar orbit theory, galactic dynamics, or satellite dynamics around irregularly shaped objects.[1]

6.1.1   Derivation of Hamiltonians from Lagrangians

In general, a Lagrangian $L = T - V$ is expressed as $L(q_i, \dot{q}_i, t)$, where q_i and $\dot{q}_i$ are the position coordinates and their time derivatives. The corresponding momenta and their time derivatives are given by

$$p_i = \frac{\partial L}{\partial \dot{q}_i}, \qquad \dot{p}_i = \frac{\partial L}{\partial q_i}. \tag{6.1}$$

The Legendre transformation gives the Hamiltonian

$$H(q_i, p_i, t) = \sum_i \dot{q}_i \mathrm{d} p_i - L(q_i, \dot{q}_i, t), \tag{6.2}$$

and the Hamilton equations of motion are then defined as

$$\dot{q}_i = \frac{\partial H}{\partial p_i}, \qquad -\dot{p}_i = \frac{\partial H}{\partial q_i}, \qquad -\frac{\partial L}{\partial t} = \frac{\partial H}{\partial t}. \tag{6.3}$$

There are a variety of ways to construct proper Lagrangians and then transform them into Hamiltonians, which in turn are arranged into the equations of motion. Perhaps the most elegant method is given by Goldstein, Poole, and Safko (p. 339). They start with a scalar version that uses implicit summation (Einstein convention)[2]:

$$L(q, \dot{q}, t) = L_o(q, t) + \dot{q}_i a_i(q, t) + \dot{q}_i^2 T_i(q, t), \tag{6.4}$$

where, if there is an ordinary potential function, $L_o = -V$, the a_i terms are coefficients of the linear velocity terms (e.g., drag), and the T_i terms are coefficients of the quadratic velocity terms (kinetic energy). This can be expressed in matrix form, where

$$\mathbf{q} = \begin{pmatrix} q_1 \\ q_2 \\ \vdots \\ q_n \end{pmatrix}, \qquad \mathbf{q}^T = \begin{pmatrix} q_1 & q_2 & \cdots & q_n \end{pmatrix}, \tag{6.5}$$

and $\mathbf{T}$ is an $n \times n$ square matrix. The Lagrangian then becomes

$$L(q, \dot{q}, t) = L_o(q, t) + \dot{\mathbf{q}}^T \mathbf{a} + \frac{1}{2} \dot{\mathbf{q}}^T \mathbf{T} \dot{\mathbf{q}}, \tag{6.6}$$

and the Legendre transformation becomes

$$H = \dot{\mathbf{q}}^T \mathbf{p} - L, \tag{6.7}$$

where $\mathbf{p}$ is a new column vector

$$\mathbf{p} = \mathbf{T} \dot{\mathbf{q}} + \mathbf{a}. \tag{6.8}$$

---

[1] See chapters 8 and 9 of Goldstein, Poole, and Safko (2002) for an extremely comprehensive discussion.
[2] For implicit summation, repeated indices imply summation. Thus $a_i b_i = \sum_i a_i b_i$.

The beauty of the matrix formulation is that multiplication by the inverse matrix of $\mathbf{T}$ produces $\dot{\mathbf{q}}$. That is,

$$\mathbf{T}^{-1}\mathbf{p} = \dot{\mathbf{q}} + \mathbf{T}^{-1}\mathbf{a}, \tag{6.9}$$

thus

$$\dot{\mathbf{q}} = \mathbf{T}^{-1}(\mathbf{p} - \mathbf{a}). \tag{6.10}$$

The Hamiltonian then becomes

$$H(q, p, t) = \frac{1}{2}\left(\mathbf{p}^T - \mathbf{a}^T\right)\mathbf{T}^{-1}(\mathbf{p} - \mathbf{a}) - L_o(q, t). \tag{6.11}$$

Starting with the matrix Lagrangian, it is a straightforward path to construct the Hamiltonian as well as the equations of motion in matrix form. Such an expression can be put into *Mathematica* format by substituting a list structure for matrices and vectors.

As an example, consider the simple projectile motion in three dimensions with gravity acting along the z-axis. We first define the velocity of our projectile,

```
vfor = {vx, vy, vz};
```

It should be noted that *Mathematica* treats a single list as a column vector, even though it looks like a row vector. Normally this means one must use the transpose operation to build the Hamiltonian in vector form. If instead the *Mathematica* "dot" notation is used, *Mathematica* understands the difference between left and right multiplication of matrices. To see its column nature, you can invoke `MatrixForm`,

```
vfor // MatrixForm
```

which yields

$$\text{vfor} = \begin{pmatrix} v_x \\ v_y \\ v_z \end{pmatrix}. \tag{6.12}$$

In the same way, we define q, p, and L_o,

```
q = {x, y, z};
p = {px, py, pz};
potentialqt = m g z;
```

Because the kinetic energy is $\frac{1}{2}mv^2$, the T matrix is simply the mass in diagonal form,

```
tee = {{m, 0, 0}, {0, m, 0}, {0, 0, m}};
```

The Hamiltonian can then be expressed as

```
hamiltonianXYZ = - (p - vfor).Inverse[tee].(p - vfor) + potentialqt
                 2
             1
```

The $\dot{p}$ equations of motion can be found from

```
∂ₓ hamiltonianXYZ
∂ᵧ hamiltonianXYZ
∂_z hamiltonianXYZ
```

while the $\dot{q}$ equations of motion are found from

∂_{px} `hamiltonianXYZ`
∂_{py} `hamiltonianXYZ`
∂_{pz} `hamiltonianXYZ`

The resulting equations can then be solved with `NDSolve`, which yields the expected result.

6.1.2 Conversion to symplectic form

One common feature of the equations of motion, whether Lagrangian or Hamiltonian, is that any variable q_i not explicitly included in either the Lagrangian or Hamiltonian is *ignorable*. This means its corresponding momentum p_i is a constant (invariant) of the motion. Quantities such as the total energy, linear momentum, or angular momentum are often invariant. In the numerical integration of such quantities there is no reason to calculate them. If you neglect this point and keep them in your equations of motion, most numerical integrators cannot recognize their invariant nature. This can accumulate errors and degrade the performance of the integrator.

Another common feature is the use of curvilinear orthogonal coordinates, which have computational advantages. If you cannot put equations into Hamiltonian form, *Mathematica* has a method within `NDSolve` called the "OrthogonalProjection" method. Basically if the equations of motion are orthogonal then the algorithm attempts to keep them that way. This is accomplished through error control of the orthogonal errors.[3] This orthogonal methodology, although it handles Hamiltonians, is a bit wasted on them. Instead `NDSolve` has a special method just for the most common Hamiltonian situations, and it is this methodology we now explore.

After deriving the matrix-vector formalism for writing the Hamiltonian form, Goldstein, Poole, and Safko (p. 339) point out that typically Hamiltonians do not produce symmetrical equations of motion without some reorganization of the variables and equations into a symplectic structure. The symplectic form is a matrix organization of the equations of motion.

To construct the symplectic form, we first define a single vector ξ (q_i, p_i) from the three position and three momentum components. The six Hamiltonian equations can then be written as

$$\dot{\xi} = \mathbf{J}\xi, \tag{6.13}$$

where

$$\mathbf{J} = \begin{pmatrix} 0 & 0 & 0 & 1 & 0 & 0 \\ 0 & 0 & 0 & 0 & 1 & 0 \\ 0 & 0 & 0 & 0 & 0 & 1 \\ -1 & 0 & 0 & 0 & 0 & 0 \\ 0 & -1 & 0 & 0 & 0 & 0 \\ 0 & 0 & -1 & 0 & 0 & 0 \end{pmatrix}. \tag{6.14}$$

We then define a canonical vector $\chi = \chi(\xi)$ such that they are related by the transformation

$$\dot{\chi} = \mathbf{M}\dot{\xi}. \tag{6.15}$$

---

[3] For equations of motion that are not orthogonal but still have known invariants, *Mathematica* provides an analogous "Projection" method in `NDSolve`.

If the matrix **M** is able to transform **J** into itself,

$$\mathbf{M}^T \mathbf{JM} = \mathbf{J}, \tag{6.16}$$

then the symplectic conditions is said to be satisfied, and the vector time dependent transformation is canonical; hence

$$\xi(t_o) = \xi(t). \tag{6.17}$$

It is at this point that NDSolve uses the Runge-Kutta numerical formalism to implement the above canonical transformation and numerically solve the differential equation.

In *Mathematica* we must state the time dependence of the Hamiltonian explicitly:

```
hamiltonianXYZ = ( 1/2 (p - vfor) . Inverse[tee] . (p - vfor) + potentialqt) /.
    {px → px[t], py → py[t], pz → pz[t], z → qz[t]}
```

We then define ξ, $\dot{\xi}$, **J**, and the equations of motion.

```
ξ = {qx[t], qy[t], qz[t], px[t], py[t], pz[t]}
ξd = {qx'[t], qy'[t], qz'[t], px'[t], py'[t], pz'[t]}
pd = {∂qx[t] hamiltonianXYZ, ∂qy[t] hamiltonianXYZ, ∂qz[t] hamiltonianXYZ,
    ∂px[t] hamiltonianXYZ, ∂py[t] hamiltonianXYZ, ∂pz[t] hamiltonianXYZ}
jay2 = ConstantArray[0, {6, 6}]
jay2[[1, 4]] = jay2[[2, 5]] = jay2[[3, 6]] = 1;
jay2[[4, 1]] = jay2[[5, 2]] = jay2[[6, 3]] = -1;
ξd2 = jay2.pd
```

To utilize NDSolve we must use == between equation sides; thus we must flatten the equations to equate them.

```
equdot1 = Flatten[Table[{ξd[[ii]] == ξd2[[ii]]}, {ii, 1, 6}]]
```

This is a symplectic arrangement that NDSolve uses to construct the algorithm for the Symplectic Partitioned Runge–Kutta (SPRK) method that we will use to solve Hamiltonian formulated equations of motion.

6.1.3 Comparison of NDSolve methods

Mathematica has several built-in equation sets and utilities to use them. The list of these sets and utilities can be seen by entering

```
$NDSolveProblems
Names["DifferentialEquations`NDSolveUtilities`*"]
```

As an example we consider the motion of a single object in a gravitational field, given by `Kepler`. We first upload the equations into *Mathematica*

```
system = GetNDSolveProblem["Kepler"];
vars = system["DependentVariables"];
```

This problem has two invariants that should remain constant (energy and angular momentum). A given numerical method may not be able to conserve either or both of these invariants, but they can be seen by

```
invs = system["Invariants"]
```

Because of the complexity of the Kepler problem, the Explicit Runge–Kutta method does not do well here. We must therefore use the Implicit Runge–Kutta method. This class of solvers has many good properties, and includes the Radau methods used by NASA JPL labs to calculate spacecraft and planetary positions. The general implicit Runge–Kutta method can be invoked by

```
NDSolve[system, time,
  Method → {"FixedStep", Method → {"ImplicitRungeKutta", "DifferenceOrder" → 10}},
  StartingStepSize → 1 / 10]
```

The conserved invariants are not held constant in this method. The deviations from constant can be seen using `InvariantErrorPlot[]`, which can be seen in figure (6.1), with energy in gray and angular momentum in black. It is clear that there is a steady increase in the errors of both as the integration proceeds. Thus the generic Implicit Runge–Kutta method should not be used.

A better variation of this method is known as the Gauss with Newton Solver. The `ImplicitSolver` method of `ImplicitRungeKutta` has the options `AccuracyGoal` and `PrecisionGoal` that specify the absolute and relative error to aim for in solving the nonlinear system of equations. However, for certain types of problems it can be useful to solve the nonlinear system up to the working precision. Such optimization is helpful here. Thus

```
NDSolve[system, time,
  Method → {"FixedStep", Method → {"ImplicitRungeKutta",
      "DifferenceOrder" → 10, "ImplicitSolver" → {"Newton",
        AccuracyGoal → MachinePrecision, PrecisionGoal →
          MachinePrecision, "IterationSafetyFactor" → 1}}},
  StartingStepSize → 1 / 10]
```

The invariant errors of this approach can again be seen in figure (6.1). Here the Hamiltonian energy is now bounded, as it should be. The second invariant is also conserved exactly (up to a round-off). This is because the Gauss Implicit Runge–Kutta is a symplectic integrator, and it conserves quadratic invariants.

Still better is to use Symplectic Partitioned Runge–Kutta method. Thus

```
NDSolve[system, time, Method → {"SymplecticPartitionedRungeKutta",
    "DifferenceOrder" → 10, "PositionVariables" → {Y₁[T], Y₂[T]}},
  StartingStepSize → 1 / 100, MaxSteps → Infinity];
```

Again plotting the invariant errors in figure (6.1), we see the SPRK method yields errors significantly smaller than the other methods. This shows the superiority using a symplectic methodology with a sophisticated error correction scheme.

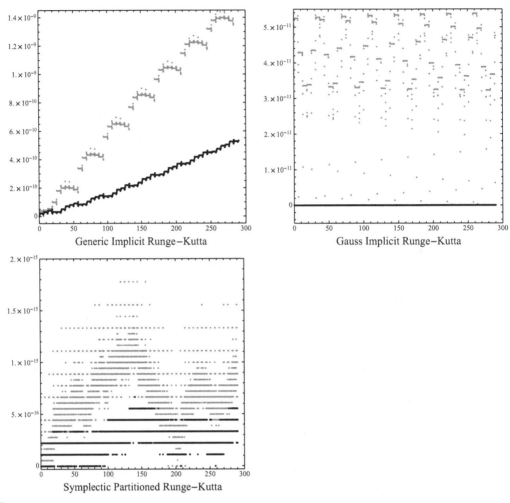

Generic Implicit Runge–Kutta

Gauss Implicit Runge–Kutta

Symplectic Partitioned Runge–Kutta

Fig. 6.1 Invariant errors for `NDSolve` methods.

6.1.4 Additional examples

A very useful guide to nonlinear differential equations including examples from the package `NDSolveProblems[]` is Davis (1966). This book is older but available in Dover reprint edition and bridges the transition from analytic to computational matters very nicely. To facilitate additional exploration of the available examples within *Mathematica* here is a list of astrophysical examples and their applications, including those illustrated in **6-1VariationalMechanics** and elsewhere. Those in bold indicate explicit use in this text's notebooks.

1. Arnold Beltrami Childress – chaotic advection flow equation – three-variable, no invariants, sinusoidal driving. (Stellar Atmospheres)
2. Arenstorf – Regularized planar three-body problem, no invariants. (Celestial Mechanics)
3. Burgers PDE – 2D, no invariants, many problems in fluids, etc. (Galactic Dynamics – Jet Motions)

4. **Henon Heiles** – Well-known chaotic system – Hamiltonian. (Galactic Dynamics-Energy invariant)
5. **Kepler** – Standard elliptical motion problem, with angular momentum and energy invariants. (Celestial Mechanics)
6. Lotka Volterra – Standard predator–prey problem; can be chaotic. Hamiltonian. Also applies to galactic cannibalism in early universe. (Galactic Dynamics)
7. **Perturbed Kepler** – GR perturbed Kepler problem, energy and angular momentum invariants. (Celestial Mechanics)
8. **Pleiades** – Seven-body planar celestial mechanics problem, second-order equations with no invariants. (Celestial Mechanics)
9. **Pollution** – A 30 constituent chemical reaction system, no invariants. (Interstellar Chemistry)
10. RigidBody – 3D Rotation Dynamics; invariants are not energy and angular momentum. (Asteroid Dynamics)

6.2 The N-body problem

One way that systems of ordinary differential equations (ODEs) arise in the physical sciences is in the description of the motion of N interacting classical particles. Newton solved this problem for the case of two particles (the two-body problem) interacting via a central force. However, for three or more particles there is no general analytic solution, and numerical techniques are of great importance in understanding the motion.

6.2.1 N-body symplectic procedure in `NDSolve`

In the numerical method known as N-body dynamics, the coupled equations of motion for the N particles are simply integrated forward (or backward) in time. Unlike the example shown by Dubin (2003) that uses Newton's law for each star, the strategy here is to use a Hamiltonian formulation so that we can take advantage of the Symplectic Partitioned Runge-Kutta (SPRK) method which uses Compensated Summation to reduce the roundoff error.

In modern times, N-body codes have been written to take advantage of massively parallel hardware structures to produce computations describing the dynamics of millions of interacting but identical particles. So why are we doing the problem over again in such a "primitive" fashion as direct Runge–Kutta numerical integration when all sorts of modern codes are already available? The reason is that these codes typically assume ensembles of equal mass particles. Of course that works fine in problems that simulate molecular flow or kinetic theory of identical particles, but real star groups (particularly those containing the brightest and most massive stars) are composed of "particles" that can be identified and labeled. The "few body" problem rather than the "many body" problem is the one we address here. There are a number of complications that are present in dynamical studies of the real clusters that we ignore for the moment that will make the problem comparatively simple and computationally tractable on personal desktop or laptop computers.

As stated in the `NDSolve` documentation, "When numerically solving Hamiltonian dynamical systems it is advantageous if the numerical method yields a symplectic map." In such a case, the phase space of the system's Hamiltonian is a symplectic manifold, which means there is a natural symplectic structure in the canonically conjugate coordinates. As a result, the Poincaré integral invariants associated with the symplectic structure are preserved.

In practical terms, if the Hamiltonian can be written in separable form,

$$H(p, q) = T(p) + V(q), \tag{6.18}$$

then *Mathematica* can use an efficient class of explicit symplectic numerical integration methods to solve the problem. A symplectic integrator can compute exactly (assuming infinite precision), the evolution of a nearby Hamiltonian, whose phase space structure is close to that of the original system. The Newtonian version of the N-body (point mass) problem is the perfect example for this method within NDSolve.

6.2.2 Deriving Lyapunov coefficients from initial conditions

In this chapter we concentrate on Hamiltonian systems that arise in the N-body problem so we can use the SPRK feature of NDSolve. The stability of such systems is most often discussed in the context of their behavior under perturbation, particularly in studies of deterministic chaos. If we limit our discussion to the information given in books that offer more than passing interest in Hamiltonian systems, our choice becomes quite narrow, consisting of only five out of some thirty works consulted. They are Goldstein, Poole, and Safko (2002); Hilborn (1994); Ott (1993); Drazin (1992); and Rasband (1990).

As a measure of chaotic behavior, the Lyapunov (variously spelled as Liapunov, Liapounov, or Ljapunov) exponent is well known.[4] Probably the most interesting of the various suggested ways to estimate the Lyapunov exponents of a system is based on the use of initial conditions. According to the Kaplan–Yorke Conjecture (Hilborn, p. 435; Ott, p. 134) there are as many Lyapunov exponents as there are state-space[5] dimensions, and that once the Lyapunov exponents are determined they may be used to estimate the fractal dimension of the attractor. A multifractal dimension determined that way is called the Lyaponov dimension.

In chaos theory, the so-called map functions are defined such that $f(x) \rightarrow x$ in an iterative fashion. For this situation, Hilborn (p. 197) and many others give an operational definition of the Liapunov exponent in terms of the average of the natural logs of the first derivatives of the map function evaluated at a succession of points. Because we are dealing with first-order differential equations in six variables[6] the Lyapunov exponents are found from the eigenvalues of the Jacobian matrix of the six first-order time derivative functions with all the initial conditions substituted into them.

If $f_1(i)$, $f_2(i)$, $f_3(i)$, $f_4(i)$, $f_5(i)$, and $f_6(i)$ are the right-hand sides of the Hamiltonian equations for $i = 1 \rightarrow n$ objects, then the Jacobian is a $6n \times 6n$ matrix of derivatives:

$$F = \begin{pmatrix} \frac{\partial f_1(1)}{\partial q_1(1)} & \cdots & \frac{\partial f_1(1)}{\partial q_6(1)} & \cdots & \frac{\partial f_1(n)}{\partial q_1(n)} & \cdots & \frac{\partial f_1(n)}{\partial q_6(n)} \\ \vdots & & \vdots & & \vdots & & \vdots \\ \frac{\partial f_6(1)}{\partial q_1(1)} & \cdots & \frac{\partial f_6(1)}{\partial q_6(1)} & \cdots & \frac{\partial f_6(n)}{\partial q_1(n)} & \cdots & \frac{\partial f_6(n)}{\partial q_6(n)} \end{pmatrix} \tag{6.19}$$

This matrix is sometimes called the Floquet matrix (Hilborn, p. 151; Ott, p. 119).

---

[4] Goldstein, Poole, and Safko, p. 491; Drazin, p. 140; Hilborn, p. 138; Ott, p. 129; and Rasband, p. 187, just to name a few.

[5] A term preferred by Hilborn, but it is equivalent to a phase space.

[6] At least three variables are normally needed for chaos from the Poincaré–Bendixson Theorem.

If one of the eigenvalues is positive then chaos is possible. In a Hamiltonian system that is conservative, the Lyapunov exponents come in positive/negative pairs with an even number of values that are 0. The sum of these for a conservative Hamiltonian is also zero (Hilborn, p. 436). Conservative Hamiltonian systems do not have phase space attractors (Hilborn, p. 372). The initial conditions in Hamiltonian systems are quite crucial whether they are chaotic or not. Some values lead to regular behavior while others lead to chaos even in the same system (Hilborn, p. 340).

6.2.3 Virial approach to the N-body problem

In Chapter 3, we applied the virial theorem to stellar pulsation, among other things. Although the virial theorem has wide application in statistical mechanics, we here focus on the virial theorem as applied to the stellar N-body problem.

Goldstein, Poole, and Safko (2002) apply the theorem to the lowest order system of point mechanics, the central force problem, which is fundamental to the consideration of two-body problems. But for explanations in the context of the classical astronomical N-body problem we rely on four somewhat older references: Chandrasekhar (1942), Pollard (1966a, 1966b), and Contopoulos (1966). The Pollard and Contopoulos papers were written during a time period when transition from analytical to computational methods was rapidly occurring, and both have details that are often omitted from more modern treatments. We concentrate on the Chandrasekhar methodology because it is given in more detail suitable for direct applicability to computation. How that early work fits into that of others is covered excellently in the summary of Contoupolis.

The virial theorem has its roots firmly in the origin of Newtonian mechanics. Lagrange applied the virial theorem to the three-body problem. Jacobi extended that to N-bodies. In every ensemble of self-gravitating point masses, there are three conservation theorems: linear momentum, angular momentum, and energy. Linear momentum is conserved such that the center of mass of the system is inertial. If we take the center of mass as the origin of the coordinate system, then

$$\sum_k m_k \mathbf{r}_k = 0, \qquad \sum_k m_k \mathbf{v}_k = 0. \tag{6.20}$$

For energy conservation, we define U as the negative of the potential energy, such that

$$U = \sum_{i \neq j} \frac{G m_i m_j}{r_{ij}}, \tag{6.21}$$

then with T as the kinetic energy,

$$T = U + E, \tag{6.22}$$

where E is the total energy.

The Lagrange–Jacobi identity gives a more useful form. If I is the moment of inertia, then

$$I = \frac{1}{2} \sum_k m_k \left(\mathbf{r}_k \cdot \mathbf{r}_k \right), \tag{6.23}$$

and

$$\ddot{I} = 2T - U. \tag{6.24}$$

While Pollard prefers the I definition of moment of inertia, Contopoulos prefers $J = 2I$, so one must be careful of factor 2 differences in the formulae. Pollard (1966a, 1966b) goes on to show some interesting inequalities of the system. If L is the total angular momentum, then Sundman's inequality is

$$L^2 \leq (2I)(2T), \qquad L^2 \leq 4I\left(\ddot{I} - E\right). \tag{6.25}$$

What this shows is that total collapse of a system ($I = 0$) can take place only if $L = 0$.

In its traditional form, the virial theorem assumes that U stays finite after $t = 0$. Then if we define the time averages as

$$\hat{T} = \lim_{t \to \infty} \frac{1}{t} \int_0^t T(\tau)\, d\tau, \qquad \hat{U} = \lim_{t \to \infty} \frac{1}{t} \int_0^t U(\tau)\, d\tau, \tag{6.26}$$

and if the time average of $\ddot{I} = 0$, then

$$2\hat{T} = \hat{U}. \tag{6.27}$$

This is the usual statement of the virial theorem, but Pollard (p. 44) shows how to justify the above average energy condition based on two limits

$$\lim_{t \to \infty} \frac{1}{t^2} I(t) = 0, \qquad \lim_{t \to \infty} \frac{1}{t}\dot{I}(t) = 0. \tag{6.28}$$

Remembering that $U = -V$ (negative of potential energy), the virial theorem is also stated as

$$2\hat{T} = -\hat{V}, \qquad 2\hat{T} + \hat{V} = 0. \tag{6.29}$$

Also note that $E = T + V$.

If $E > 0$, Contopoulos (p. 175) shows that if $I \to \infty$ as $t \to \infty$, then if m_l is the largest mass of any star in the ensemble we have

$$m_l \sum_{i-1}^{n} r_i^2 \geq 2I, \tag{6.30}$$

so that the ejection of a single star is possible because its distance can tend to infinity without having other stars do the same. This is known as evaporation from a cluster.

Contopoulos (p. 178) also shows how to analyze a cluster of equal mass objects (a model globular cluster) to obtain a total mass estimate using the virial theorem.[7] But as shown by Chandrasekhar (1942), the virial system for an isolated system is valid only if it is in a relaxed state where original orbits are obliterated by stellar encounters. Collins (p. 18) has pointed out that consideration of the ergodic (or its modern version, the quasi-ergodic) hypothesis is more important to the validity of the virial theorem than simple considerations of relaxation time, particularly in stellar dynamics. We will assume that any system to which we are applying the virial is one that is completely relaxed, and that some sort of collisional equilibrium such as that encountered in an isothermal ideal gas has been achieved. That type of state is what Contopoulos refers to as stationary, which means the statistical distribution function of the velocities is steady state and not evolving in time.

1. If the system is stationary, the total kinetic energy T is defined as

$$T = \frac{M}{2}\langle v^2 \rangle, \tag{6.31}$$

where M is the total mass and $\langle v^2 \rangle$ is the RMS velocity of the entire cluster. The mass M can then be estimated with the virial theorem.

[7] For modern studies of globular clusters see Heggie and Hut (2003), http://www.maths.ed.ac.uk/~douglas/gmbp/gmbp.html http://www.ids.ias.edu/~piet/act/astro/million/index.html.

2. A second needed assumption is that the stars in the system are of equal mass.

3. An auxiliary quantity is the effective cluster radius r_{eff}, shown by Chandarsekhar to be

$$\frac{n(n-1)}{r_{eff}} = \frac{1}{2}\sum_{i=1}^{n}\sum_{j=1}^{n}\frac{1}{r_{ij}}, \qquad i \neq j. \tag{6.32}$$

4. Invoking the virial theorem we have

$$M\langle v^2 \rangle = \frac{G}{2}\sum_{i=1}^{n}\sum_{j=1}^{n}\frac{m_i M_j}{r_{ij}} = \frac{Gn(n-1)m^2}{r_{eff}} \sim \frac{GM^2}{r_{eff}}, \tag{6.33}$$

   or

$$\langle v^2 \rangle = \frac{GM}{r_{eff}}. \tag{6.34}$$

5. Introducing yet another assumption from statistical mechanics, we assume equipartition of energy so that the 3D kinetic energy can be estimated from observations of the speeds along the line of sight; thus

$$\langle v_r^2 \rangle = \frac{GM}{r_{eff}}, \tag{6.35}$$

   where v_r is the observed radial velocity.

6. A practical method of obtaining the effective radius is given by Schwartzchild (1954) and quoted by Contopoulos. It involves doing star counts in projected volumes, and assumes the cluster is spherical.

6.2.4 Limitations of the virial approach

Collins (1978), being a great proponent of the virial theorem, took its failure in specific cases very seriously and describes the fundamental assumption that leads to difficulties in astrophysical applications compared to, for example, kinetic theory in statistical mechanics. Similarly Chadrasekhar (1942) knew well the same problem and couched the difficulty in somewhat different terms.

The virial theorem is always presented in the form of time averages of the quantities involved. In instances of systems having very short relaxation times (the time required to come to collisional equilibrium), most of the time the system is found in equilibrium so that the time averages refer to that state. Astrophysical relaxation times are much too long to calculate time averages over many relaxation times, so we must use phase (or space) averages, that is, the averages of all quantities over the whole system in a relatively short time period. But how does one know that the statistical state of the system is stationary over such a short time? The answer is that you don't. You must make some assumption, and doing that is what the ergodic hypothesis is all about.

The ergodic (or quasi-ergodic) hypothesis basically allows one to use phase averages instead of time averages in the virial theorem. The theorem's roots go back to Boltzmann with some refinement by Gibbs and Maxwell (Collins, p. 19). The word ergodic implies that a particle will eventually reach every point in phase space, but that is not correct. The quasi-ergodic hypothesis requires only that a trajectory pass arbitrarily close to any point in a finite amount of time. In modern times, the ergodic hypothesis boils down to the issue of whether phase averages can be substituted for time averages or not. There are now several facets of dynamical theory that have some bearing on the question. In particular, if the system has isolating integrals of the equations of motion that restrict the coverage of space space, the use of the virial theorem with replacement of the time averages by phase averages should be treated with caution.

6.3　The solar system

We will now use *Mathematica* to examine the following important question: Is the solar system stable? How do we know that planetary orbits do not have a non-zero Lyapunov exponent, so that they may eventually fly off their present courses, possibly colliding with one another or with the sun? There has naturally been a considerable amount of very advanced work on this fundamental problem in celestial mechanics. Here, we simply use our N-body approach to solve for the orbits of the planets and that of an intruder, proving that the system is stable over the next 300 years. This is not very long compared to the age of the solar system, but it is about the best we can do using *Mathematica* unless we are willing to wait for long times for the code to complete. More advanced numerical integrators, running on mainframe computers, have evaluated the orbits over much longer time periods.

6.3.1　Effect of an intruder mass

To test the stability of the solar system we will consider the effect of a $0.1 M_\odot$ mass intruder star with a closest approach of about 10 000 AU. This distance is about $1/20$ of a parsec. Since one parsec is the average distance between stars in the solar neighborhood it is likely that the sun has had many such approaches of similarly massive stars in the last 4 billion years.

```
tint = 300; impact = 10 000; starmass = 0.1; vel = 30; incl = 30; Δt = 1; Δtg = 0.5;

vxS = 0; vyS = -vel × vcl[[1]] × Cos[Degree × incl];
vzS = -vel × vcl[[1]] × Sin[Degree × incl];

star = {starmass, -impact, -vyS × tint / 2, -vzS × tint / 2, vxS, vyS, vzS};
```

The data for the solar system is based on a JPL table of mass in solar masses, position (x, y, z) in AU, and velocity (v_x, v_y, v_z) in AU/year.[8] For example,

```
sun = {1.9891*^30 × mc[[1]], dc[[1]] × -7.0299*^8, dc[[1]] × -7.5415*^8,
    dc[[1]] × 2.38988*^7, vc[[1]] × 14.1931, vc[[1]] × -6.9255,
    vc[[1]] × -0.31676};
```

The planetary data is then combined into a single table for the solar system and intruder.

```
solarsys = {sun, mercury, venus, earth, mars, jupiter, saturn, uranus,
    neptune, pluto, star};

planets0 = Table[Table[solarsys[[n]][[j]], {j, 2, 7}], {n, 2, npart}];
```

---

[8] See **6-2Starsun** for the complete table.

To solve the problem using SPRK and NDSolve we first create a mass list

```
mass = Join[{sun[[1]]}, Table[solarsys[[n]][[1]], {n, 2, Length[solarsys]}]]
```

We then define the initial conditions for the Hamiltonian:

```
G = gc; gg = Table[Table[-G×mass[[ii]] mass[[jj]], {ii, 1, npart}],
   {jj, 1, npart}];
initqx = Join[{sun[[2]]}, Table[planets0[[n]][[1]], {n, 1, npart - 1}]];
...
initpz = Join[{sun[[1]] * sun[[7]]},
   Table[mass[[n + 1]] * planets0[[n]][[6]], {n, 1, npart - 1}]];
```

The positions and momenta are given relative to the center of mass of the system. The Hamiltonian is then given by

$$H := \sum_{k=1}^{npart} \left(\frac{(px[k][t])^2 + (py[k][t])^2 + (pz[k][t])^2}{2 * mass[[k]]} + \sum_{ii=1}^{k-1} gg[[ii, k]] \Big/ \right.$$
$$\left. \sqrt{\left((qx[k][t] - qx[ii][t])^2 + (qy[k][t] - qy[ii][t])^2 + (qz[k][t] - qz[ii][t])^2\right)} \right);$$

It is a deferred evaluation mode function because of the :=. The Hamiltonian for the initial values follows a similar form:

$$initH = \sum_{k=1}^{npart} \left(\left((initpx[[k]])^2 + (initpy[[k]])^2 + (initpz[[k]])^2\right) \Big/ (2 * mass[[k]]) + \right.$$
$$\sum_{ii=1}^{k-1} gg[[ii, k]] \Big/ \sqrt{\left((initqx[[k]] - initqx[[ii]])^2 + \right.}$$
$$\left. (initqy[[k]] - initqy[[ii]])^2 + (initqz[[k]] - initqz[[ii]])^2\right) \right)$$

The initial conditions must be put into a flattened list so NDSolve[] can read them:

```
icqx = Table[qx[jj][t] == initqx[[jj]] /. t → 0, {jj, 1, npart}];
...
ics1 = Join[icqx, icqy, icqz, icpx, icpy, icpz];
```

Likewise the Hamilton equations are flattened into a list

```
eqqx = Table[qx[jj]'[t] == D[H, px[jj][t]], {jj, 1, npart}];
...
eqpz = Table[pz[jj]'[t] == -D[H, qz[jj][t]], {jj, 1, npart}];
ODEs = Join[eqqx, eqqy, eqqz, eqpx, eqpy, eqpz];
```

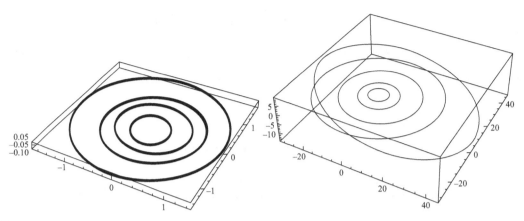

Fig. 6.2 Orbital motion of inner (left) and outer (right) planets over 300 years.

Finally we create a flattened list for the variables:

```
varqx = Table[qx[jj][t], {jj, 1, npart}];
. . .
varpz = Table[pz[jj][t], {jj, 1, npart}];
vars = Join[varqx, varqy, varqz, varpx, varpy, varpz];
```

In the SPRK method, the position variables must be identified separately from the momenta. We can then run the N-body code for the planet positions forward in time.

```
eqns = Join[ODEs, ics1];
sol = NDSolve[eqns, vars, {t, tstart, tint},
    Method → {"SymplecticPartitionedRungeKutta",
      "DifferenceOrder" → 10, "PositionVariables" → posvar},
    StartingStepSize → startstep, InterpolationOrder → All, MaxSteps → 10^7];
```

This can take considerable time to run, even on a fast machine. Once the calculation is complete we can calculate the positions of the sun, planets, and intruder relative to the center of mass:

```
orbitfunctions1 = Table[{qx[n][t], qy[n][t], qz[n][t]} /. sol[[1]], {n, 1, npart}];
```

The resulting graphs seen in figure (6.2) show that both the inner and outer planets are highly stable over 300 years.

6.3.2 Calculating Lyapunov exponents

Now that we have calculated the solution to our N-body problem, we can use the Hamiltonian to calculate the Lyapunov exponents by the Jacobian method. We begin by determining the generating functions of the canonical equations of motion as partial derivatives:

```
nBJacobi[ii_] := {D[Flatten[...
    Table[{D[H, px_jj[t]], D[H, py_jj[t]], D[H, py_jj[t]], -D[H, qx_jj[t]],
        -D[H, qy_jj[t]], -D[H, qz_jj[t]]}, {jj, 1, npart}]], pz_ii[t]]}
```

From these we can calculate the Jacobian matrix:

```
matrixX = Table[nBJacobi[ Ceiling[kk - 5 Chop[kk / 6]]]
    [[ Ceiling[Chop[kk / npart]]]], {kk, 1, 6 npart}];
matrixI = matrixX /. t → 0;
```

Setting $t \to 0$ indicates the initial conditions are to be substituted. The result is a 66×66 matrix ($6\times$ the number of bodies). Because we have calculated the initial conditions, we can substitute them directly:

```
matrixY = matrixI /. ics1;
```

The conditions at $t = 0$ are determined by the characteristics of the Jacobian matrix. By analogy with a 2D state space, if the trace of the matrix is positive then the point is a repellor point. If the trace is negative it is a node. The type of node or repellor point depends on the determinant. If the determinant is positive, and if

$$\frac{(\mathrm{Tr}\mathbf{J})^2}{n^2} > \mathrm{Det}\mathbf{J}, \tag{6.36}$$

where n is the dimension of the matrix, then it is just a standard node or repellor. If

$$\frac{(\mathrm{Tr}\mathbf{J})^2}{n^2} < \mathrm{Det}\mathbf{J}, \tag{6.37}$$

then is is a spiral node or repellor. The determinant and trace can be calculated by

```
Det[matrixY]; Tr[matrixY];
```

which gives a determinant of 0 and a positive trace.

The index of an evaluation point is the number of eigenvalues whose real parts are positive. For a node the index is 0. For a pure repellor point the index is equal to the dimension of the state space $6n$. For a saddle point the index is either n (a curve) or $2n$ (a surface). The eigenvalues can be found from

```
λ = Eigenvalues[matrixY]
```

and the index is found by

```
reλ = Re[λ]; nre = Length[reλ];
ii = 0; Do[If[ reλ[[jj]] > 0, ii = ii + 1], {jj, 1, nre}]
```

This yields an index of 13. Because the index in this case is greater than $n = 11$, and less than $2n = 22$, it must be some sort of hyperspace repellor saddle point.

The sum of all the eigenvalues is greater than 0, the dimensionality of the problem is the whole phase space. Because the sum is not 0, the Hamiltonian seems not to be conserved. For a dissipative system the sum should be less than 0, so the system is also not dissipative. This may mean the system is chaotic, but Fourier and correlation properties of the trajectories is necessary to verify this claim. It is often claimed that the orbit of Pluto is chaotic even without an intruder, so Pluto might be the one contributing most to the chaos of the system.

An explanation of these results can be found in discussions give by Hilborn (p. 326) and Ott (p. 22). If our results had shown that the sum of the Lyapunov exponents were actually 0, then there would have been constants of motion, including H as the total energy. Such systems are called integrable Hamiltonian systems.

For integrable systems, one can go on to find canonical transformations of the equations of motion that relate to the action variables J, the constants of motion, in terms of the p's and q's. Associated with the action variables are the so-called angle variables θ. The existence of the J's indicates that all trajectories are confined to an N-dimensional surface in state space. The canonical transformations we are interested in results in a Hamiltonian that has only J's and no θ's. Most integrable systems do not need numerical techniques for a solution. Hilborn goes on to list properties that distinguish Hamiltonian systems that are integrable.

Integrable systems are always periodic or quasi-periodic, and cannot be chaotic. Trajectories in state space are confined to nested N-dimensional tori. But because all the signs point to our N-body Hamiltonian as being nonintegrable, there is one property such systems have that will be very important in our discussion of galactic dynamics in Chapter 8: nonintegrable systems have more degrees of freedom than they have constants of motion. If a system is (or becomes) nonintegrable then the lack of constraints allow the trajectories to wander through state space. If one turns an integrable system into a nonintegrable one through a perturbation of the original system (such as our perturbing star), then the result in state space is described by the famous Kolmogorov–Arnold–Moser (KAM) Theorem (Goldstein, Poole, and Safko, p. 487; Ott, p. 224; Hilborn, p. 337; Rasband, p. 174), which describes the infinitesimal evolution of the integrable Hamiltonian into a nonintegrable one.

6.3.3 Chaotic behavior and time series

So far we have ignored practical methods of determining the behavior of a chaotic system from actual data. One of the most important is known as Attractor Reconstruction (Strogatz, 1994; Baker and Gollub, 1996), or Time Series Analysis (Hilborn, 1994; Drazin, 1992). This method uses the extraction of multidimensional descriptions of state space dynamics from the time series data of a single dynamical variable. This is made possible through the use of what is called an embedding scheme.[9] The use of embedding schemes is often used in medical fields, and is described in Krasner (1990). Medical research has developed singular value decomposition (SVD) methods for reducing noise, which is of particular interest to astrophysics. Strogatz (p. 440) has given some good advice on the limitations of the process. Although some of the factors given by Strogatz limit our ability to extract the attractors from any one object in our system, we nevertheless pursue two of the factors needed before performing such extractions, correlation time and its related data set, the power spectrum of the distances of each object from the center of mass.

For each object in our system, the position has been determined as a data function of discrete points at discrete times. Therefore we must use the discrete Fourier transform, Fourier[], rather than the

---

[9] See Hilborn, pp. 436–443; Rasband, pp. 200–203; Baker and Gollub, pp. 137–150.

continuous or interpolated `FourierTransform[]`. Thus, for a given data list $u[t]$ of length n,

$$F[s] = \frac{1}{\sqrt{n}} \sum_{t=1}^{n} u[t]e^{2\pi i(t-1)(s-1)/n}, \tag{6.38}$$

and the power spectrum is then

$$P[s] = |F[s]|^2 . \tag{6.39}$$

The autocorrelation function can be determined from `InverseFourier`; thus

$$C[t] = \left| F^{-1}[P] \right|, \tag{6.40}$$

where

$$F^{-1}[P] = \frac{1}{\sqrt{n}} \sum_{s=1}^{n} P[s]e^{-2\pi i(s-1)(t-1)/n}. \tag{6.41}$$

The autocorrelation function can then be fit to a periodic exponential function to determine the correlation time. The correlation time is a measure of time range of predictability and is a measure of chaos or stability. Stable objects have longer correlation times, whereas chaotic objects have shorter times.

A full analysis of the system requires that the correlation time be calculated for every object in our system, but here we calculate it for Mercury. Calculation of the others will follow the same form.[10]

Earlier we calculated the positions of the sun, planets, and intruder star relative to the center of mass. Rather than repeat this calculation, the positions have been exported as a data table. In the case of Mercury, this table is `mercuryxyzdata`, which gives the cartesian position as a function of time. We wish to calculate the power spectrum of the radial distance of Mercury from the center of mass, so our table must be converted to radial distance.

```
mercuryraddata = Table[√(mercuryxyzdata[[ii, 1]]² +
        mercuryxyzdata[[ii, 2]]² + mercuryxyzdata[[ii, 3]]²),
    {ii, 1, dimpoints[[1]]}];
```

The power spectrum can then be calculated as

```
fftradmercury = Abs[Fourier[mercuryraddata]]²;
```

As seen in figure (6.3), there are many different periodicities present in the power spectrum. The autocorrelation is then found from

```
corradmercury = Abs[InverseFourier[fftradmercury]];
listcorradmercury =
  Table[{(ii - 1) Δtg, corradmercury[[ii]]}, {ii, 1, Ceiling[dimpoints[[1]] / 2]}];
```

---

[10] For the full calculation, see **6-4TimeSeriesSS**.

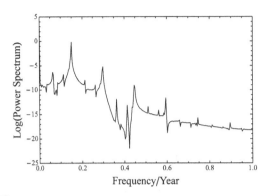 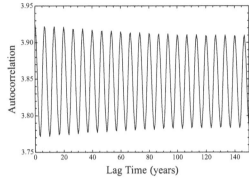

Fig. 6.3   Power spectrum and autocorrelation of Mercury.

The autocorrelation appears to be a decaying harmonic function, as figure (6.3) shows, so we define a function form to fit

```
form = a e^-k1 x (1 + b Cos[k2 x]); params = {a, b, k1, k2};
ymercury =
  NonlinearModelFit[listcorradmercury, form, params, x, MaxIterations → 10 000]
ymercury1 = Normal[ymercury]
```

This yields a correlation function for Mercury as

$$C(t) = 1.4 \, e^{-1.5 \times 10^{-6} t} \left(1 + 1.7 \cos\left(1.5 \times 10^{-10} t\right)\right). \tag{6.42}$$

The correlation time is found by the zero crossing point (to within roundoff level). It must be found by trial and error using the log of the autocorrelation function. The dominant root should be real if possible, and obtained to be the same value with several trial roots covering a few decades on numbers.

```
tsolmercury = FindRoot[Log[ymercury1] == -35, {x, 10^7}]
solmercury = Re[tsolmercury[[1, 2]]]
```

The prediction error is estimated as the lack of regression accuracy of the log function used to find the root:

```
solmercuryerr = √(1 - ymercury["RSquared"]) solmercury
```

This gives the correlation time for Mercury to be $2.3142 \pm 0.0291 \times 10^7$ years. When we calculate the correlation times for all the planets, we find they range from a maximum of $4.5872 \pm 0.0124 \times 10^9$ years for Venus to a minimum of $1.1734 \pm 0.0187 \times 10^5$ years for Pluto. With the exception of Pluto, the correlation times are on the order of 10^9 to 10^7, which again points to Pluto as a possible source of chaos within the system.

In comparing the correlation times of the planets, there seems to be a loose relationship between the eccentricity of a planet's orbit and its correlation time, as seen in figure (6.4). A more detailed

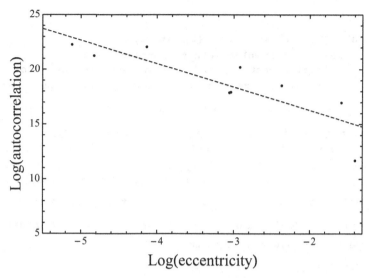

Fig. 6.4 Correlation time vs. orbital eccentricity.

analysis[11] seems to indicate stability depends mostly on eccentricity, followed by distance from the sun, then planetary mass. These dependencies agree with dynamical expectations of a system dominated by self-perturbation rather than the influence of the intruder. They also indicate why massive close binary systems in orbits of small eccentricity are so numerous in multiple star systems.

6.3.4 Test of the virial theorem

We now test the virial theorem for our system assuming no intruder is present. The Hamiltonian and related variables are calculated just as before, but excluding the stellar intruder.[12] The initial virial $2T + V$ is calculated as

$$
\mathtt{initV} = \sum_{\mathtt{k=1}}^{\mathtt{npart}} \Bigg(\big((\mathtt{initpx[[k]]})^2 + (\mathtt{initpy[[k]]})^2 + (\mathtt{initpz[[k]]})^2 \big) \big/ (\mathtt{mass[[k]]}) +
$$

$$
\sum_{\mathtt{ii=1}}^{\mathtt{k-1}} \mathtt{gg[[ii, k]]} \big/ \sqrt{ \big((\mathtt{initqx[[k]]} - \mathtt{initqx[[ii]]})^2 +
$$

$$
(\mathtt{initqy[[k]]} - \mathtt{initqy[[ii]]})^2 + (\mathtt{initqz[[k]]} - \mathtt{initqz[[ii]]})^2 \big) } \Bigg)
$$

The virial theorem states that this should be zero. In fact it is slightly positive (3×10^{-4}).

---

[11] See **6-4TimeSeriesSS**.
[12] See **6-5PointMassVirial** for details.

The equations of motion are again solved through the SPRK method:

```
sol = NDSolve[eqns, vars, {t, tstart, tint},
    Method → {"SymplecticPartitionedRungeKutta",
        "DifferenceOrder" → 10, "PositionVariables" → posvar},
    StartingStepSize → startstep, InterpolationOrder → All, MaxSteps → 10^7];
```

We can then calculate the virial for the unperturbed solar system over 300 years:

```
rcomponents[jj_, tt_] := {qx[jj][t] /. sol[[1, jj]],
    qy[jj][t] /. sol[[1, jj + npart]], qz[jj][t] /. sol[[1, jj + 2 npart]]} /. t → tt;
pcomponents[jj_, tt_] := {px[jj][t] /. sol[[1, jj + 3 npart]],
    py[jj][t] /. sol[[1, jj + 4 npart]], pz[jj][t] /. sol[[1, jj + 5 npart]]} /. t → tt;
mv2fun[jj_, tt_] := pcomponents[jj, tt].pcomponents[jj, tt];
ke[t_] := Sum[mv2fun[i, t] / (2 mass[[i]]), {i, 1, npart}];
pe[t_] := Module[. . . pesum];
virial[t_] = 2 ke[t] + pe[t];
```

If we then put this in a data table and make a linear fit

```
virialdata = Table[virial[t], {t, 0, tint, Δt}];
avH1 = Mean[virialdata]
σH1 = StandardDeviation[virialdata]
```

we find the virial has a value of 2.88×10^{-4} with a standard deviation of 1.36×10^{-18}. Thus the virial remains positive. The solar system only approximately satisfies the virial. It is clear then that the solar system fails to satisfy the ergodic hypothesis. There are several possible causes for this:

1. The phase averages cannot properly replace the time averages. Hence while $T + V$ is negative, $2T + V$ is slightly positive.
2. The system shows some signs of having motions that are chaotic through Lyapunov exponents being unpaired and positive.
3. Because the system is nearly coplanar, there may be third integrals of the motion that prevent portions of phase space from being covered properly.
4. The solar system moves in a galactic orbit that is under the influence of the galactic potential.
5. No attempt to take into account mutual tidal dissipation forces or rotational motions has been made.
6. The existence of the correlation time power law relationship with mass, distance, and eccentricity for the system seems, to indicate some sort of ergodic constraint.

6.4 The Trapezium system

We now look at the four brightest stars in the Orion Nebula, known as Trapezium. The positions of these stars in the sky and their velocities are known from observation, as is the general distance of the cluster. These must be converted to coordinates in terms of the center of mass[13] in order to calculate the N-body

---

[13] See 6-6NBInitialconditions.

problem. What is not known are the distances of the stars along the line of sight (x-distances), which cannot be determined from observation. We must therefore generate a set of x-distances at random subject to a physical constraint such as the virial theorem or the equipartition of energy.

6.4.1 Statistical methods of estimating line of sight distances

Selecting stellar distances along the line of sight relative to the central star is a fundamental difficulty of the N-body approach to stellar dynamics. One can devise, for example, a number of Monte Carlo schemes based on the equipartition of energy, the virial,[14] or some combination of both. The coplanar situation (all x-distances are zero) is that of minimum energy, but such a configuration is not likely to be the minimum energy configuration that actually exists. We must therefore try various tricks to estimate the x-distances whose parameter space will be explored computationally.

When generating x-distances, one must be careful not to introduce bias into the system. For example, we experimented with a Monte Carlo selection of two component distances, then used the virial theorem to find the potential energy of the third component. This could then be used to determine the third distance. However it was found that this procedure did not sample the available phase space in an unbiased manner.

Here we will look at a method that does a good job of selection. We assume the shape of the system is roughly spherical and then use the moments of a uniform cosine distribution around the line of sight. Here we define the x-axis as the line of sight distance relative to the center of mass, while `initqy` and `initqz` are the observed positions relative to the center of mass. We first define a scale for the random line of sight positions from `initqy` and `initqz`,

$$\mathtt{avcos} = \frac{\int_0^{\pi/2} \mathtt{Cos}[\theta]\, d\theta}{\int_0^{\pi/2} 1\, d\theta};$$

```
dist2D = Table[0, {ii, 1, nB}];
Do[dist2D[[ii]] = √(initqy[[ii]]² + initqz[[ii]]²), {ii, 1, nB}];
```

$$\mathtt{avls} = \frac{\sqrt{\sum_{ii=1}^{nB} (\mathtt{dist2D[[ii]]})^2}}{8 \times \mathtt{avcos}};$$

We can then estimate the distances from a random distribution using `avls`:

```
Do[xdepths[[ii]] = avls × Random[] × Sign[1 - 2 × Random[]], {ii, 1, nB}];
```

$$\mathtt{cmxdepth} = \frac{\sum_{i=1}^{4} \mathtt{xdepths[[i]]} \times \mathtt{mas[[i]]}}{\sum_{i=1}^{4} \mathtt{mas[[i]]}};$$

```
xxdepths = xdepths - cmxdepth;
initqx[[1]] = xxdepths[[1]]; initqx[[2]] = xxdepths[[2]];
initqx[[3]] = xxdepths[[3]] initqx[[4]] = xxdepths[[4]];
```

---

[14] See **6-5PointMassVirial**.

Because we have estimated the `initqx` values, a full analysis should examine multiple generated sets to see if the evolution of the system depends critically on initial conditions. For brevity we examine only one set of values.[15]

6.4.2   Is Trapezium unstable?

Direct calculation of the Hamiltonian and the virial show that the virial is about 10 times more positive than the Hamiltonian is negative; therefore the virial is not satisfied for our choice of `initqx`, which would seem to indicate Trapezium is unstable. This brings up an extremely important point in celestial mechanics. When one calculates the total energy and the virial based on the *observed* parameters of a system, often the results indicate the visible components form an unstable system. Yet like the solar system we know such systems have existed at least long enough for the components to survive as individual entities with independently determined ages.

It is well known that many types of galaxies seem to be stabilized by "dark" matter, as we will see in Chapter 8. Many clusters of galaxies seem to be stabilized by a combination of dark matter and intergalactic hot gas emitting only x-rays.[16] The Orion Trapezium is one of the few very young stellar clusters that can be seen by optical means. Spitzer Space Telescope has discovered many more such young systems in the infrared that are completely shrouded in the dust/gas cloud from which they formed. Images of the Orion Trapezium show that the star cluster itself is partially embedded in dust from the Orion A radio source, and so we investigate whether there is enough mass in these clouds to allow the high initial velocities.

The Trapezium and the rest of the Orion Nebula Cluster (ONC) is unusual in that there are well over 100,000 solar masses of molecular cloud material within 100 parsecs of Trapezium itself (Wilson et al., 2005), and at least one incidence of 70,000 solar masses within 10 parsecs. To make calculations faster we will again assume the ONC is spherical around Trapezium and so satisfies Gauss' law. But this assumption cannot be made for the molecular clouds OriA and OriB. Although the clouds are extended objects, each one has several major condensations of CO radiation where the entire mass of that part of the cloud can be placed as a crude "multipole" expansion of the extended objects. Of course these clouds are moving relative to each other and relative to the Trapezium, but in order to not have to solve three differential equations for each cloud's straight line motion (and thus slow down the computation) we have opted to keep the clouds stationary at least for this exploratory computation. Then the only variations in the potentials are the "tiny" changes of distances caused by the motions of the stars themselves.

The cloud data is taken from Wilson et al. and converted to Right Ascension (RA) and Declination (Dec) in degrees:

```
cloudlist = {{"trap", 82.85, -5.45, 470, 0}, {"OriA1", 82.375, -3.98, 521, 12 300},
    {"OriA2", 84.3, -6.18, 465, 69 500}, {"OriA3", 85.75, -8.58, 412, 23 300},
    {"OriB1/2", 85.5, -1.7, 422, 65 300}, {"OriB3", 89, 1.68, 383, 18 000}};
```

The cloud masses form the last column of our data, so this is extracted separately:

```
masscL = Table[cloudlist[[ii + 1, 5]], {ii, 1, 5}]
```

---

[15] Multiple generated sets can be found in **filenum.csv**, and can be analyzed by appropriate modifications of **6 6NBInitialconditions**.

[16] In the case of the solar system, it may be that Chandrasekhar's dynamical friction from the surrounding stars may be responsible

We next transform these to the *xyz* coordinate system centered on Trapezium. The *x*-coordinate is simply the line of sight distance converted to AU:

```
xAUc = Table[206 265 (cloudlist[[ii + 1, 4]] - cloudlist[[1, 4]]) , {ii, 1, 5}]
```

For the *y* and *z* coordinates we must transform the observations using spherical trigonometry:

```
dec = Table[cloudlist[[ii, 3]] °, {ii, 1, 6}];
ra = Table[cloudlist[[ii, 2]] °, {ii, 1, 6}];
cosdec = Cos[dec]; sindec = Sin[dec]; Δra = ra - ra[[1]];
cosphi = Table[sindec[[1]] sindec[[ii + 1]] +
    cosdec[[1]] cosdec[[ii + 1]] Cos[Δra[[ii + 1]]], {ii, 1, 5}]
phi = ArcCos[cosphi] / Degree
```

These angular positions are then converted to linear distances in AU.

```
difδ = Table[dec[[1]] - dec[[ii + 1]], {ii, 1, 5}]; cosΔδ = Cos[difδ];
zAUc = 470 Sin[difδ] 206 265; yAUc = -470 Sin[Δra] 206 265;
```

The cloud data can then be included in the Hamiltionan and virial calculations:

$$
\text{initH} = \sum_{k=1}^{\text{npart}} (\ldots) +
$$

$$
\sum_{jj=1}^{5} \sum_{k=1}^{nB} -\text{gc mass}[[k]] \text{ masscL}[[jj]] / \sqrt{((\text{xAUc}[[jj]] - \text{initqx}[[k]])^2 +}
$$

$$
(\text{yAUc}[[jj]] - \text{initqy}[[k]])^2 + (\text{zAUc}[[jj]] - \text{initqy}[[k]])^2)
$$

$$
\text{initV} = \sum_{k=1}^{\text{npart}} (\ldots) +
$$

$$
\sum_{jj=1}^{5} \sum_{k=1}^{nB} -\text{gc mass}[[k]] \text{ masscL}[[jj]] / \sqrt{((\text{xAUc}[[jj]] - \text{initqx}[[k]])^2 +}
$$

$$
(\text{yAUc}[[jj]] - \text{initqy}[[k]])^2 + (\text{zAUc}[[jj]] - \text{initqy}[[k]])^2)
$$

Direct calculation shows that for Trapezium both the virial and the Hamiltonian are negative when the Orion Molecular Clouds are taken into account. This is not the case if they are left out. Thus we might conclude that the Orion Trapezium is a stable cluster. It should be noted, however, that the intense light from the Trapezium and other massive stars will eventually dissipate the gas and dust, and the cluster will then be much less stable. By then the highest mass young stars will have disappeared along with the clouds. We have also not taken galactic rotation into account, but this will be explored in Chapter 8.

6.4.3 Calculating orbital motions

We are now ready to calculate the evolution of Trapezium both forward and backward in time. The process follows that of Section 6.3.1 with only minor changes. The forward and backward solutions must be calculated separately starting with the initial ($t = 0$) positions and momenta of the cluster stars.

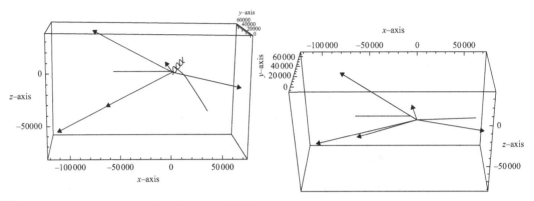

Fig. 6.5   Motion of Trapezium backward (left) and forward (right) 1000 centuries.

For the backward calculation, the signs of the momenta must be defined as the negative of the initial values; thus

```
icpx = Table[px_jj[t] == -initpx[[jj]] /. t → tstart, {jj, 1, nB}];
icqy = Table[qy_jj[t] == initqy[[jj]] /. t → tstart, {jj, 1, nB}];
icqz = Table[qz_jj[t] == initqz[[jj]] /. t → tstart, {jj, 1, nB}];
```

The calculation forward in time uses the initial values directly.

The Hamiltonian follows the same form as before:

$$H := \sum_{k=1}^{nB} (\ldots) + \sum_{jj=1}^{5} (\ldots);$$

from which the equations of motion are derived as before. The equations are then solved using the SPRK method:

```
sol1 = NDSolve[{eqs, ics1}, vars, time,
    Method -> {"SymplecticPartitionedRungeKutta",
       "DifferenceOrder" → 10, "PositionVariables" → posvar},
    StartingStepSize → startstep, InterpolationOrder → All,
    MaxSteps → 1 000 000 000];
```

The results can be seen in figure (6.5), with the directions to the surrounding clouds indicated as well.

It should be noted that Mathematica presents a particular challenge when plotting astronomical graphs. The 3D *Mathematica* plot routines use rotation matrices to achieve their results, and all assume they are dealing with right-handed orthogonal coordinate systems. This means that even in 2D plots, the x-axis goes to the right with the y-axis vertical. However, most astronomical angular systems such as RA-Dec and galactic latitude-longitude without special treatment will be plotted with x reversed to follow astronomical convention.

When it comes to the 3D situation, *Mathematica* generally has the z-axis perpendicular to the x–y plane with positive z determined by the right-hand rule. If we make $-x$ be the RA and y the Dec, then z points toward the observer rather than outward as required by astronomical convention. Unless the data sets are specially treated before plotting, *Mathematica* will plot the z information backwards to the order seen in the sky. It should also be noted that, to save computation time, *Mathematica* will plot only the "region of interest." To plot all the data you must use the option `PlotRange -> All`.[17]

As a check of the accuracy of these calculations, one can calculate the initial and final Hamiltonians. The initial Hamiltonian is given by

$$
\mathtt{initH} = \sum_{k=1}^{\mathtt{nB}} \left(\frac{1}{2\,\mathtt{mass[[k]]}} \left((\mathtt{initpx[[k]]})^2 + (\mathtt{initpy[[k]]})^2 + (\mathtt{initpz[[k]]})^2 \right) + \ldots \right) +
$$

$$
\sum_{\mathtt{jj}=1}^{5} \sum_{k=1}^{\mathtt{nB}} -\mathtt{gc\,mass[[k]]\,masscL[[jj]]} \Big/ \sqrt{\Big((\mathtt{xAUc[[jj]]} - \mathtt{initqx[[k]]})^2 + }
$$

$$
(\mathtt{yAUc[[jj]]} - \mathtt{initqy[[k]]})^2 + (\mathtt{zAUc[[jj]]} - \mathtt{initqy[[k]]})^2 \Big)
$$

For the final Hamiltonian we must determine the final position and momenta of the cluster stars:

```
qqx = Table[0, {ii, 1, nB}]; qqy = Table[0, {ii, 1, nB}]; qqz = Table[0, {ii, 1, nB}];
ppx = Table[0, {ii, 1, nB}]; ppy = Table[0, {ii, 1, nB}]; ppz = Table[0, {ii, 1, nB}];

Do[qqx[[i]] = sol1[[1, i, 2]] /. t → tint, {i, 1, nB}];
...;
Do[ppz[[i]] = sol1[[1, 5 nB + i, 2]] /. t → tint, {i, 1, nB}];
```

From these we can calculate the fractional error of the Hamiltonian:

$$
\Delta H = \frac{2\,\mathtt{Abs[finalH - initH]}}{\mathtt{Abs[finalH + initH]}}
$$

The resulting fractional error is on the order of 10^{-12} to 10^{-14}, which is quite reasonable.

6.4.4 Power spectrum and autocorrelation properties

The power spectrum and autocorrelation can be calculated for each member of the cluster, both forward and backward in time.[18] Again we will consider just one star[19] as an example.

From our earlier calculations we have the position of the star relative to the center of mass as a data table.[20] From this we calculate the radial distance from the center of mass:

```
starDraddatar = Table[√(starDxyzdata[[ii, 1]]² + starDxyzdata[[ii, 2]]² +
    starDxyzdata[[ii, 3]]²), {ii, 1, dimpoints[[2]]}];
```

From this we calculate the power spectrum,

```
fftradDx = Abs[Fourier[starDraddatax]]²;
```

---

[17] See **6-7Nbodybk** and **6-8Nbodyfwd** for more details. For these notebooks, when starting the N-body calculations it was realized that a convenient right-hand coordinate system could be obtained if the standard *xyz* coordinate system were redefined such that the *y*-axis became the RA, the *z*-axis as Dec, and *x*-axis as the positive radial direction. Although convenient for computation and algorithmic visualization, this did not solve the display problem. Thus we had to reverse the *y*-axis values before putting points into any of the 3D plot programs. Otherwise a coordinate inversion will occur.

[18] See **6-12MotionNBbk** and **6-13MotionNBfw**.

[19] The star we consider is known as Theta Orionis D.

[20] See **6-12MotionNBbk**.

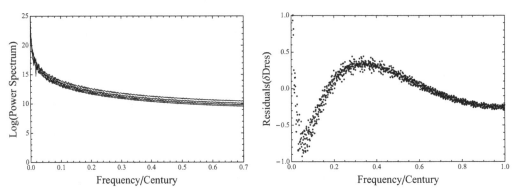

Fig. 6.6   Power spectra (left) and residuals (right) for Trapezium distances.

The result can be seen in figure (6.6). In this case the power spectra are quite smooth, without discrete frequencies. In the figure, the curves from high to low are Dec direction, radial, then RA, then line of site. The fact that all the lines are distinct indicates nonplanar motion. The small variations within the overall curve indicate quasi-periodic motion with periods less than integration time step (in this case 100 years). We then fit this curve to a function

$$\texttt{form} = a e^{-k_1 t} + c. \tag{6.43}$$

Thus

```
form = a e^{-k1 u} + c ; params = {a, k1, c};

yDr = NonlinearModelFit[listfftradDr, form, params, u, MaxIterations → 100 000];

yD0r = yDr["BestFitParameters"]
```

which yields

$$P(t) = 7.84 e^{-6.82t} + 10.10. \tag{6.44}$$

If discrete frequencies are present in the power spectrum, we need a semi-empirical quantitative way to characterize them. Here we use FitResiduals.

```
yDres = yDr["FitResiduals"];

δDres = Table[{ (ii - 1) / (Δt dimpoints[[2]]) , yDres[[ii]]},

    {ii, 1, Ceiling[dimpoints[[2]] / 2]}];

plotresD = ListPlot[δDres]
```

The result, seen in figure (6.6), represents the residual spectral powers. Visually quasi-periodicities, if present, may be seen in the graph. An estimate of the standard deviation of the quasi-periodic fluctuation in the power spectrum is obtained from the residuals from a power series fit. The larger this number the stronger the quasi-periodic power and the nearer to chaotic motion.

```
linDres = LinearModelFit[δDres, {1, x, x², x³, x⁴, x⁵, x⁶, x⁷}, x];

linDres1 = Normal[linDres];

ΔlinD = linDres["FitResiduals"];

resDdev = StandardDeviation[ΔlinD]
```

This gives a residual deviation of about 0.03, which is reasonably small.
   The autocorrelation is found by

```
corradDr = Abs[InverseFourier[fftradDr]];

listcorradDr = Table[{(ii - 1) Δt, Log[corradDr[[ii]]]},
   {ii, 1, Ceiling[dimpoints[[2]] / 2]}];
```

This is then fit to an exponential function to determine the correlation time:

```
form = a e⁻ᵏ¹ˣ ; params = {a, k1};

xDr = NonlinearModelFit[listcorradDr, form, params, x, MaxIterations → 100 000];

xD0r = xDr["BestFitParameters"]; xD2r = xDr["ParameterErrors"];

xder = 10² / xD0r[[2, 2]]; xderrorr = xder xD2r[[2]] / xD0r[[2, 2]];
```

The resulting correlation time is $(3.14 \pm 0.02) \times 10^6$ years. Calculation of the correlation times for the other trapezium stars gives a mean $1/e$ correlation time for the Trapezium as 2.9×10^6 years.

6.5 Chaotic systems

Over the years, experimental or observational ways to determine the presence of chaos have become more sophisticated, but they also have become more quantitative and less "visual." We have so far focused on these more sophisticated methods, though we retained one "visual" technique that involves the Fourier power spectrum analysis. We did not utilize the most famous visual techniques, known as the Poincaré section, because it is difficult to apply to the N-body problem.

Chaos determinations are especially important in Hamiltonian dynamics, as we have seen. In Section 6.1 we recommended that students invest in Braun (1993), Hubbard and West (1991), and Hubbard and West (1995) as guides to the dynamical uses of differential equations. Here we present a basic overview of chaos from the historical viewpoint, following Baker and Gollub (1996). In the early days of computing, certain types of computational instability were observed that covered up the "real" dynamical behavior. We examine a case where that happened.

6.5.1   When is numerical instability not chaos?

Mathematica includes an example called "Pleiades" in the tutorial NDSolveExtrapolation, which we examine here. The problem is called "Pleiades" because it has seven bodies in it, but there the resemblance to the real cluster ends. Here the example is coplanar, the masses are arbitrarily assigned,

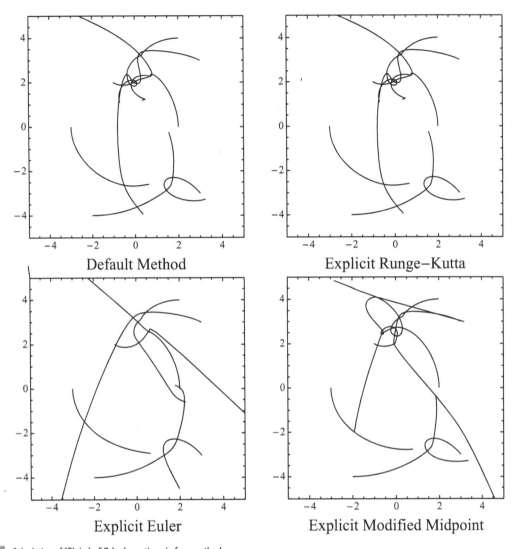

Fig. 6.7   Calculation of "Pleiades" 7-body motion via four methods.

and no invariants are given. It should be a conservative system with both energy and angular momentum conserved.

While the N-body problem is an ideal problem to attack using Hamiltonian methods, the *Mathematica* equation set provided for Pleiades is not in a form suitable for SPRK calculation. Rather it is a set of second-order differential equations provided by Newton's laws, with no velocity equations. The reason is that this example is used to illustrate NDSolve's capability at solving a system of second-order equations with no velocity equations. This is why no energy calculation is done. We first import the Pleiades equations:

```
system = GetNDSolveProblem["Pleiades"];
```

Because this is planar motion, the 2D components are the dependent variables. The default solution is to simply invoke NDSolve[],

```
eDsol = NDSolve[system];
```

the result of which can be seen in figure (6.7). It is well known that such a set of equations can be solved numerically using the Nystrom form of the Runge–Kutta. However, the Nystrom form is not discussed in the *Mathematica* documentation separately. If the Nystrom method is embedded in NDSolve[], then it is most likely found within ExplicitRungeKutta, as the Nystrom method is indeed a Runge–Kutta method:

```
erksol = NDSolve[system, Method → "ExplicitRungeKutta", StartingStepSize → 1 / 50];
```

As seen in figure (6.7) the result is nearly identical to the default method. However, if the Nystrom method is embedded in NDSolve, only as an Explicit Runge–Kutta, then the Euler call should not work.

```
eEsol = NDSolve[system, Method → "ExplicitEuler", StartingStepSize → 1 / 50];
```

Despite the second-order equations, a solution is still obtained. This means the method employed by *Mathematica* is more general than the Nystrom method and works with a variety of methods. The second-order formulation offers no computational superiority. It seems likely that NDSolve[] reduces all second-order systems to the equivalent first-order differential equations before doing a numerical solution.

The Euler method produces a result somewhat different from the first two methods. The Euler result is similar to that achieved by the Modified Midpoint method,

```
emmsol = NDSolve[system, Method → "ExplicitModifiedMidpoint",
    StartingStepSize → 1 / 50];
```

Although all four example methods show qualitative similarities, the last two agree better with themselves than with the Runge–Kutta methods. Because there are no invariants specified for this problem, it is not possible to determine the errors as we have done previously. With no error analysis it is difficult to determine which solutions are best. Therefore it is recommended that the Hamiltonian SPRK formulations or some other symplectic or projection method be used whenever possible, as error control is a natural byproduct of those numerical methods.

In the aforementioned Pleiades example, with no constraints to keep the numerical integration "on track," the solution one gets often depends on the numerical method and step size used. Almost any method including Euler's will perform correctly if the step size is small enough, but a small step size means slow progress and lots of computation time. This was particularly true when computer word lengths were only 4-bit or 8-bit.

In the early days of computing, whenever a solution would wander away from expectation it was always blamed on word length. Never was any blame ascribed to the equations themselves. Over the years, as methods improved and word lengths got longer, it was realized which methods were suspect and should be avoided. In addition, it was discovered there were ranges of parameters for certain differential equations where the behavior was not "classical." This type of motion was called deterministic chaos, or simply chaos. Chaos and random in modern times are not the same thing. Random (stochastic) is implied in statistical and quantum mechanics to be a system's property regardless of the phase space in which the system operates. Chaos is often observed when systems are operated out of their normal parameter space.

Certain differential equations are more easily driven to chaos than others, but there are certain necessary conditions that must exist:

1. The system must be nonlinear.
2. The system must have at least three "dimensions" (often provided by second-order differential equations in a single variable).
3. A small change in initial conditions produces a profound change in the behavior of the system.

There are many examples of chaotic systems that can be explored in Braun (1993), Hubbard and West (1991), and Hubbard and West (1995). It should be noted, however, that there are far too many books on Chaos and Fractals to provide a comprehensive list.

6.5.2 Using EquationTrekker to study differential equations

When learning about the chaotic behavior of differential equations it is often instructive to be able to visualize the behavior of the solutions. *Mathematica* actually provides a tool to do just that, known as EquationTrekker. It is a built-in package that interacts directly with a *Mathematica* notebook in real time. There is also a nonmodal version called EquationTrekkerNonModal.

To use Trekker you must first activate the package; thus

Needs["EquationTrekker`"]

Within the package is a function called EquationTrekker[]. If you use this function in a particular place in your notebook, *Mathematica* will pause at that place, produce a new window interface, and stay paused there until you close the external window. At that point the actions taken in the external window get incorporated graphically in your notebook.

As an example, consider a simple harmonic oscillator:

$$\textbf{EquationTrekker}\left[\textbf{y''[x] + y[x] == 0, y, }\left\{\textbf{x, }\frac{\pi}{8}\textbf{, 2}\pi\right\}\right]$$

When this command is activated, the Trekker window appears waiting for your interaction. The window that comes up is fairly blank, but the harmonic equation is placed at the upper right. Initially the plot is a blank phase diagram with axes for velocity $y'(x)$ plotted versus $y(x)$. At the top left of the window are several icons that stand for various plotting options. The option in current use is shaded. The cursor within the plot is a cross and if (after setting the drawing option [double click second button from the left] to line or points) you click on a position then that sets an initial condition (which appears in the conditions windows). From left to right, the first option button is a dotted square (dragging the mouse selects certain points and lines), the second if clicked will allow the line of "trek" associated with the initial point to be either solid (lines) or dotted (points). This choice has to be made or trek will not plot either your point or the associated phase path. (If you do not like the automatic color assignment it can be changed by clicking on the color patch and a palette of color choices will come up and you can pick another.) The magnifying glass is third and clicking it will give two zoom choices, zoom out

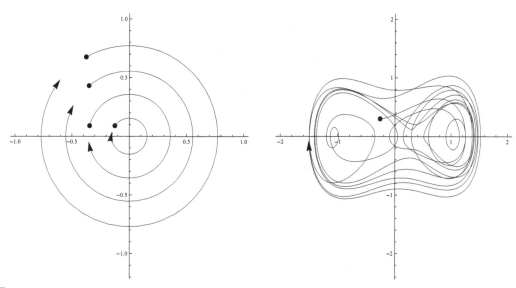

Fig. 6.8 Phase diagrams for simple harmonic motion (left) and Duffing's equation (right).

or return to zoom to fit. The crossed arrows icon, if clicked, allows one to move the plot around in the window. The cursor itself locates the point of interest (POI) you want in your plot. When you are satisfied with your plot, click away the window in the usual way for your computer and control will be passed back to your sheet while printing your graph there.

An example result can be seen in figure (6.8). The motion of a simple harmonic oscillator is not chaotic, so in the phase space the treks are circles. As an example of a chaotic system, consider the function known as Duffing's equation. This function involves three parameters, but only one initial condition has been chosen:

```
EquationTrekker[x''[t] + x[t] == α Cos[ω x[t]], x, {t, -10, 10},
  PlotRange → {{-10, 10}, {-10, 10}}, TrekParameters → {α → 3, ω → 1}]
```

The resulting phase diagram, seen in figure (6.8), is quite complicated, and plotting more than one trek will just make it harder to interpret.

6.5.3 The Poincaré section

The calculated solutions representing the continuous time function of a differential equations are known as "flows." For a flow the errors in the direction tangent to the trajectory are marginally stable, and thus there is at least one Lyapunov exponent that is 0. Any perturbation in this direction can be compensated by a shift in time. The effect of such a time shift is not particularly interesting, so its effect is often eliminated by utilizing the Poincaré section.

A Poincaré section is created by choosing an $n - 1$ hyperplane within the n-dimensional phase space. One then calculates the intersection of the flow with the hyperplane. Typically a 2-dimensional Poincaré

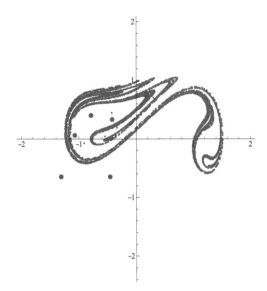

Fig. 6.9   Poincaré section for Duffing's equation.

surface is chosen, for which the flow intersection becomes a collection of data points. As an example, consider a Poincaré section of Duffing's equation,

```
EquationTrekker[{ x''[t] + γ x'[t] - x[t] + x[t]^3 == ε Cos[ω t]},
 x, {t, 1000, 10 000}, PlotRange → {{-2, 2}, {-2, 2}},
 TrekParameters → {γ → .15, ε → .3, ω → 1.},
 TrekGenerator → {PoincareSection, "SectionCondition" → Mod[ω t, 2 π],
    "SectionVariables" → {x, x'}, MaxSteps → ∞}]
```

When the Trekker window comes up, you must select the plot option of points, or else the diagram will make no sense. The result can be seen in figure (6.9).

The usefulness of the resulting graph depends highly upon the placement of the Poincaré surface. The best surface is one that maximizes the number of intersections. Another aspect for the choice of the surface of section is to try to maximize the variance of the data within the section, as their absolute noise level is independent of the section.

Exercises

6.1   Consider a simple model of a point mass orbiting a gravitational central field.
  1. Derive the Lagrangian and Hamiltonian equations for this model.
  2. From these equations determine the constants of motion.
  3. Show that the orbital motion of the mass is bounded by a function of these constants, and derive these boundaries.

6.2   From *Wikipedia*, find the mass and orbital properties of the 10 most massive asteroids.
  1. Calculate the virial for these asteroids.
  2. What does your result say about the stability of these objects?

6.3 In **6-5PointMassViral** a relation between eccentricity and autocorrelation is derived from a fit of our planetary data. Using this fit as a model, estimate the correlation times of the asteroids above.

6.4 Using **StarSun** and its associated notebooks, analyze the stability of the solar system for an intruder with a closest approach of 100 AU. (The other parameters should retain their default values.)

6.5 Study the effects of altering either the mass or initial velocity for the default closest approach, and compare with the results obtained in that case.

6.6 If you have sufficient computer power, run the N-body notebooks assuming a different distance along the line of sight and investigate what effect this has on the estimates of stability of the Trapezium. (There are four examples provided in the "put into user home" folder of Chapter 6, which can be used as a starting point.)

6.7 Use **6-10VClipNBbk** and **6-11VClipNBfw** to make sections of a video clip to produce a demonstration video using one of the initial condition sets (or your own) from the above exercise.

6.8 Find a differential equation that is believed to be chaotic for certain values.

1. Verify using `EquationTrekker[]` that the chaotic domain is indeed correct.

2. Using the Poincare section classify what type of attractor or repeller is present.

6.9 The N-body notebooks have a built-in expandability so that more than four bodies can be included in the analysis. Of course the computer requirements are correspondingly increased. We have found three additional objects that could be put into the N-body calculation. They are the B-N object at $9.3M_\odot$, radio source I at $5M_\odot$, and radio source N at $2M_\odot$ (Rodriguez et al., 2008). Here are the respective parameters:

ravel={14.06,14.503,14.34}
decal={-22.8+60,-30.5+60,-32.9+60}
xvels={21,27,27}
pevalsra={-5.3,4.5,0}
pevalsdc={9.4,-5.7,-13}

7

Astrophysical plasmas

It is evident from observation that most of the interstellar medium is permeated with charged particles and permanent magnetic fields. If the ionization in a particular region is complete (no neutral particles) the gas is called a plasma. Thus, interstellar space is always a low density, nearly collisionless, environment where particles may go centuries without encountering a particle of the same kind. It is most certainly dominated by a plasma or at least a highly ionized gas. In this chapter we consider the behavior of light and electric charges in such plasmas in a variety of situations. We will typically use protons and electrons as test particles.

Many treatments of plasmas in astrophysics consider only "cold," virtually collisionless plasmas, but there are a number of instances, particularly in the vicinity of stars and protostars, where one must examine higher density, higher temperature situations. A cold plasma is one in which the kinetic motion of the protons and electrons generally can be ignored. For warm plasmas, electron and ion temperature becomes a contributing factor and we must take kinetic theory into account via the Maxwell–Boltzmann equation (MBE). Finally we look at two diverse, but actually closely related, applications of plasma theory in astrophysics, the first using pulsars to map the electron density and magnetic field within the Milky Way galaxy, the second being a model of solar wind.

7.1 Charges in cold plasmas

The nearest places showing interstellar-like conditions are found in Earth's or other planetary magnetospheres. In addition, the sun emits streams of charged particles (as do many other stars) called solar (or stellar) wind. The solar wind being a plasma is itself fairly strongly magnetized, and so it behaves most of the time not as streams of independent particles such as those presented here, but as collective plasmas whose motions are described by the laws of magnetohydrodynamics (MHD). We discuss these in Section 7.5. These streams are quite effective in keeping the real interstellar medium away from the sun. Therefore most of our information about the interstellar medium (ISM) is obtained via remote sensing. The exception to this are high-energy cosmic rays that are able to penetrate the solar wind and sometimes even Earth's magnetic field, atmosphere, and crust. They are true samples of the interstellar medium, though biased because of their energies. However, their paths can be described by the equations presented here. Because our equations can be directly compared with magnetospheric observations, we present calculations in the context of Earth's magnetospheric/ionospheric conditions with the understanding that we can easily scale results to interstellar conditions when needed.

7.1.1 Magnetic fields

In astrophysical contexts, only gravity and magnetism are the really important long-range forces because the permanent charge separation necessary for electrical forces is rarely achieved for long times and over immense distances. The exception to this is the vicinities of individual stars and planets where electric fields do play a role. Here we will focus on the effects of a magnetic field.

An electric charge moving through a uniform magnetic field $\mathbf{B}$ experiences a Lorentz force

$$\mathbf{F} = \frac{q\mathbf{v} \times \mathbf{B}}{c}, \tag{7.1}$$

where q is the charge and $\mathbf{v}$ is its velocity. Because the resulting force is perpendicular to the velocity, the charge spirals along the magnetic field lines at constant speed. The radius of motion for the charge is known as the Larmor radius. Setting the magnitude of the magnetic force equal to the centripetal force, we find

$$R_L = \frac{p_c c}{qB}, \tag{7.2}$$

where p_c is the transverse momentum of the charge. The frequency of the spiral motion is known as the cyclotron frequency, and is given by

$$\omega_c = \frac{v_c}{R_L} = \frac{qBv_c}{p_c c}, \tag{7.3}$$

where v_c is the transverse velocity. Because the Larmor radius is small compared to the scale of interstellar magnetic fields, charged particles become tied to magnetic field lines. The only way charged particles can escape magnetic field lines is through collisions. The circular motion of a charge about a magnetic field is known as the gyro motion.

Typically a charged particle will spiral along a magnetic field. If v_z is the component of velocity along the field line then a charge will spiral with a pitch angle given by

$$\theta = \tan^{-1}\left(\frac{v_c}{v_z}\right). \tag{7.4}$$

If the charge encounters a strengthening magnetic field, the Larmor radius will decrease and the transverse velocity will increase due to the conservation of the charge's angular momentum. Because the magnetic field does no work on the charge, an increase in the transverse velocity v_c must correspond to a decrease in v_z due to conservation of kinetic energy. If the magnetic field strengthens sufficiently, v_z will vanish and the charge will be reflected in the opposite direction.

Because angular momentum is conserved, the magnetic moment

$$\mathbf{M} = \frac{q}{2c}\mathbf{v} \times \mathbf{r} \tag{7.5}$$

is also conserved. The magnitude of this moment can be written in terms of the Larmor radius; thus

$$M = \frac{v_c p_c}{2B}, \tag{7.6}$$

and therefore

$$v_c = \frac{2BM}{p_c}. \tag{7.7}$$

If the pitch angle of a charge is initially θ_o, in a magnetic field of initial strength B_o, then $v_z \to 0$ when the magnetic field reaches

$$B = \frac{B_o}{\sin^2 \theta_o}. \tag{7.8}$$

In this way an increasing magnetic field acts as a magnetic mirror that reflects charges. If a charge is trapped between two magnetic mirrors it is reflected back and forth without a possibility of escape (barring a rare collision). Such motion is known as bounce motion.

In general, the motion of charged particles in a diffuse plasma are composites of simpler motions:

1. Gyro motion – around the magnetic field
2. Bounce motion – occurs when the magnetic field lines diverge in one direction and converge in the opposite (and vice versa), trapping the particle.
3. Drift motion – motion of an entire orbital pattern at constant velocity

7.1.2 The Hamiltonian viewpoint

Gurnett and Bhattacharje (2005) is one of the more up-to-date plasma books, and has topics other texts do not have. In particular, they discuss some applications of Hamiltonian mechanics[1] to some well known single-particle magnetic field configurations. The Hamiltonian approach has not only analytical advantages, but when using *Mathematica* there is a distinct computational advantage as well.

The Hamiltonian H is an expression for the total energy which for conservative potentials is a constant of motion. Hence

$$H = T + V, \tag{7.9}$$

where T is the kinetic energy and V is the potential energy. In Hamiltonian theory the kinetic energy is expressed in terms of the momenta rather than the velocities, while the potential can be a function of coordinates and/or momenta. The potential can even be an explicit function of time if that is required. For simple situations the H function can be written by inspection, but in complex situations the transition from Lagrangian to Hamiltonian should be followed carefully to keep from making mistakes or omissions.

Unlike the Lagrange equations of motion, which are usually second order, the Hamiltonian equations are stated as systems of first-order equations expressed in terms of the generalize coordinates q_i and their conjugate momenta p_i. Thus $H = H(q_i, p_i)$ and the equations of motion are

$$\dot{q}_i = \frac{\partial H}{\partial p_i}, \tag{7.10}$$

$$\dot{p}_i = -\frac{\partial H}{\partial q_i}. \tag{7.11}$$

Gurnett and Bhattacharjee comment (p. 62) that although Hamiltonian formulations are often useful for qualitative investigations "it is difficult, if not impossible, to obtain closed-form analytical solutions." They also state that although the Hamiltonian approach does rely on symmetry, it does not assume adiabaticity (i.e., linearity). Indeed because Hamilton's equations are nonlinear they hold great interest in modern mechanics because such equations can become deterministically chaotic.

Over the years there have been many schemes devised to make the Hamilton equations of motion symmetric in structure by changing the nomenclature of the variables. The one that enables efficient numerical computations of numerical solutions to them is a special matrix formulation called symplectic structure (Goldstein, Poole and Safko, p. 343). It fortunately happens that *Mathematica*'s function NDSolve has a special option that uses symplectic methods to solve Hamiltonian equations of motion

---

[1] For a full modern treatment of Hamiltonians, see Goldstein, Poole, and Safko (2002).

regardless of the number of coordinate dimensions. Because the codes we present below use the *Mathematica* built-in matrix solution routines, the computations are extremely fast and efficient.[2]

Here we examine two computational examples. The first is that of an axially symmetric magnetic mirror. In this case the magnetic field is given by

$$\mathbf{B}_\rho = -\frac{\partial A_\phi}{\partial z}, \qquad \mathbf{B}_z = \frac{1}{\rho}\frac{\partial}{\partial \rho}\left(\rho A_\phi\right), \tag{7.12}$$

where $A_\phi\left(\rho, z\right)$ is the magnetic vector potential. The Hamiltonian is then

$$H = \frac{p_\rho^2}{2m} + \frac{p_z^2}{2m} + \frac{1}{2m\rho^2}\left(p_\phi - q\rho A_\phi\right)^2. \tag{7.13}$$

The vector potential can in principle be found directly from a specification of the B-field components. For simplicity we consider the magnetic field of two widely separated coils, which produces a relatively weak field at their center with a strong symmetric gradient of the field strength at their ends. In other words, we use the coil fringe field superpositions to create the mirror region. This is in contrast to Helmholtz coils, which have very large radii compared to their separation so that they produce a uniform field at the mid plane between the coils.

Although it looks from the equations above that a differential solution might be possible, the vector potential we seek is not separable. Thus with *Mathematica* to help us we try instead to derive the vector potential by direct integration over the current. We assume we have two parallel (in the x–y plane) coaxial circular coils of identical properties separated by λ in the z-direction. The potentials for these coils are found by

$$\texttt{aAL} = \int \frac{1}{\sqrt{\texttt{r1}^2 + \left(\frac{\lambda}{2} + \texttt{z}\right)^2 + \rho^2 - 2\,\texttt{r1}\,\rho\,\texttt{Cos}[\phi 1]}}\,\texttt{d}\phi 1$$

$$\texttt{aAR} = \int \frac{1}{\sqrt{\texttt{r2}^2 + \left(\frac{\lambda}{2} - \texttt{z}\right)^2 + \rho^2 - 2\,\texttt{r2}\,\rho\,\texttt{Cos}[\phi 2]}}\,\texttt{d}\phi 2$$

It can be shown that both of these integrals are 0 for $\phi = 0$ so only the values at $\phi = 2\pi$ need be evaluated. The vector potential then becomes

```
vectA[ρ1_, z1_, λsep_, radloop_] = (r1 aAL - r2 aAR) /.
    {φ1 → 2 π, λ → λsep, z → z1, ρ → ρ1, r1 → radloop, φ2 → 2 π, r2 → radloop};
```

From this one can calculate and plot the resulting magnetic field.[3] The resulting Hamiltonian is given by

$$\texttt{hmirror} = \frac{\texttt{p}\rho^2}{2\,\texttt{mM}} + \frac{\texttt{pz}^2}{2\,\texttt{mM}} + \frac{1}{2\,\texttt{mM}\,\rho\rho^2}\,(\texttt{p}\phi - \texttt{qQ}\,\rho\rho\,\texttt{constM}\,\texttt{vectA}[\rho\rho, \texttt{zz}, \texttt{sep}, \texttt{rad}])^2$$

---

[2] For documentation, see the online tutorial Help > NDSolve > Advanced Numerical Differential Equation solving in *Mathematica* > ODE Integration Methods > Methods > Symplectic Partitioned Runge Kutta (SPRK).

[3] See **7-1ChargePaths**.

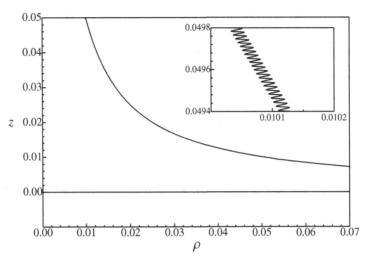

Fig. 7.1 Electron motion in a simple axial magnetic field.

where mM and qQ are the electron mass and charge respectively. There are six first-order equations of motion:

```
p1d = - (∂zz hmirror); p2d = - (∂ρρ hmirror); p3d = - (∂ϕϕ hmirror);
q1d = (∂pz hmirror); q2d = (∂pρ hmirror); q3d = (∂pϕ hmirror);
```

The resulting equations are very nonlinear, so it is difficult to pick initial conditions that result in stability. The step size must be very small; otherwise NDSolve[] will hit singularities. Due to the nonseparable nature of the Hamiltonian, if you attempt a solution by the SPRK method, *Mathematica* will reply with an error. We can, however, use the Implicit Runge-Kutta method with Gauss coefficients. The nonlinearity of these equations means the method can be touchy and time consuming. It does, however, produce a reasonable result:

```
system2 = NDSolveProblem[...];

solnD = NDSolve[system2, {t, 0, tend},
  Method → {"FixedStep", Method → {"ImplicitRungeKutta", "DifferenceOrder" → 10}},
  StartingStepSize → step1, MaxSteps → 10^6]
```

The result is not, however, very useful. As can be seen in figure (7.1), our two-loop system is not an effective magnetic mirror system that can contain a plasma. The potential depends on

$$\left(p_\phi - q\rho A_\phi\right)^2, \tag{7.14}$$

and for electrons with q negative the potential is always positive unless A is always negative. With the Hamiltonian always positive, no bound trajectories can exist and electrons will always escape. To have bound electron orbits there must be an additional force with sufficient negative potential.

For a second example, consider a dipole magnetic field with magnetic moment M. In spherical coordinates the magnetic potential is given by

$$A_\phi = \frac{\mu_0 M}{4\pi} \frac{\sin \theta}{r^2}. \tag{7.15}$$

The Hamiltonian is then

$$H = \frac{1}{2m} \left(p_r^2 + \frac{p_\theta^2}{r^2} + \frac{p_\phi^2}{r^2 \sin^2 \theta} \right) + \frac{1}{2mr^2 \sin^2 \theta} \left(p_\phi - \frac{q\mu_0 M \sin^2 \theta}{4\pi r} \right)^2. \tag{7.16}$$

Here the spherical coordinate kinetic energy expression follows Goldstein, Poole, and Safko (p. 341) while the potential function follows Gurnett and Bhattacharjee (p. 66).

Again, the SPRK approach fails in NDSolve[]. To use SPRK in Hamiltonian problems it is essential that the kinetic energy be a quadratic form of the momenta while the potential energy is a function of the coordinates. Because this is not the case we must use the Implicit Gauss Runge–Kutta method:

$$\mathbf{hdipole} = \frac{1}{2\,\mathbf{mM}} \left(\mathbf{pr}^2 + \frac{\mathbf{p\theta}^2}{\mathbf{rR}^2} + \frac{\mathbf{p\phi}^2}{\mathbf{rR}^2\,\mathbf{Sin[\theta 2]}^2} \right) + \frac{1}{2\,\mathbf{mM}\,\mathbf{rR}^2\,\mathbf{Sin[\theta 2]}^2} \left(\mathbf{p\phi} - \frac{\mathbf{qQ}\,\mathbf{mu0}\,\mathbf{dimom}\,\mathbf{Sin[\theta 2]}^2}{4\,\pi\,\mathbf{rR}} \right)^2 ;$$

Expressing the Hamilton equations as before, we can then calculate the solution

```
system3 = NDSolveProblem[. . .];

solnG = NDSolve[system3, {t, 0, tend1},
   Method → {"FixedStep", Method → {"ImplicitRungeKutta", "DifferenceOrder" → 10}},
   StartingStepSize → 0.00001, MaxSteps → 10^7]
```

In the magnetic dipole case the potential depends on

$$\left(p_\phi - \frac{q\mu_0 M \sin^2 \theta}{4\pi r} \right)^2, \tag{7.17}$$

so for electrons it is always positive unless p_ϕ is more negative than the other term is positive. Again, if the Hamiltonian is positive then no bound trajectories can exist. To have bound electron orbits one needs an additional force with a potential that is more negative than the other terms are positive.

7.2 Photons in cold plasmas

In the plasma physics literature, the topic of this section is usually called E–M wave propagation in a cold plasma. The references on this subject are too numerous to quote, so we focus on approaches that are most conducive to computational treatments.[4] A good example of a traditional treatment is seen in Davies (1966), where the first approach to propagation is to study the refractive index in a uniform

---

[4] See, for example Davies (1966), Gurnett and Bhattacharjee (2005), Kivelson and Russell (1995), Shu (1991, 1992), Sturrock (1994), and Tanenbaum (1967).

plasma. This treatment similar to that found in undergraduate E–M courses. To make the treatment accessible it is generally assumed:

1. Simple harmonic progressive waves
2. Steady-state solutions
3. Plane waves with a given polarization
4. Electrically neutral
5. Charge distribution has statistical uniformity
6. Uniform external magnetic field
7. Only electrons are effective
8. Electron collisions
9. Thermal motions of the electrons are unimportant (i.e., a cold plasma)
10. The magnetic properties are those of free space

7.2.1 The Appleton–Hartree equation

The electromagnetic waves are governed by Maxwell's equations:

$$\nabla \cdot \mathbf{D} = 4\pi\rho, \quad \nabla \times \mathbf{E} = -\frac{1}{c}\frac{\partial \mathbf{B}}{\partial t},$$

$$\nabla \cdot \mathbf{B} = 0, \qquad \nabla \times \mathbf{H} = \frac{4\pi}{c}\mathbf{j}, \tag{7.18}$$

where ρ is the charge density and $\mathbf{j}$ is the current density, and

$$\mathbf{D} = \epsilon\mathbf{E}, \quad \mathbf{B} = \mu\mathbf{H}, \tag{7.19}$$

where ϵ and μ are the permitivity and permeability respectively.

Following Davies, we apply these equations to the wave and then impose the properties of the medium through "constituitive relations." Davies points out that there are two ways to interpret the medium. One is as a conductor with currents. The other is as a polarizable medium in which the ion movements are included. The description cannot be both of these at the same time, so Davies chooses the polarizability viewpoint.

Using a coordinate system with axes 1, 2, and 3, where the direction of the 3-axis is determined by the right-hand rule $1 \times 2 = 3$, the geometry chosen for the wave is to set the direction of propagation along the 1-axis. The external magnetic field, B_0, is taken to be in the 1–2 plane, making an angle θ with the 1-axis. For the external magnetic field, the component along the direction of propagation is called the longitudinal B-field, B_L, and the component along the 2-axis is called the transverse B-field, B_T. As shown by Davies, the final results can then be expressed by four parameters:

$$X = \frac{Ne^2}{\epsilon_o m\omega^2}, \qquad Y_L = \frac{eB_L}{m\omega}, \qquad Y_T = \frac{eB_T}{m\omega}, \qquad Z = \frac{\nu}{\omega}, \tag{7.20}$$

where N is the electron density (number/m^3), m is the electron mass, ω is the angular frequency of the wave, and ν is the collision frequency.

The magnitude and orientation of the external magnetic field can be found from its transverse and longitudinal components such that

$$\frac{B_T}{B_L} = \tan\theta, \qquad B_0 = \sqrt{B_T^2 + B_L^2}. \tag{7.21}$$

The wave polarization R is expressed through the ratio of the electric fields, magnetic fields, or polarizabilities:

$$R = \frac{P_3}{P_2} = \frac{E_3}{E_2} = -\frac{H_2}{H_3}. \tag{7.22}$$

If R is real then the E, D, and Poynting vectors are linearly polarized at an angle $\arctan(R)$ with respect to the 2-axis. The magnetic vector H makes the same angle with respect to the 3-axis. When R is complex the H and D vectors describe ellipses in the 2–3 plane.

The refractive index of the medium is defined as

$$n = \frac{c}{V} = \sqrt{\frac{\mu\epsilon}{\mu_o\epsilon_o}}, \tag{7.23}$$

where V is the wave velocity of the medium. If there are no special magnetic properties then $\mu = \mu_o$ and the refractive index is

$$n = \sqrt{\frac{\epsilon}{\epsilon_o}}, \tag{7.24}$$

which is the square root of the dielectric constant of the medium.

To calculate the index of refraction for a cold plasma we must first find a relationship between the four plasma parameters and the E-field ratio R. Without going into details, there is a quadratic relation:

$$Y_L R^2 - \frac{iY_T^2}{1 - X - iZ} R + Y_L = 0. \tag{7.25}$$

Unfortunately *Mathematica* does not factor these roots very well, so the solutions must be optimized by hand to resemble the form given in Davies. There are two expressions for the polarization ratios, given by

$$R_2 = \frac{i}{2Y_L}\left(\frac{Y_T^2}{(-1 + X + iZ)} - \sqrt{\frac{Y_T^4}{(1 - X - iZ)^2} + 4Y_L^2}\right), \tag{7.26}$$

$$R_4 = \frac{i}{2Y_L}\left(\frac{Y_T^2}{(-1 + X + iZ)} + \sqrt{\frac{Y_T^4}{(1 - X - iZ)^2} + 4Y_L^2}\right). \tag{7.27}$$

As a result of these two roots, the complex refractive index has two distinct forms, known as the Appleton–Hartree formula:

$$n_1 = \sqrt{1 - \frac{X}{(1 - iZ + iY_L R_2)}}, \tag{7.28}$$

$$n_2 = \sqrt{1 - \frac{X}{(1 - iZ + iY_L R_4)}}. \tag{7.29}$$

This means there are two "characteristic" waves in a magnetoionic plasma. These waves are known as ordinary and extraordinary rays because they correspond to the rays of the same names in birefringent crystals such as Iceland spar. The real part of the reflective index (sometimes denoted by μ) indicates the phase speed of the wave. The imaginary part indicates the amount of energy loss by absorption. In free space the phase speed and c are the same; hence $\mu = 1$. If $\mu = 0$ then there is reflection at that point.

7.2.2 Ordinary versus extraordinary rays

So how does one tell which index of refraction corresponds to which type of ray? In n_1 notice that if $Y_L = 0$ but $Y_T \neq 0$, the terms containing Y_T still cancel, giving the refractive index of the nonmagnetic case. This is what is expected for an ordinary ray. It means the ordinary ray is reflected from a height where the nonmagnetic plasma would have its reflection, even if $Y_T \neq 0$. This also means the ordinary ray has a polarization negative sign. In n_2 notice that if $Y_L = 0$ but $Y_T \neq 0$, the terms containing Y_T never reduce to that of the nonmagnetic case. This is what is expected for an extraordinary ray. It means the extraordinary ray is reflected from a height different from the nonmagnetic plasma unless there is no magnetic field. It also means the extraordinary ray has a polarization positive sign.

If the magnetic field is 0, then the two expressions reduce to the same form:

$$n_{B=0} = \sqrt{1 - \frac{X}{1 - iZ}}. \tag{7.30}$$

Thus there is only one wave in a nonmagnetic plasma, and it is the ordinary ray. If in addition there is no aborption ($Z = 0$) the refractive index depends only on the electron density:

$$n_0 = \sqrt{1 - X}. \tag{7.31}$$

Perhaps the most confusing aspect of the distinction between ordinary and extraordinary waves is their respective polarization states. Even Davies (p. 211), who is normally very careful about such things, has a diagram showing the supposed relative orientations of the circular polarization states in quasi-longitudinal propagation, but the note with the figure says only that the magnetic field is perpendicular to the paper, not whether the field is into (or out of) the page. The diagram shows that the + component is left handed with respect to a normal out of the page and the − component is right handed.

Sturrock (p. 82), however, makes it very clear how to tell the handedness of the rays. In his consideration of the propagation parallel to the magnetic field vector he makes the following astute observation, relative to the magnetic field direction. Because of the $\mathbf{v} \times \mathbf{B}$ rule of magnetism, electrons (being negative ions) will circulate in a right-handed manner around the B-field vector. This means that electrons can preferentially absorb right-hand circular polarized waves and thus will have a singularity in the dispersion relationship. That also means that extraordinary waves must be identified with right-hand polarization relative to the propagation of the B-field. That leaves ordinary waves to be left-handed.

So why is this important? In terrestrial radio communications within Earth's geomagnetic field frequent use is made of circularly polarized antennas to increase the efficiency of detection of faint signals. To avoid preferential absorption of the extraordinary ray, one desires to transmit and receive only the ordinary ray.

So how is that arranged? Earth's magnetic field convention is that the magnetic dipole field at present has a "south" magnetic pole in the Northern Hemisphere, and a "north" magnetic pole in the Southern Hemisphere. This means the magnetic force lines originate in the Southern Hemisphere and descend into the Northern Hemisphere. This is why the north magnetic end of a bar magnet points north. East–west transmission is essentially across magnetic lines, and does not have an extraordinary mode. On the other hand, north–south transmissions do have an extraordinary mode that can be absorbed. Because northward transmissions are "parallel" to the magnetic field lines, the extraordinary ray will be right-handed relative to the field direction, so transmission should be left-handed. Reception of the transmission on a "southward" looking antenna (even if pointing upward) will be looking at the

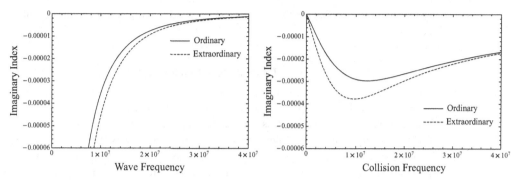

Absorption dependence on wave frequency and collision frequency.

extraordinary ray coming down (perhaps after reflection from the ionosphere or magnetosphere), but from the receiver side the polarization will look right-handed so the receiving antenna should use right-hand polarization. The same antenna transmitting southward should use the same polarization to transmit back, because relative to reversed transmission direction the direction of the ordinary ray should be right-handed, not left-handed.

7.2.3 Wave propagation with absorption

The calculation burden of the Appleton–Hartree equation is considerable, so often the full theory is not confronted. Here we consider the basic effects of absorption within this model before considering a more general approach in the next section.[5]

In the Appleton–Hartree equation absorption is dependent on the average collisional frequency of the ions, ν. It is common to plot the refractive index versus the X parameter in what are called dispersion curves. For our purpose, the same thing is accomplished by plotting the refractive index versus the wave frequency and angle.

In *Mathematica* it is useful to re-express the parameters of equation (7.20) in the form

$$X = \frac{80.6164N}{f^2}, \qquad Y_L = \frac{f_L}{f}, \qquad Y_T = \frac{f_T}{f}, \qquad Z = \frac{\nu}{f}, \tag{7.32}$$

where N is the electron density per cubic meter, $80.6164N$ is the square of the electron plasma frequency, and f is the wave frequency in Hertz. The electron gyro frequency can be defined as f_H, and thus $f_L = f_H \cos\theta$ and $f_T = f_H \sin\theta$. In this way various 3D plots of the refractive index can be plotted.[6] Although the refractive index does depend on the angle of incidence, the effect can be subtle. We will therefore look at the dependence of collision frequency for a fixed incident angle (in this case $20°$).

It is the imaginary component of the refractive index that acts as the attenuation component; therefore a plot of the imaginary component serves as a measure of the amount of absorption of the electromagnetic wave. As seen in figure (7.2), the absorption depends on both the wave and collision frequencies.[7]

---

[5] For a more general study of Appleton-Hartree, see **7-2EMWaves**.

[6] See **7-2EMWaves**.

[7] Here wave frequency is plotted with a fixed collision frequency of 4×10^5. The collision frequency is plotted with a fixed wave frequency of 1.1×10^7. For a general 3D plot, see **7-2EMWaves**.

In particular it is clear that for both ordinary and extraordinary rays absorption is mainly a low frequency phenomenon.

7.2.4 Polarization and absorption

When dealing with the complex refractive index, the real component is taken as the periodic part while the imaginary component is taken as the attenuation part. We passed over the fact that the E-field ratio is purely imaginary. When substituted into the refractive index it was also multiplied by i, thus rendering the index (except for the Z term) real. What this means is that the E_2 and E_3 field components are orthogonal to each other (as are H_2 and H_3). Here we return to the original Appleton–Hartree polarization equations (7.26) and (7.27). In this case the 2-axis is parallel to the transverse magnetic field component B_T while the 3-axis is perpendicular to the 2 direction and the 1 direction. Unlike refractive index, the imaginary parts for the E-field ratio are the "normal" situation as these are transverse waves. The real ratios are those fields that are parallel.

The E-field vector rotates such that when aligned along the 2-axis it has the length E_2 and when aligned along the 3-axis it has the length E_3. That means the vector end point traces out an ellipse unless R is unity, in which case the shape is a circle. To generate our ellipses on the same center point we use the eccentric anomaly version of the orbit formula (Danby, 1988), where x and y from the ellipse center are given by

$$x_o = a\sqrt{\left(1 - e^2\right)} \sin E, \qquad y_o = a \cos E, \tag{7.33}$$

for the ordinary wave, and

$$x_x = a\sqrt{\left(1 - e^2\right)} \cos E, \qquad y_x = a \sin E, \tag{7.34}$$

for the extraordinary wave. Here a is the semi-major axis of the ellipse and e is the eccentricity. In terms of the magnitude of the ratio R of the E-fields, the eccentricity of the ellipse is obtained from the ratio

$$R = \frac{1 - e}{1 + e} \quad \text{or} \quad R = \frac{1 + e}{1 - e} \tag{7.35}$$

depending on whether the ratio is greater or less than 1 respectively. The results can then be plotted with ParametricPlot[].

Without absorption, in the case of the ordinary ray, $R > 1$; thus

```
rr0 = Abs[
    rR7 /. {yT6 -> fH1 Sin[θ4] / freq2, yL6 → fH1 Cos[θ4] / freq2, xX6 → cnst2 nN2 / freq2²}]
```

We can then solve for the eccentricity:

```
e0solve = If[Abs[Re[rr0]] < 1,
    Solve[Abs[Re[rr0]] ⩵ (1 - e) / (1 + e), e], Solve[Abs[Re[rr0]] ⩵ (1 + e) / (1 - e), e]]
```

Since $R > 1$ for the extraordinary ray, we can follow the same process to calculate its eccentricity. The resulting plot of these polarizations can be seen in figure (7.3), with the semi-major axis $a = 1$.

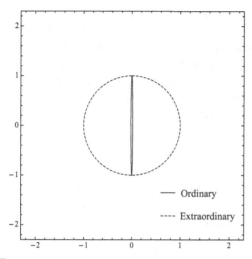 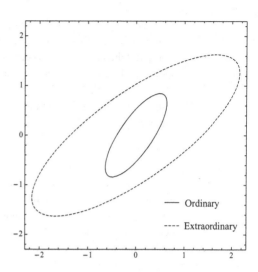

Fig. 7.3 E-field polarization without (left) and with (right) absorption.

To see the effect of absorption we must recalculate R_1 and R_2 with a typical absorption term.[8] For the ordinary ray we can calculate as before:

```
rrO1 = rR6 /. {zZ6 → v1 / freq3, yT6 -> fH1 Sin[θ5] / freq3,
    yL6 → fH1 Cos[θ5] / freq3, xX6 → cnst2 nN3 / freq3²}
```

This yields a complex value; however the magnitude of $R > 1$, so the eccentricity can be calculated just as before:

```
eOsolve1 = If[Abs[Re[rrO1]] < 1,
    Solve[Abs[Re[rrO1]] == (1 - e) / (1 + e), e], Solve[Abs[Re[rrO1]] == (1 + e) / (1 - e), e]]
```

For the extraordinary ray the magnitude of $R < 1$; thus

```
rrX1 = rR8 /. {zZ6 → v1 / freq3, yT6 -> fH1 Sin[θ5] / freq3,
    yL6 → fH1 Cos[θ5] / freq3, xX6 → cnst2 nN3 / freq3²}

eXsolve1 = If[Abs[Re[rrX1]] < 1,
    Solve[Abs[Re[rrX1]] == (1 - e) / (1 + e), e], Solve[Abs[Re[rrX1]] == (1 + e) / (1 - e), e]]
```

The results can be plotted as before; however, since R is complex the semi-major axis must be multiplied by the magnitude of the imaginary factor. This reduces or enhances the ellipse sizes in comparison to the noncollisional cases. As seen in figure (7.3), the extraordinary ray experiences a collisional enhancement at the expense of the ordinary ray.

7.2.5 Propagation in an anisotropic media

An isotropic medium is one where the phase velocity is independent of the direction of propagation. A magnetoionic medium therefore is said to be anisotropic because the phase velocity is dependent on

---

[8] Here we have chosen $\theta = 10°$, $f = 1 \times 10^3$, $N = 10,000$, and $\nu = 3 \times 10^6$.

the propagation angle with respect to the magnetic field. In an isotropic medium a wave originating at a point has a spherical wave front and the wave direction (perpendicular to the tangent to the wave front) and the ray direction (vector to origin) coincide. In an anisotropic medium, the wavefront is not spherical. In the case of a magnetoionic medium the wavefront is elongated in the direction of the external magnetic field. Thus there is an angle α between the ray direction that makes an angle ξ with respect to the magnetic field direction and the wave normal that makes an angle θ with respect to the same magnetic field direction. As shown by Davies (p. 95), if μ denotes the refractive index then

$$\tan \alpha = \frac{1}{\mu} \frac{d\mu}{d\theta}, \tag{7.36}$$

where α is measured in the convention of Davies where angles from the wave normal (axis 1) to the magnetic vector are considered positive. The degree of wave front distortion is most pronounced at low frequencies and illustrates the complexity of VLF signal propagation under terrestrial, solar corona, or heliospheric conditions. Of course these effects scale to higher wave frequencies in high B-field regions such as pulsar magnetospheres.

7.2.6 Time of arrival distortion

Just as there is wave front distortion in magnetoactive media, so too there is time of arrival distortion. In Chapter 5 we mentioned the relativistic excess time delay for signals passing near the sun. There was no mention of solar corona effects because it was assumed the radar measurements were at high enough frequencies that the solar corona would produce no measurable effect. However, early Crab Nebula pulsar pulse delay measurements were often made at frequencies so low that the coronal delays completely swamped the relativistic effects (Goldstein and Meisel, 1969). Removal of the plasma induced delays can be accomplished today because the coronal electron density profiles obtained from solar X-ray satellite images give not only the electron densities, but also the magnetic field strengths. Here we complete the Appleton–Hartree formalism by evaluating the magnetoionic phase and time delays. We cannot actually evaluate the integrals involved (Davies, 1966) because we would need to have actual electron density and magnetic field maps available, but we can show their form and calculate the kernels.

The transit time for the front of constant phase is

$$\Delta T_{ph} = \int \mu \cos \alpha \, ds, \tag{7.37}$$

and for the group path we have

$$\Delta T_{gr} = \int \frac{\cos \alpha}{\mu} ds. \tag{7.38}$$

The real parts of μ and α are associated with the times while the imaginary parts are associated with the path absorption losses. Plots of these for a typical angle ($\theta = 70°$) can be seen in figure (7.4).

It is clear the degree of time of arrival distortion is most pronounced at low frequencies. These effects scale to higher wave frequencies in high B-field regions such as pulsar magnetospheres. Along with time of arrival fluctuations due to electron density changes along the path, there will also be frequency fluctuations that mimic the effects of physical Doppler motions. Hence

$$\frac{\Delta f}{f} \propto -\frac{dT_{ph}}{dt}. \tag{7.39}$$

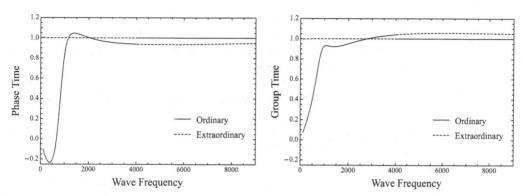

Fig. 7.4   Time of arrival distortion for waves with absorption.

The main source of such T_{ph} fluctuations is turbulence in the medium. Many media have turbulence as an additional complication to the propagation of waves. Such fluctuations are not unlike light fluctuations in starlight that even amateur astronomers are familiar with called scintillation. Radio astronomers also refer to such behavior as phase fluctuations. These are an important source of error in interferometric observations.

7.3 Generalized magnetoionic radio wave propagation

In a widely quoted paper, Sen and Wyller (1960) generalized the Appleton–Hartree expressions by considering collisions to be governed by Maxwell–Boltzmann statistics for the electrically neutral particles that collide with the electrons, and the departure of the electrons from a Maxwellian distribution caused by the electromagnetic fields through the Boltzmann Transport Equation.[9] Sen and Wyller were guided by experimental work[10] with microwaves in nitrogen and air where the momentum collision frequency was found to be proportional to electron energy. A reinterpretation of the Sen–Wyller work using modern collision cross sections has been published by Freidrich, Finsterbusch, Tokar and Spöcker (1991).

To start a generalized collisional theory, one usually assumes a slightly ionized Lorentz gas, that is, a gas where one constituent is much more massive than another. Essentially in this case it is assumed that the neutrals are quite heavy compared with the electrons so that the neutrals maintained a Maxwellian distribution independent of the electrons. The electrons may have had a Maxwellian distribution without the impressed fields on, but once the sinusoidal E-field and constant B-field are applied the electrons take on non-Maxwellian steady state solutions. The electron velocity distribution is obtained by a perturbation from the velocity distribution of the neutrals. Such a process is described by the Boltzmann Transport Equation using the Chapman–Enskog method. What is needed are generalized expressions for the conductivity followed by the complex refractive index and polarization valid for any velocity dependence of the collision frequency. Sen and Wyller arduously progress through the details in their paper, constantly pausing to discuss how their expressions are related to the

---

[9] See the *Mathematica* notebooks in the appendix for a detailed discussion of transport phenomena.
[10] Phelps and Pack (1959) and Phelps (1960).

Appleton–Hartree functions where ν is constant, and providing graphs that display numerical results from which quantitative comparisons can be drawn. It would be a monstrous undertaking to repeat the elegant results of Sen and Wyller in *Mathematica* so we will not even try. Instead we will concentrate on the computational results their theory enables.

7.3.1 The Sen–Wyller algorithm in computational form

For computational purposes, in cases in which the collision frequency is expressible with μ constant (Appleton-Hartree), or μ proportional to velocity or velocity squared, the Sen-Wyller results involve the so-called C script integrals. These integrals are a member of the integral family discussed by Dingle, Arndt and Roy (1957). The integrals as named in the original papers are not listed in Weisstein (2003), nor can they be found through the Wolfram web site *MathLink*. But in spite of the lack of a formally named function available for computation that generates the results directly, it turns out that *Mathematica* does solve the C script integral directly. Therefore we can easily create a suitable function for finishing the generalized theory.

It should be noted that *Mathematica* generates a conditional function for the C integral directly. Thus the result for the integral

$$cC = \frac{1}{p!} \int_0^\infty \frac{\epsilon^P e^{-\epsilon}}{\epsilon^2 + x^2} d\epsilon \tag{7.40}$$

will have to be cut and paste to create a new function:

```
cScript[p1_, x1_, prec_] :=
```

$$N\left[\frac{1}{p!}\left(\text{Gamma}[-1+p]\ \text{HypergeometricPFQ}\left[\{1\}, \left\{1-\frac{p}{2}, \frac{3}{2}-\frac{p}{2}\right\}, -\frac{x^2}{4}\right] + \pi\left(\frac{1}{x^2}\right)^{\frac{1}{2}-\frac{p}{2}}\text{Csc}[p\ \pi]\ \text{Sin}\left[\frac{p\ \pi}{2}+\sqrt{x^2}\right]\right)\ /.\ \{p \to p1,\ x \to x1\},\ prec\right]$$

This function either requires three arguments ($p, x, prec$) or two arguments (p, x), the former stating a numerical precision to the integral. Thus

```
cScript[3 / 2, 1]
```

yields the function in analytical form, while

```
cScript[3 / 2, 1, 20]
```

yields 0.25396602433678820751.

We can now use the equations of Sen–Wyller to construct a *Mathematica* function that calculates the refractive index for the several angular frequencies that are specified for a plasma plus the angle that the propagation path makes with the magnetic field. Because there are many operations involved we will use the Module construct. The arguments defined are the propagation angle ϕ, the wave frequency ω, the plasma frequency ωp, the gyromagnetic frequency ωb, the collision frequency ν_M as defined by Sen–Wyller, and the numerical precision $pr1$.

For the ordinary ray,

$$\text{refr0}[\phi\_, \omega\_, \omega p\_, \omega b\_, \nu M\_, \text{pr1}\_] := \text{Module}\Bigg[\ldots$$

$$\sqrt{\frac{\text{aA} + \text{bB Sin}[\phi]^2 + \sqrt{\text{bB}^2 \text{Sin}[\phi]^4 - \text{cC}^2 \text{Cos}[\phi]^2}}{\text{dD} + \text{eE Sin}[\phi]^2}}\Bigg]$$

and for the extraordinary ray,

$$\text{refrX}[\phi\_, \omega\_, \omega p\_, \omega b\_, \nu M\_, \text{pr1}\_] := \text{Module}\Bigg[\ldots$$

$$\sqrt{\frac{\text{aA} + \text{bB Sin}[\phi]^2 - \sqrt{\text{bB}^2 \text{Sin}[\phi]^4 - \text{cC}^2 \text{Cos}[\phi]^2}}{\text{dD} + \text{eE Sin}[\phi]^2}}\Bigg]$$

We can also modify the refractive index functions to determine the polarization ratio under the Sen–Wyller formalism. This modification involves only changing the last expression in each function. Thus for the ordinary and extraordinary rays respectively,

$$\text{rRO}[\phi\_, \omega\_, \omega p\_, \omega b\_, \nu M\_, \text{pr1}\_] := \text{Module}\Bigg[\ldots$$

$$-\frac{\text{bB Sin}[\phi]^2 - \sqrt{\text{bB}^2 \text{Sin}[\phi]^4 - \text{cC}^2 \text{Cos}[\phi]^2}}{\text{cC Cos}[\phi]}\Bigg]$$

$$\text{rRX}[\phi\_, \omega\_, \omega p\_, \omega b\_, \nu M\_, \text{pr1}\_] := \text{Module}\Bigg[\ldots$$

$$-\frac{\text{bB Sin}[\phi]^2 + \sqrt{\text{bB}^2 \text{Sin}[\phi]^4 - \text{cC}^2 \text{Cos}[\phi]^2}}{\text{cC Cos}[\phi]}\Bigg]$$

7.3.2 Estimating collision frequencies from a Maxwellian distribution

Because Sen and Wyller did not have modern computers available when their paper was written, much of their effort was spent finding under which circumstances the old Appleton–Hartree formula could be used using a suitable definition of collision frequency. Because in the Sen–Wyller case the collision frequency has an electron speed dependence, it is useful to associate the mean collision frequency with one of the two mean speeds of the standard Boltzmann distribution, the most probable speed or the RMS speed. If one chooses the most probable speed, then $\nu_{AH} = \nu_m$. If one chooses the root mean square speed as the reference (as does Sen and Wyller) then $\nu = (3/2)\nu_m$. They establish that if the exponent of the electron velocity dependence of the collision frequency is $2n$ then the relationship between the AH frequency and the most probable collision frequency becomes

$$\frac{\nu}{\nu_{AH}} = \frac{2n+3}{3}. \tag{7.41}$$

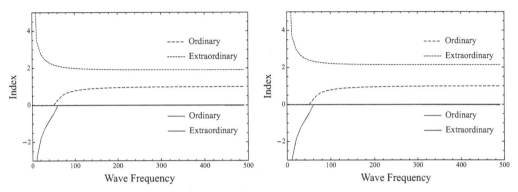

Fig. 7.5 Quasi-longitudinal (left) vs. quasi-transverse (right) VLF waves.

For evaluation of the collision frequency for a particular Maxwellian gas, we resort to using the mean free path theory for an ideal gas (even if the real gas is molecular). From elementary kinetic theory (Laurendeau, p. 301) the binary collision rate per unit volume is given by

$$\nu_M = n_1 n_2 d_{12}^2 \sqrt{\frac{8\pi kT}{\mu_{12}}}, \tag{7.42}$$

where n_1 is the electron density, n_2 is the neutral density, d_{12} is the hard sphere approximation center-to-center distance, k is the Boltzmann constant, T is the gas temperature, and μ_{12} is the reduced mass between the neutrals and electrons (essentially the mass of the neutrals). For a dilute ionized gas, as is usually assumed for the Appleton–Hartree situation, we have $n_1 \ll n_2$. Only for a fully ionized gas would $n_1 n_2 = n^2$. For d_{12}, the electron contributes nothing to the distance between colliding particles, so this is essentially the neutral particle hard sphere radius. In a model atmosphere whether terrestrial, planetary, or stellar, the parameters in this expression can usually be obtained fairly easily. In the interstellar medium we can even use this formula to estimate collision frequencies that are exceedingly small.

7.3.3 Propagation of very low frequency waves

When there is no collisional term in the Appleton–Hartree equations, the polarization and refractive index behavior is fairly well determined. When the real part of the refractive index reaches 0, wave reflection occurs. Correspondingly the E-field transverse components along the dimensions along the transverse axis do the same. When absorption occurs these two "rules" are broken.

If the propagation is exactly along the magnetic field lines, it is called longitudinal propagation, and the angle is 0 degrees. If the propagation is perpendicular to the magnetic field lines it is called transverse, and the angle is 90 degrees. In Sen-Wyller theory three cases are derived: longitudinal, transverse, and in-between. Here we are not sure about convergence of the Sen–Wyller expressions when $\theta = 0$ and $\theta = 90°$. Therefore for the former we shall let $\theta = 2°$ and call the situation quasi-longitudinal. For the latter we shall let $\theta = 88°$ and call it quasi-transverse.

In figure (7.5) we have plotted the refractive indices and absorption indices for both the ordinary and extraordinary rays. The case presented here corresponds to very low frequency (VLF) transmission of the D layer of Earth's ionosphere, known as whistlers. It is clear that the ordinary wave is reflected back

($n = 0$) until an angular frequency of about 50 Hz, where partial reflection occurs until about 200 Hz as $n \to 1$. The absorption factor e^{-n} of the ordinary ray goes to unity at about 60 Hz. The extraordinary ray has a large refractive index at small frequencies, then levels off at $n = 2$ rather than $n = 1$. This means the phase velocity of the extraordinary ray exceeds the speed of light by a factor of 2. Although this seems like a violation of special relativity, it is not, as we shall see. The absorption index for the extraordinary ray is 0 for all frequencies, so the extraordinary ray is unattenuated through the layer.

The ordinary and extraordinary rays "heard" at your locality becomes the whistler that can be heard at sunrise and sunset when the electron density in the lower ionosphere is rapidly changing. The whistler pulses of such low radio frequencies are created by lightning strokes at the opposite geomagnetic point of Earth's dipole field. The electron density of the outer ionosphere then guides the waves along magnetic lines back down to be received at the other locality. Of course your local lightning produces whistlers that are heard at the other magnetic conjugate point. It works both ways. Whistlers have been detected from other planets with magnetic fields. Whistlers of extremely high frequencies are also generated in pulsar atmospheres. Solar radio bursts are between these extremes. Thus the radio propagation theory we have outlined has astrophysical as well as geophysical importance.

It is clear from figure (7.5) that there is not much difference between the quasi-longitudinal and quasi-transverse results. In the comparable cases without absorption the Appleton–Hartree theory usually shows profound effects. So why is that not true when collisions are included? The parameters given are quite appropriate for whistlers in the terrestrial ionosphere and magnetosphere. Whistlers are easily observed at high geomagnetic latitudes where vertically projected VLF waves take on a quasi-longitudinal behavior as they are trapped between magnetic lines of force. On the other hand, horizontally projected VLF waves at the same latitudes behave in the quasi-transverse mode and get trapped between the lower ionosphere and the ground in "waveguide" mode. The properties of these two propagation modes are quite similar in accord with Sen–Wyller theory.

7.3.4 Phase velocity and group velocity

Our discussion so far has been concerned with the propagation of single monochromatic waves that can be described as waves that are functions of $\omega t - \mathbf{k} \cdot \mathbf{r}$, where the angular frequency ω is a position vector with some origin and $\mathbf{k}$ is a vector in the propagation direction (the ray) known as the wave vector, and its magnitude is called the propagation factor, wave factor, or wave number. The magnitude of $k = 2\pi/\lambda$, where λ is the wavelength. The speed of the wave pattern is the phase velocity, which in free space is equal to the speed of light c. As we have seen, for whistlers the extraordinary wave has a refractive index that indicated its phase velocity is at least $2c$. Does this not violate special relativity? Actually no.

Remember that the phase velocity is the speed at one frequency, and a plasma is a dispersive medium that by definition can have a different phase speed and polarization for each frequency. Although the wave pattern at one frequency has an amplitude, it alone does not carry energy. Instead there is an energy propagation velocity know as the group velocity. Sturrock (1994, Appendix B) presents a nice discussion of group velocity that is much more general than is usually found. Essentially Sturrock proves that given a conserved quantity (energy or momentum for example) with an ambient density and associated flux it is the group velocity (not the phase velocity) that is the proportionality factor between them. On the other hand the phase velocity and group velocity for waves are related to each other by

$$V_{phase} V_{group} = c^2. \tag{7.43}$$

So when the extraordinary ray has a phase velocity of $2c$, it has a group velocity of $0.5c$, which is of course less than c. Hence group velocity is always less than c as long as the phase velocity never gets less than 1.

7.4 Pulsar signals as probes of the interstellar medium

After a long excursion through the details of electromagnetic propagation through ionized media, we are ready to deal with issues of electromagnetic wave propagation through a very dilute ionized medium, the interstellar medium. The discovery of pulsars not only revolutionized the studies of white dwarfs and neutron stars, but it also influenced studies of the interstellar medium.

7.4.1 Faraday rotation

The observed rotation of the plane of linear polarization of light has been known for a long time thanks to the pioneering studies of Michael Faraday. Of course this predates considerably the observation of Faraday rotation in the radio domain. In quasi-longitudinal propagation, Davies (p. 210) indicates that the ordinary ray and the extraordinary ray are circularly polarized in opposite directions, but that the ordinary ray rotates in its direction more rapidly than the extraordinary rotates in its direction. Of course which sense is observed as right-handed and which as left-handed depends on which direction the magnetic field is pointed with respect to the down coming wave. The net effect is that the two circular polarizations combine to form a linear polarization that itself rotates, producing the Faraday effect. Because the linear polarization rotates in the sense of the ordinary ray, for electrons the linear polarization rotates left-handed relative to the magnetic field direction.

In terrestrial ionospheric studies, observations of artificial satellite signals or radar transmissions in "moon-bounce" mode have shown Faraday rotation in the received signal. Such observations give an estimate of the total electron content of the ionosphere. The amount of rotation depends on the path length, with the total observed angle of rotation given by

$$\Omega = \frac{1}{2}\left(K_+ - K_-\right), \tag{7.44}$$

$$K_+ = \frac{2\pi}{\lambda_o}\int \mu_+ \cos\alpha \, ds, \tag{7.45}$$

$$K_- = \frac{2\pi}{\lambda_o}\int \mu_- \cos\alpha \, ds, \tag{7.46}$$

where λ_o is the free space wavelength of the wave.

In the moon bounce experiments (as well as interstellar observations), it is usually true that the frequencies used are always high enough to be well above any plasma frequency so that an approximation of $\cos\alpha = 1$ can be made. Thus for quasi-longitudinal propagation we have

$$\Omega_{QL} = \frac{\pi}{\lambda_o}\int (\mu_+ - \mu_-) \, ds. \tag{7.47}$$

For QL waves we have no significant absorption on the path; thus

$$\Delta\mu = (\mu_+ - \mu_-) = \sqrt{1 - \frac{X}{1 + Y_L}} - \sqrt{1 - \frac{X}{1 - Y_L}}. \tag{7.48}$$

To second order, this reduces to

$$\Delta\mu = \frac{X Y_L}{\sqrt{1 - X}}. \tag{7.49}$$

If the wave frequency is much larger than the plasma frequency then the X in the denominator can be ignored. To render $d\Omega$ unitless it must be multiplied by a factor of constants. In *Mathematica* this can be done by running the expression without adjustment, then cutting and pasting to obtain the dimensionless part:

```
δΩ = ((  π
       ----- FullSimplify[xX yL /. {xX →   nN e²    , yL ->  e B₀ Cos[θ] }]) /.
      (c / f)                            e₀ mE ω²             mE ω

    {ω → 2 π f, e → qe[[1]], mE → m[[1]], e₀ → ε0[[1]],

    c → UnitConvert[Quantity[1, "SpeedOfLight"]][[1]]})

UnitConvert[Quantity[1, "VacuumPermeability"]][[1]]
```

which yields

$$d\Omega = 0.0297 \frac{N \cos\theta B_o}{f^2}, \tag{7.50}$$

where N is the number of electrons per cubic meter, f is the wave frequency in Hertz, and B_o is in gauss. The wave frequency can be expressed in terms of the free space wavelength $f = c/\lambda_o$; thus

```
δΩλ2 = (δΩ /. f → c / λ₀) /. c → UnitConvert[Quantity[1, "SpeedOfLight"]][[1]]
```

The full angle Ω divided by λ_o^2 is called the rotation measure (RM). This definition is observationally grounded but the value of the integral and the conversion constant depend on the definition of the electron density. Here we assume electrons per cubic meter; thus

```
cnst4 = δΩλ2[[1]]; rM = cnst4 Hold [∫nN Cos[θ] B₀ ds]
```

Here the Hold[] operator prevents *Mathematica* from evaluating the integral.

It is necessary to have measurements at several frequencies to resolve the direction of the rotation and hence the orientation of B as well as to obtain the value of the integral. The main difficulty is that measurements of the Faraday rotation are mod 2π (actually mod $\pm\pi$) and so the rotation measure must be obtained at high enough frequencies that the RM for the most distant objects does not exceed the observable range of angle. Just to significantly overcome the RM of the terrestrial magnetosphere and the solar wind produced heliosphere appears to require frequencies in the low microwave range. This had to be checked from other studies, but appears to be true for interstellar Faraday measurements at 1 to 2 GHz, a range now achievable but always suffering from cell phone interference. Hence we define the full rotation Ω using the frequency in MHz, and define the integral by `integralnB`.

cnst3 = $\delta\Omega$[[1]]; Ω = $\dfrac{\texttt{cnst3 integralnB}}{\texttt{fMHz}^2\ \texttt{10}^{12}}$

One way to circumvent the $\pm\pi$ limitation in the observational determination of the Faraday rotation is to use a filter bank and observe the differential change in Ω between successive filter frequencies. We assume that the separation of the filters is a constant $\Delta fMHz \ll freqMHz$, where $freqMHz$ is the frequency at the center of the filter bank. As long as measurements are consistent with each other without any jumps, then the average can be taken and set equal to the derivative of the Ω expression. If the filter channels are noisy then a least squares procedure must be devised. Therefore

dΩdf = $\partial_{\texttt{fMHz}}$ Ω

and introducing the observational average $d\Omega/dfMHz$ as $avd\Omega dfMHz$,

integral2 = Solve[dΩdf == avdΩdfMHz, integralnB]

One can eliminate the integral and just use the average rotation rate as a function of frequency to get RM directly, which we define as `rM1`. We define `rM2` for RM as a function of the integral:

rM1 = rM /. Hold$\left[\int$ nN Cos[θ] B$_0$ d s$\right]$ $\rightarrow$ integral2[[1, 1, 2]];

rM2 = rM /. Hold $\left[\int$ nN Cos[θ] B$_0$ d s $\right]$ -> integralnB;

The integral can be solved as a function of RM by

intsolve = Solve[rM0 == rM2, integralnB]

In pulsars, the polarization varies over a pulse, as well as pulse to pulse, so appropriate pulse averages must be taken (Manchester and Taylor, 1977). This is a tedious procedure.

7.4.2 Dispersion estimates of the electron density

Because the polarization of pulsars is so highly variable, it is difficult to get just the interstellar contribution to the Faraday rotation from their pulses. On the other hand, the electron density integrated along the same line of sight is relatively easy to obtain through the frequency dispersion observed for the pulsar pulse arrival times. This is possible because the pulsar pulses are so brief and the pulse frequency content is so high that the group velocity is an appropriate measure of the pulse dispersion. Because the K vectors used to obtain the Faraday rotation are so close to each other we can define the average K for the propagation. Hence

$$K_{ave} = \frac{1}{2}\left(K_+ + K_-\right) \tag{7.51}$$

$$K_+ = \frac{2\pi}{\lambda_o}\int \mu_+ \cos\alpha\ ds, \tag{7.52}$$

$$K_- = \frac{2\pi}{\lambda_o}\int \mu_- \cos\alpha\ ds, \tag{7.53}$$

The group velocity is obtained from

$$\frac{dK}{d\omega} = \frac{1}{V_{gr}},\tag{7.54}$$

and again it is assumed that $\cos\alpha = 1$.

In *Mathematica*, we can express the effective dispersion relation including the magnetic field contribution for quasi-transverse propagation. Because neither X nor Y_L have a distance dependence, the integral over ds is converted to s.

```
kave = π/λ₀ (√(1 - xX/(1 + yL)) + √(1 - xX/(1 - yL))) /.

    {xX → cnst2 nN 4 π² / ωD², yL → 2 π cnst1 bB Cos[θb] / ωD}

kav = kave /. λ₀ → 2 π c / ωD;
```

From this we find the group velocity grV

```
grV = 1 / ∂_ωD kav
```

For the time delay, we want the reciprocal of the velocity, i.e., time per distance. Since the interstellar B-fields from Faraday rotation measurements are found to be microgauss we can make the substitution for that as well:

```
dtperds = Simplify[1 / grV] /. bB → 10⁻⁶ bBμ
```

Both the B-field and the electron density N are small for interstellar media, so we can expand this to second order in $bB\mu$ and nN:

```
dtds1 = Collect[Normal[Series[dtperds, {bBμ, 0, 2}]], bBμ];
dtds2 = Collect[Simplify[Normal[Series[dtds1, {nN, 0, 2}]]], nN];
```

Pulsar pulse measurements are made at MHz frequencies so that Earth's ionosphere and the heliosphere add little to the delays. Therefore we convert our result to MHz

```
dtds3 = Collect[FullSimplify[dtds2 /. ωD → 2 π fMHz 10⁶], nN];
dtds4 = dtds3 /. bBμ → 0;
```

This yields

$$\frac{dt}{ds} = \frac{1}{c}\left(1 + 4.03082 \times 10^{-11}\frac{N}{f^2} + 2.43713 \times 10^{-21}\frac{N^2}{f^4}\right)\tag{7.55}$$

Multiplying by ds and integrating, we then find the time delay:

$$t = \frac{s}{c} + 1.34454 \times 10^{-19}\int\frac{N ds}{f^2} + 8.12938 \times 10^{-30}\int\frac{N^2 ds}{f^4}.\tag{7.56}$$

The first term is the nondispersed transit time for free space. It is the same for every observation, and will subtract out if we compare the time of pulse arrival at one frequency with that at another. The latter terms are the first- and second-order dispersion terms. Because the second-order term is a factor of $10^{-21}/f^2$ smaller than the first-order term, it is usually dropped.

When comparing transit times for two frequencies, we will assume $f_1 > f_2$; thus

$$\Delta t = 1.34454 \times 10^{-19} \int N ds \left(\frac{1}{f_2^2} - \frac{1}{f_1^2} \right). \tag{7.57}$$

The dispersion constant D is an observational quantity defined as

$$D = \Delta t \left(\frac{1}{f_2^2} - \frac{1}{f_1^2} \right)^{-1}, \tag{7.58}$$

where the frequencies are in Hz rather than MHz; thus we have

$$D = 1.34454 \times 10^{-7} \int N ds. \tag{7.59}$$

If this situation were analogous to the Faraday rotation case, where rotation measure refers directly to the observationally measured quantity leaving the associated electromagnetic parameters with the integral over ds, the dispersion constant would also be called the dispersion measure. But alas that is not so. Instead the integral term itself is called the dispersion measure (DM). This can be confusing because the various E–M parameters are now associated with D. Thus DM is still proportional to D, but the parameters in the transformation depend on the units of the integral. Because there is no problem of angular mod transformations in DM, it is customary to determine it by simply considering observations at two or more frequencies using D, the dispersion constant.

7.4.3 An empirical model of the galactic electron density

The online ATNF pulsar catalog[11] contains DM values for nearly 2000 pulsars. The list of these pulsars is given as `PulsarDMR.csv`, and includes the AFTN number, J2000 (RA,dec) designation, galactic longitude, galactic latitude, DM, and DM error, in that order. To plot these as a map of electron density, we must first import the data into *Mathematica*:

```
dmp = Import["PulsarDMR.csv"];
```

The file contains a header row, which we must ignore. We can then create a data matrix from the list:

```
dMT = Table[{0, 0, 0, 0, 0, 0}, {ii, 1, numd - 1}];
. . . ;
Do[dMT[[ii, 6]] = dmp[[ii + 1, 6]], {ii, 1, numd - 1}];
```

Because we only want to plot DM with respect to galactic coordinates, we create a new table using only those values, which we can then plot as a `3DListPlot[]`:

```
dmglatlong = Table[{dMT[[ii, 3]], dMT[[ii, 4]], dMT[[ii, 5]]}, {ii, 1, numd - 1}];
ListPlot3D[dmglatlong, PlotRange → {{0, 360}, {-50, 50}, {0, 2500}},
  AxesLabel → {"glongitude", "glatitude", "DM"}]
```

---

[11] http://www.atnf.csiro.au/research/pulsar/psrcat

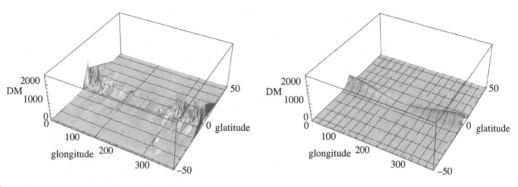

Fig. 7.6   Raw DM (left) vs. empirical fit (right) for pulsar data.

The result can be seen in figure (7.6). It is clear that most of the plasma lies along the galactic plane. Because these data appear well organized, we can make an empirical fit to our data:

```
latlongfit =
    NonlinearModelFit[dmglatlong1, b Abs[Cos[x1°/2]ᵃ] e^(c Abs[Sin[x2°]]), {a, b, c}, {x1, x2}]
```

which can also be seen in figure (7.6).

It was recognized quite early in pulsar studies that the DM values should be correlated with object distance. But except for those pulsars associated with supernova remnants, the only direct distance indicators are those pulsars in whose directions the 21 cm hydrogen absorption is seen. Using the list given in Manchester and Taylor (p. 127), we produce a very approximate DM-distance calibration using 18 objects. The list order here is mean distance in parsecs, galactic longitude, galactic latitude, and DM:

```
n2data = Table[{distlist[[ii, 4]], distlist[[ii, 2]]°,
        distlist[[ii, 3]]°, distlist[[ii, 1]]}, {ii, 1, numdist}];

nlm2 = NonlinearModelFit[n2data, a x1 Abs[Cos[ x2/2 ]]ᶜ e^(b Abs[Sin[x3]]),

    {a, b, c}, {x1, x2, x3}, MaxIterations → 1000]
```

The coefficient corresponding to the first distance-DM fit with galactic coordinate correlations is 0.295 ± 0.006 parsecs per DM, which is essentially the same value as that assumed in Manchester and Taylor. The correlation is about 0.87, which is relatively high. It is therefore clear that the standard practice of using DM as a distance indicator is a reasonable approach.

7.4.4 Estimates of the interstellar magnetic field

From the integral definitions of RM and DM we can obtain an estimate of the B-field by simple division of the integrals

$$\langle B\cos\theta\rangle = \frac{\int NB\cos\theta\, ds}{\int N ds} = 4.0664 \times 10^{11} \frac{RM}{DM}. \qquad (7.60)$$

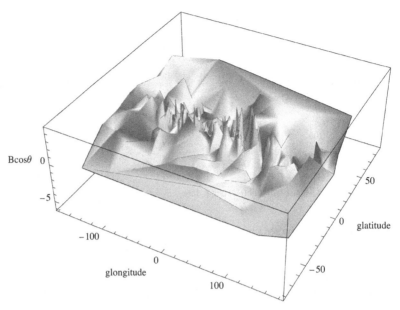

Fig. 7.7   Map of the galactic magnetic field.

Both RM and DM are standard observational quantities. Typically RM is given in radians/m^2, while DM is in parsec/cm^3, so one must be careful about units. Again pulsar data including DM and RM can be found at the online ATNF pulsar catalog[12]. A graph can be generated from this data as before, and is seen in figure (7.7).

Unlike the electron density map, the galactic mangetic field map is very much less well organized. A similar result was obtained from early pulsar $\langle B \cos \theta \rangle$ measurements. In those studies, however, a cluster of negative $\langle B \cos \theta \rangle$ values around galactic longitude 90° and a similar positive cluster around 270° was taken as an indication of a general longitudinal B-field. This also seems present in the equatorial plane of figure (7.7), but there are many local field reversals along the way. The change from + to − appears to be in the region between 220° to 240° longitude and −20° to 0° latitude. A crude estimate of the tilt is at about 20°. Because the solar equatorial ascending node is located at galactic longitude of 200° and latitude −23°, it seems this feature is not galactic but rather the result of solar wind blowing plasma into interstellar space.

7.5   Warm plasmas

Previously we explored the trajectories of single charged particles in external electromagnetic fields. Then we explored the propagation of electromagnetic waves in "cold" ionized media. We assumed the primary influence was the more mobile electrons rather than the more massive ions. Here we shift from the properties of "cold" plasmas to "warm" ones and illustrate the differences between them. In warm

---

[12] http://www.atnf.csiro.au/research/pulsar/psrcat

plasmas the Boltzmann Transport Equation (BTE) plays a central role. Although we will not delve into the details of this equation here, we have provided several *Mathematica* notebooks as introductions.[13] Although both Sturrock (1994) and Gurnett and Bhattacharjee (2005) provide excellent and somewhat equivalent narratives of the problems encountered in warm plasmas, it is easier to follow Gurnett and Bhattacharjee because they do their derivations in SI units, which is more in keeping with modern EM treatments as well as space physics texts such as Kivelson and Russell (1995).

7.5.1 Debye shielding

By definition a plasma is a nearly fully ionized gas that has electrons and ions distributed in such a way that there is strict average charge neutrality. As we noted in Section 7.1, this means any electric fields that may occur are not permanent, although because ions and electrons have different mobilities charge separation can occur.

If we have a pure hydrogen plasma, it is a "simple" gas of electrons and protons. With no net charge there is no net E-field, but the interstellar medium has electron density fluctuations that indicate there can be local departures from strict charge neutrality. As shown by Gurnett and Bhattacharjee (p. 7), Poisson's equation describes the electrostatic potential created with the addition of just one test charge Q to a plasma of electron number density n_e and ion number density n_o,

$$\nabla^2 \Phi = -\frac{e}{\epsilon_o}(n_o - n_e). \tag{7.61}$$

Because the ions (protons) are much more massive than the electrons they do not recoil at the addition of the test charge. If there is a Maxwellian speed distribution for the electrons before the appearance of the potential, there will be a modified distribution afterwards

$$f_e(v) = n_o \left(\frac{m_e}{2\pi k T_e}\right)^{3/2} e^{\frac{\frac{1}{2}m_e v^2 + e\Phi}{kT_e}}. \tag{7.62}$$

Integrating over the speed gives the electron density caused by the potential,

$$n_e = 4\pi \int_0^\infty f_e(v)\, dv = n_o e^{e\Phi/kT_e}. \tag{7.63}$$

Substituting this into the Poisson equation,

$$\nabla^2 \Phi = -\frac{n_o e}{\epsilon_o}\left(1 - e^{e\Phi/kT_e}\right). \tag{7.64}$$

We assume the potential Φ is only a function of r, but even with this simplification there is not a general analytical solution. We must therefore further simplify the equation by expanding the exponential term to first order,

$$\nabla^2 \Phi = \frac{d^2\Phi}{dr^2} + \frac{2}{r}\frac{d\Phi}{dr} = \frac{n_o e^2}{kT_e}\Phi(r). \tag{7.65}$$

This is known as the Debye–Hückle equation, and represents the first-order potential function caused by a swarm of electrons clustered around a single positive ion. For its solution we assume the potential is 0 at infinity and asymptotic to the Coulomb potential as $r \to 0$; thus

$$\Phi(r) = \frac{Q}{4\pi r\epsilon_o}e^{-r/\lambda_D}, \tag{7.66}$$

---

[13] See **7-5Maxwell**, **7-6Boltzmann** and **7-7CollisionB**.

where Q is the charge of the positive ion, and

$$\lambda_D = \sqrt{\frac{n_o e^2}{k T_e \epsilon_o}}, \tag{7.67}$$

is known as the Debye length.

The number of electrons within a Debye cube (λ_D^3) is an important plasma parameter. Gurnett and Bhattacharjee stress the importance of $N_D \doteq n_o \lambda_D^3$ as an indicator of plasma behavior:

1. In the foregoing derivation, the Debye–Hückle equation is valid only if $e\Phi/kT_e \ll 1$. The approximate volume of one electron is the inverse of the density. This means the minimum radius is $1/\sqrt[3]{n_o}$. This means

$$\frac{e\Phi}{kT_e} = \frac{1}{4\pi N_D^{2/3}}, \tag{7.68}$$

   and therefore Debye–Hückle is valid for $N_D \gg 1$.

2. In the section on electromagnetic wave propagation we characterized the plasma by average quantities without justification. The plasma can be considered a continuum fluid as long as $N_D \gg 1$, or the same as the Debye–Hückle condition.

3. One can estimate the ratio of the kinetic energy to the electrostatic potential in the same way using

$$\langle r \rangle = \frac{1}{\sqrt[3]{n_o}}, \tag{7.69}$$

   with the result that

$$\frac{KE}{PE} = 6\pi N_D^{2/3}. \tag{7.70}$$

4. If one compares the collective oscillation frequency of electrons in the plasma (ω_{pe}) with the individual collisions of the electrons with ions (ν_{ei}), then as shown by Gurnett and Bhattacharjee,

$$\frac{\omega_{pe}}{\nu_{ei}} = \sqrt{\frac{\pi}{2}} \frac{128 N_D}{\ln(12\pi N_D)}, \tag{7.71}$$

   and once again a large value of N_D implies the validity of the continuum assumption. If it turns out $N_D \ll 1$, one may safely perform single particle analysis.

7.5.2   Kinetic theory in plasma physics

There is a tendency in modern plasma theory to unify all oscillations (electromagnetic, plasma, and acoustical) into one combined theory because they operate in that way within the plasma itself. However, undergraduate students need to have a theoretical treatment of each phenomenon separately. Previously we discussed electromagnetic oscillations and here we introduce ion-electron mass motions that are coupled together as primarily an electrostatic interaction in a cold plasma where the thermal motions can be ignored for the moment.

Plasma oscillations

In such a restricted and idealized situation we assume a slab of electrons set into motion relative to the (stationary) ions. Electrons at postion x are now at position $x + \xi$. It is assumed that the shift occurs without disturbing the surrounding layers, particularly those at $\pm\infty$. The E-field that appears is

$$E = n_e Q \frac{\eta}{\epsilon_o}, \tag{7.72}$$

where n_e is the electron density and Q is the electric charge. Thus the equation becomes

$$m_e \frac{d^2\eta}{dt^2} = -QE. \tag{7.73}$$

Substitution of the slab field gives a sinusoidal solution,

$$\eta(t) = C_1 \cos\left(\frac{\sqrt{n_1}Qt}{\sqrt{m_e\epsilon_o}}\right) + C_2 \sin\left(\frac{\sqrt{n_1}Qt}{\sqrt{m_e\epsilon_o}}\right), \tag{7.74}$$

and the coefficient can be recognized as the square of the electron angular plasma frequency we encountered earlier. That is no accident, because electromagnetic waves traveling through a plasma set up electrostatic oscillations in the electrons as they pass.

If we include more massive positive ions displaced in a direction opposite to the electrons about the center of mass, there will be two equations of motion. The center of mass equation requires that

$$m_p \eta_p + m_e \eta_e = 0. \tag{7.75}$$

If the slabs are the same lengths, then the potential is symmetric in the distance difference between the relative slab positions. If we use the center of mass conditions to eliminate the distance not in the corresponding equation, then the equations become

$$m_e \frac{d^2\eta_e}{dt^2} = -\frac{n_1 Q^2}{\epsilon_o} \eta_e \left(1 + \frac{m_e}{m_p}\right), \tag{7.76}$$

$$m_p \frac{d^2\eta_p}{dt^2} = -\frac{n_1 Q^2}{\epsilon_o} \eta_p \left(1 + \frac{m_p}{m_e}\right). \tag{7.77}$$

The solutions are again sinusoidal, but the plasma frequency coefficients are a bit different being the square root of the sum of the squares of the individual angular frequencies. Each species has a separate plasma frequency and the effective frequency of the combination is simply the RMS sum of the squares of the individual angular frequencies. For electrons and equally charged positive ions with the same number density, this reduces to

$$\eta = \sqrt{\frac{(m_e + m_p)\, n_1 Q^2}{m_e m_p \epsilon_o}}. \tag{7.78}$$

Sound speed

We have assumed a Maxwellian distribution of electrons and ions in our plasma. It is not clear, however, whether the gas will have one temperature or many. In kinetic theory temperature is controlled via collisions. If the mass density is reasonably high in the gas, whereby collisions can maintain thermal equilibrium and equipartition of energy among different constituents, then the average kinetic energy for all species together should be $3kT/2$. But in plasmas the temperature of electrons and ions can be perturbed by internal electric fields or external magnetic fields. In a tenuous plasma the charged particle species will equilibrate between themselves faster than between different species. Thus one often encounters nonequilibrium situations where each species has a different temperature. This is particularly true for electrons compared with everything else. Hence each species satisfies equipartition with its own kind. As a result the sound speeds are also different. The formula for the speed of sound is

$$C_s = \sqrt{\frac{kT_s}{m_s}}, \tag{7.79}$$

where T, k, and m have their usual thermodynamic meaning. Because the speed of sound has the same parameters included in the plasma frequency and Debye length of each species, it should not be surprising that these parameters are simply related, such that

$$C_s = \omega_{ps}\lambda_{D_S}. \tag{7.80}$$

Cyclotron frequency

When a cold plasma is in a magnetic field, electromagnetic propagation tells us that the electrons and ions take up circular motions due to the $\mathbf{v} \times \mathbf{B}$ force. As before, the angular frequency of this oscillation is

$$\omega_{cs} = \frac{e_s B}{m_s}, \tag{7.81}$$

where B is the magnetic field, m is the species mass, and e is the respective charge (positive or negative). These motions are harder to keep synchronized than longitudinal electrostatic oscillations, so they are damped more highly. In a hot plasma this damping is called Landau damping, and results in energy being transferred from electrons to ions.

7.5.3 Maxwellian distributions and their evolution

Our cold plasma theory only modeled effects that depended on electron density. Here we show how temperature affects not only the electron behavior, but the ions as well, resulting in the Debye length and volume for each species. Key to obtaining the Debye parameters is averaging over the Maxwellian distribution. But it was argued how conditions might force the plasma to have different electron and ion temperatures. One such mechanism was the perturbation of electrons and ions by electrostatic oscillations. Suppose you wanted to know how a perturbed gas returns to its equilibrium temperature T. That information is given through the Boltzmann Transport Equation (BTE) or one of its many derivative forms (Vlasov, Fokker-Planck, Chapman-Enskog, Navier-Stokes, etc.).[14]

The power of the BTE approach is that it involves only the Maxwell–Boltzmann function in a single vector equation. Once the time evolution of the function is found then it may be averaged at each point to give the measurable properties of the gas such as number density, average velocity, kinetic energy density, and momentum density (or pressure tensor) by direct integration of the Maxwell–Boltzmann

---

[14] A general discussion of the BTE is given in **7-6Boltzmann** and **7-7Collisions**. Other facets of kinetic and transport theory are found in **17Transport** and **7-5Maxwell**.

function over velocity. Of these properties only the pressure tensor should be unfamiliar. All four are examples of "moments" of the Maxwell–Boltzmann distribution over velocity space:

1. Number density = zero-order moment
2. Average velocity = first-order moment
3. Average kinetic energy = $\frac{1}{2}mv^2$, where $v^2 = \mathbf{v} \cdot \mathbf{v}$ is the second-order moment
4. Pressure tensor $P = \langle m (v - \langle v \rangle) (v - \langle v \rangle) \rangle$ = second-order moment where v and $\langle v \rangle$ are vectors.

In the last example, the product is a dyadic (second order tensor) that can be expanded into a 3×3 matrix. It represents the flow of momentum relative to a coordinate system that itself is moving with velocity $\langle v \rangle$. Dividing this by the mass that gives the velocity "dispersion" of the distribution. If the velocity dispersion is isotropic then the off diagonal elements will be 0 and the same pressure $P = n_o k T$ along the diagonal. The Maxwell–Boltzmann distribution in this case is

$$f(v) = n_o \left(\frac{m}{2\pi kT} \right)^{3/2} e^{-m(v-U)^2/2kT}, \tag{7.82}$$

where U is the vector average velocity and v is the velocity vector. If there is some reason the pressure is anisotropic (e.g., because of a magnetic field), one obtains a bi-Maxwellian distribution

$$f(v) = n_o \left(\frac{m}{2\pi kT_\perp} \right) \left(\frac{m}{2\pi kT_\parallel} \right)^{1/2} e^{-mv_\perp^2/2kT_\perp} e^{-mv_\parallel^2/2kT_\parallel}. \tag{7.83}$$

The standard Boltzmann Transport Equation is

$$\left[\frac{\partial}{\partial t} + \bar{v} \cdot \nabla_r + F_{ext} \cdot \nabla_p \right] f(\bar{r}, \bar{p}, t) = \left(\frac{\partial f}{\partial t} \right)_{coll}, \tag{7.84}$$

where $\mathbf{r}$ is the position vector, $\mathbf{v}$ is the velocity vector, F the external force, and $f(\mathbf{r}, \mathbf{p}, t)$ is the distribution function. If the right-hand side is 0, the BTE is said to be collisionless. If the equation is collisionless and F is the Lorentz force, the equation is known as the Vlasov equation. In the Vlasov equation, the left side is $df/dt = 0$. If all the constants of motion $C_i(q, p)$ are obtained from the Hamiltonian of the individual particle motion, or $C_i(r, v)$ from the adiabatic invariants, then any function $f(C)$ is a solution of the Vlasov equation provided the correct boundary conditions can be applied. That means if

$$H = \frac{1}{2} \frac{p_s^2}{m_s} + Q_s \Phi(q), \tag{7.85}$$

then the solution of the Vlasov equation is

$$f(v) = n_o \left(\frac{m}{2\pi kT} \right)^{3/2} e^{(p^2/2m + Q\Phi)kT}. \tag{7.86}$$

It is this equation that was used in the derivation of the Debye–Hückle law.

7.5.4 Warm plasmas with no magnetic field

We have sketched the foundations for developing a theory of a warm plasma, but do not actually repeat the steps taken by Sturrock (1994) and Gurnett and Bhattacharjee (2005), as that would lead us too far astray into a very complex process. Suffice it to say that the moment equations lead to a charge density continuity equation and a momentum equation including a pressure term. This, combined with adiabatic equations of state of the form

$$P_s = P_o n_s^{\gamma_s}, \tag{7.87}$$

where n_s is the density of each charge component, finally yields a dispersion relationship (a function of angular frequency and have number k) for longitudinal modes. These modes do not exist in a cold plasma. The transverse mode in this instance is identical to the cold case and is not affected by the pressure. Each charge (electron or ion) has a dispersion relation. For electrons these are known as the Langmuir mode. For ions they are called the ion acoustical modes. These both have the speed of the electrons in them, but we know that the sound speed of the electrons is related to the Debye length so that

$$Ce^2 = \omega_{pe}^2 D_\lambda^2. \tag{7.88}$$

In *Mathematica* the Langmuir mode is found from

```
solLang = Solve[[(1 - (ωpe² / (ω² - γe cCe² knum²))) /. cCe → (ωpe λDD)] == 0, ω]
```

Although there are two roots, there is really only one mode, so that the frequency of oscillation increases with temperature or Debye length and the adiabatic constant of the electrons. If $k \to 0$ (long wave limit) the cold plasma result is obtained regardless of the temperature. Because the phase velocity is ω/k and group velocity is $d\omega/dk$ we have the solutions

```
phaseV = solLang[[2, 1, 2]] / knum; groupV = ∂knum solLang[[2, 1, 2]];
```

The plots in figure (7.8) represent a zero-order approximation to the solutions in a warm plasma obtained by the method of moments, but are these solutions valid? Subsequent to the derivations of the dispersion relations using the moments of the BTE, Gurnett and Bhattacharjee reconsidered the electrostatic wave situation in greater detail and arrived as did others at the conclusion that there are many instabilities that the moment approach misses. In most instances where a dispersion relation is derived, a Fourier approach is used (including just substituting a plane wave solution and solving). In their Chapter 8, Gurnett and Bhattacharjee return to use a full Vlasov analysis, but they point out the problems with a Fourier solution for the dispersion as used by Vlasov himself. The only real new success is the Langmuir solution for electrons (known as the Bohm–Gross dispersion) where it is shown that the constant $\gamma = 3$, and not 5/3 as might be assumed from the usual monoatomic gas situation. The several instabilities that show up are

1. Cold beam instabilities such as the two stream instability often discussed in the "cold" situation
2. Buneman instability – a cold electron bream through positive ions at rest
3. Ion acoustic instability – due to finite temperature in ions

In the Fourier solution all these lead to an exponentially growing situation. Landau realized that the problem was the singularity that occurs when the actual velocity is equal to the phase velocity. This situation can be resolved by treating the problem as an initial value problem using Laplace transforms rather than Fourier transforms. The Laplace transform approach is not without its price, as frequencies

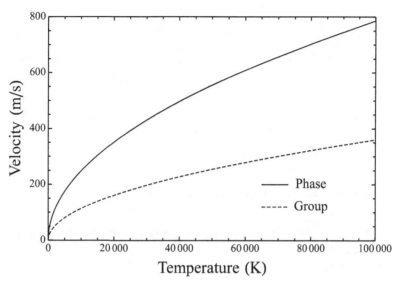

Fig. 7.8 Ion acoustical phase and group velocities.

can be complex numbers and the solutions are generally done in the complex plane via the method of residues. If a Cauchy distribution of velocities is assumed instead of a Maxwellian one, the solutions are analytic and it can be shown (Gurnett and Bhattacharjee, p. 300) that the time dependence of the electrostatic potential becomes

$$\Phi\left(t, \omega_p, \gamma_g\right) = \Phi(t=0)e^{\pm i\omega_p t}e^{-\gamma_g t}. \tag{7.89}$$

When a Maxwellian distribution is assumed, the same type of Laplace transform analysis shows (Gurnett and Bhattacharjee, p. 308) that the real frequency expression (with $\gamma = 3$) is preserved, but in addition it gives a damping factor. The result is the same as the weak growth rate approximation; hence

$$\gamma_{Landau} = -\sqrt{\frac{\pi}{8}} \frac{\omega_{pe}}{kp\lambda_D^3} \exp\left[-\frac{1}{2kp\lambda_D^3} - \frac{3}{2}\right]. \tag{7.90}$$

The ion acoustic result is a bit trickier to handle. Central to the Laplace transform approach is the integral

$$Z(\zeta) = \int_{-\infty}^{\infty} \frac{e^{-z^2}}{z - \zeta} dz, \tag{7.91}$$

where, following Gurnett and Bhattacharjee,

$$z = \sqrt{\frac{m}{2k_B T}} v_z, \tag{7.92}$$

$$\zeta = \sqrt{\frac{m}{2k_B T}} \left(\frac{i\gamma + \omega}{k_{\text{num}}}\right), \tag{7.93}$$

$$\omega_p \lambda_D = \sqrt{\frac{k_B T}{m}}. \tag{7.94}$$

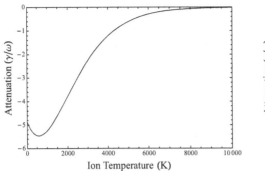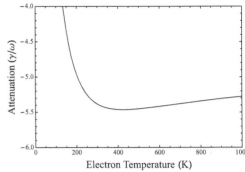

Fig. 7.9 Attenuation constant (γ / ω) for Landau damping at low temperatures.

Gurnett and Bhattacharjee do a series expansion before integrating, but *Mathematica* is able to do the integral directly:

```
int1 = Integrate[ e^(-z²) / (z - ζ1), {z, -∞, ∞}, Assumptions → Im[ζ1] ≠ 0, PrincipalValue → True]
```

The dispersion equation for electrons plus one species of ion becomes

$$D(k, \gamma - i\omega) = 1 - \frac{1}{(k_{num}\lambda_D)^2} \frac{1}{2} \left(Z'(\zeta_e) + \frac{T_e}{T_i} Z'(\zeta_i) \right) = 0 \qquad (7.95)$$

From this we can plot the attenuation constant (γ / ω) for the various cases, as seen in figure (7.9).

7.5.5 Warm plasmas with a magnetic field

To consider the effect of an external magnetic field on a warm plasma we must utilize magnetohydrodynamics (MHD). Like the Saha equation, MHD has its roots in astrophysics because early astrophysicists needed to develop operational analytical models long before sophisticated theories were available in the disciplines themselves. Essentially MHD is a blend of "ordinary" fluid dynamics and Maxwell's equations. In spite of its somewhat ad hoc nature, some surprising concepts and results were found that are still in use today. These include frozen in magnetic fields, Alfvén waves, MHD shocks, and magnetic reconnection, all concepts encountered in modern astrophysical texts and research papers. We cover only the most basic aspects of MHD here.[15]

To utilize MHD we must first define the mass density as the summation of the mass densities ρ_m. The fluid velocity is then the mass weighted average U. Then several modified, but borrowed, equations from kinetic theory come into play:

1. Mass continuity equation (1st moment of BTE)
2. Momentum equation (2nd moment of BTE)
3. Generalized Ohm's law
4. The equation of state

[15] For a wider coverage of the computational aspects of MHD, see **7-9MHD**.

These are combined with Maxwell's equations. Finally, charge neutrality is formalized by Ampére's law, $\nabla \cdot J = 0$.

There are two forms of Ohm's law. When $\mathbf{E} + \mathbf{U} \times \mathbf{B} = 0$, it is called an ideal MHD plasma. For that it is assumed that the internal collision frequency is 0, and as a result the conductivity becomes infinite. For an ideal MHD fluid:

1. Fluid velocity perpendicular to B is the single particle drift velocity, $U_\perp = \mathbf{E} \times \mathbf{B}/B^2$.
2. The E parallel to B is 0.
3. Magnetic lines are equipotentials if $\partial A/\partial t = 0$.

The resistive MHD fluid is more widely used because it is the simplest deviation from the ideal MHD fluid. Its Ohm's law is

$$\mathbf{E} + \mathbf{U} \times \mathbf{B} = \frac{\mathbf{J}}{\sigma}, \tag{7.96}$$

where σ is the conductivity. If the MHD is resistive then the pressure is a scalar. The fluid changes are slower than the kinetic theory changes so that the conduction current dominates over the displacement current. If the pressure is isotropic then the ordinary adiabatic equation of state is valid.

For resistive MHD, the evolution of the magnetic field is given by the equation

$$\frac{\partial \mathbf{B}}{\partial t} = \nabla \times (\mathbf{U} \times \mathbf{B}) + \frac{1}{\sigma \mu_o} \nabla^2 \mathbf{B}, \tag{7.97}$$

where the first term is the "convective" term and the second is the "diffusive" term. The extremes that can be assumed is whether convection dominates or diffusion dominates. The magnetic Reynolds number (which is analogous to the ordinary Reynolds number) determines these two extremes:

$$R_m = \frac{|\nabla \times (\mathbf{U} \times \mathbf{B})|}{\left|\frac{1}{\mu_o \sigma} \nabla^2 \mathbf{B}\right|} \tag{7.98}$$

For $R_m \gg 1$ then equation (7.97) leads to a flux constancy condition

$$\frac{\partial \mathbf{B}}{\partial t} = \nabla \times (\mathbf{U} \times \mathbf{B}), \tag{7.99}$$

$$\frac{d\Phi}{dt} = 0. \tag{7.100}$$

This means the magnetic field is "frozen" into the plasma. The geometry of "frozen-in" B-fields leads to the notion of magnetic flux tubes along which the plasma flows while the tubes themselves move through space. However, one must be careful when using this concept, as it is only valid if $\mathbf{E} + \mathbf{U} \times \mathbf{B} = 0$.

If $R_m \ll 1$ then the equation reduces to

$$\frac{\partial \mathbf{B}}{\partial t} = \frac{1}{\sigma \mu_o} \nabla^2 \mathbf{B}. \tag{7.101}$$

This is a diffusion equation whose speed depends on the inverse of the conductivity.

The notion of magnetic field pressure is also obtained from the fact that

$$(\nabla \times \mathbf{B}) \times \mathbf{B} = -\nabla \left(\frac{B^2}{2\mu_o}\right) + \frac{1}{\mu_o} (\mathbf{B} \cdot \nabla) \mathbf{B}. \tag{7.102}$$

The scalar gradient can be grouped with the gas pressure p. The second term can be resolved into two components. The one aligned along the B-field cancels the magnetic pressure in that direction. That means only the magnetic pressure perpendicular to the B-field can exert a force on the plasma. The second component arises from a gradient of the B-field, and thus represents a curvature force, one of the ways to exert a force on a single charge as was discussed in Section 7.1. The important thing about this instance of curvature is that it acts perpendicular to B and therefore is similar in nature (and function) to a mechanical string tension.

7.6 Solar wind

The sun is the nearest star to Earth, and as such is it often defined as just an "average" star. However, it is "average" in a logarithmic, not linear, way in mass, radius, temperature, composition, and even age. In the past we used to think the sun was unique with respect to its magnetism and the richness of its planetary companions, but even those views have changed and continue to change. Exoplanets continue to be discovered around other stars in interesting patterns that have similarities to many of the objects in our solar system.

When it comes to magnetic stars like our sun, it is the observable manifestations of magnetic interaction called solar activity (sunspots, solar flares, and coronal mass ejections for example) that command the most attention because they are relatively easy to "see." In other stars such activity is largely inferred, but is present in less than a majority of stars like the sun. One key property of this activity is how highly variable it is at all spatial and temporal scales. Because any energy-producing process on the sun should have some detectable result on the Earth, the study of solar-terrestrial relationships is a thriving and vital subject. Correlations between solar activity and variations in the upper atmosphere of Earth (aurorae for example) have been known for more than a century. But it is only relatively recently that some connection with the lower terrestrial atmospheric behavior ("real" weather and climate) could be convincingly demonstrated (Eddy, 1978; Herman and Goldberg, 1978).

In the 1970s and 1980s, as outlined in the references, the sun presented astronomy and astrophysics with four main frontier subjects:

1. The solar neutrino problem
2. The nearly steady-state transport of solar material into space
3. Seismic sounding of the sun
4. Solar modulation of the climate/weather pattern

Of these four, we consider only the second one. In particular, the origin and evolution that is present whether the sun is in an active state or not, the so-called solar wind.

7.6.1 History of the solar wind

The view of the outer atmosphere of the sun prior to the solar wind's inclusion in the sun's structure was a bit puzzling, not the least of which was that the corona by all indications was extremely hot and that the lower solar atmosphere actually reached a minimum temperature above which the temperature increased rapidly with radial distance without any indication of a major decrease out at least to many solar radii. All calculations indicated that the corona required for staying reasonably isothermal, a profound amount of energy transfer from below just to stay hot with cooling times on the order of hours without such a supply. Several coronal static models were devised but none worked satisfactorily in one way or the other. As higher resolution views were obtained by balloon-borne telescopes, it became evident that the solar disk was covered by small convection cells in constant motion. Above these granules, features called spicules were observed in the chromosphere. These eventually were connected with the coronal fine structure seen during eclipses of the sun. MHD models of these features indicated that there was a net flow of acoustical energy from near or below the photosphere. These waves, upon reaching the low density (but hot in regions of the chromosphere), became shock waves[16] that traveled along magnetic lines. Sometimes these disturbances went across magnetic lines to deposit their shock energy into the corona.

So what powers the granulation and by implication the corona? At some distance below the photosphere the temperature increases as one goes to the core. Not too far below the photosphere, the temperature reaches the point where the hydrogen is nearly completely ionized. Below that point the hydrogen is a fully ionized plasma. In the hydrogen ionization zone, the temperature rises faster than would be the case in radiative equilibrium, and that causes the convection that eventually overshoots into the photosphere. So how do we know this is a true scenario? According to best solar models, all solar type stars with a composition like that of our sun should have these hydrogen convective zones and if there is a connection of the convection with a corona, then all solar type stars should have coronae. It was discovered that like most one million degree objects, the solar corona emits x-rays that have been observed since early sounding rocket experiments in the 1960s. Thirty years later when x-ray satellites were launched with enough sensitivity to see x-rays coming from stars other than the sun, a number of nearby solar type stars indeed were found with x-ray emitting coronae. This confirmed the convective heating mechanism of the solar corona. Because the net heat flow was outward it was inescapable that the sun would have to emit particles to keep it at a fairly constant observed temperature. Of course this is an oversimplified picture, but qualitatively correct. Once the x-ray corona could be observed continuously, it was found not only that the sun could produce auroral and geomagnetic disturbances via chromospheric explosions (solar flares) but also the corona was observed to produce violent CMEs (coronal mass ejections) independently of the solar flares, and when these occurred toward Earth, disturbances similar to solar flare effects would be observed.

Until the solar wind provided an explanation of where the corona could release energy in a nonradiative way, the thermal adjustment mechanism was a bit obscure. But although the solar flares and CMEs accounted for many geophysical disturbances during high solar activity, attempts to account for disturbances at sunspot minimum failed badly because when one followed such disturbances back to the sun the trail ended up at coronal heights in the atmosphere where no coronal emissions were visible. These were called coronal "holes." The only manifestation of heating was enhanced helium (not hydrogen because it was already too hot for neutral hydrogen to exist) emission in the upper chromosphere/lower corona boundary. If the helium emission in the "hole" represented the "visible" form of an energy transport mechanism, then the form was "invisible" particle acceleration that would become the solar wind before normal x-ray emission could be excited. Of course, once the properties of the solar wind were established, all of the connections cited earlier became obvious and the concept of the solar wind became an established scientific reality with little further controversy (Parker, 1978).

The orbital flight of Skylab in 1975 marked the beginning of the modern solar wind period, and the progress of our knowledge about the wind can be obtained by comparing our two primary reference papers by Hundhausen (1978 and 1995). For pre-Skylab viewpoints see book accounts by (Parker, 1963; Brandt, 1970; and Hundhausen, 1972).

---

[16] See **7-10Shock**.

7.6.2 MHD as applied to the solar wind

Because the solar wind is an extension of the hot solar corona where hydrogen is completely ionized, its dynamics in the steady state is well described by the MHD equations. In fact these same equations not only have steady dynamical solutions, but they can also be shown to satisfy hydrostatic equilibrium equations in spherical symmetry, as usually assumed to simplify the solution.

We start with the MHD conservation equations for mass and momentum in a spherical corona (Hundhausen, p. 97). The mass conservation and momentum equations are particularly simple:

$$\frac{1}{r^2}\frac{d}{dr}\left(\rho u r^2\right) = 0, \tag{7.103}$$

$$\rho u \frac{du}{dr} = -\frac{dp}{dr} - \rho \frac{GM_\odot}{r^2}, \tag{7.104}$$

where ρ is the density, $u(r)$ the velocity, and p the pressure.

If one assumes $u = 0$, the first equation drops out and the second equation becomes trivial. Following MHD as before, we choose the ideal gas law:

$$p = nk_B\left(T_i + T_e\right), \tag{7.105}$$

where n is the number density. The mass density is $\rho = n(m_e + m_i)$, which becomes

$$\rho = \frac{(m_e + m_i)\,p}{k_B\left(T_i + T_e\right)}. \tag{7.106}$$

If we take p_o and r_o as a reference pressure at a given radius, then

$$p = e^{\frac{G(m_e + m + i)M_\odot(r_o - r)}{k_B r_o r(T_e + T_i)}}. \tag{7.107}$$

But unlike the parallel atmosphere assumption we have used before, the spherical case does not go to 0 when the distance goes to infinity. This creates a problem for the static case. The only solution is to make the reference distance infinite; otherwise interstellar space would be flooded with high-pressure gas from nearby stars, and this excess pressure cannot be reconciled with estimates from interstellar gas emissions.

This impasse for the static case is what caused E. N. Parker to explore dynamical situations in the 1950s. His early work is summarized in Parker (1963). We return to equations (7.103) and (7.104), but this time we assume an isothermal equation of state

$$p = 2nk_B T = \frac{2\rho k_B T}{m}, \tag{7.108}$$

where $2T = T_i + T_e$ is constant and we define the mass density $\rho = nm$. Equation (7.103) can then be integrated to yield

$$I = r^2 \rho u, \tag{7.109}$$

where I is constant. This means the mass flux per solid angle is constant. Equation (7.104) then becomes

$$\frac{1}{u}\frac{du}{dr}\left(u^2 - \frac{2k_B T}{m}\right) = \frac{4k_B T}{mr} - \frac{GM_\odot}{r^2}. \tag{7.110}$$

Although this is a first-order differential equation, the solution is not trivial. There is a critical radius

$$r_c = \frac{GM_\odot m}{4k_B T},$$ (7.111)

where the right-hand side of the equation vanishes. Likewise there is a critical velocity

$$u_c = \sqrt{\frac{2k_B T}{m}},$$ (7.112)

where the left-hand side vanishes. It should be noted that the critical velocity u_c is also the coronal sound speed. Following Hundhausen (p. 102), we set $u(r_c) = u_c$. This condition makes sense because the critical velocity is that reached with the kinetic energy of the ions equal to $k_B T$. The critical radius is that distance where a quarter of the gravitational potential energy is also $k_B T$. With this constraint, equation (7.110) becomes

$$\frac{\mathrm{d}u^2}{\mathrm{d}r}\left(1 - \frac{u_c^2}{u^2}\right) = \frac{4u_c^2}{r}\left(1 - \frac{r_c}{r}\right).$$ (7.113)

This can be integrated to yield

$$\left(\frac{u}{u_c}\right)^2 - \ln\left(\frac{u}{u_c}\right)^2 = 4\ln r + 4\frac{r_c}{r} + C,$$ (7.114)

where C is a constant of integration. There are two roots to this equation:

1. The subsonic solution, where $u/u_c < 1$ and the solution monotonically decreases as $r \to \infty$
2. The supersonic solution, which monotonically increases, and where $u/u_c > 1$ as $r \to \infty$

In the first case, for large r

$$\ln\frac{u}{u_c} \simeq -2\ln r;$$ (7.115)

thus $u \propto 1/r^2$. From equation (7.109), this means $\rho \to I$, and is therefore a (nonvanishing) finite value. Likewise the pressure at infinity is nonvanishing. As with the static case, this is not a physical solution. We are therefore left with the latter case as the dynamic solution for the solar wind. The behavior of the velocity can be seen in figure (7.10).

These calculations verify that the solar wind is a supersonic flow of electrons and ions out from the sun. But we ignored the fact that in an MHD plasma the magnetic field is frozen in. Thus, while the model wind itself flows nearly radially outward, the magnetic configuration remains to be determined.

7.6.3 MHD as applied to the solar wind magnetic field

Near sunspots, the strong magnetic fields have lines that are in a closed loop form as they go from a spot of one polarity to a spot of another. Because there is transport of energy from both spots, the spot areas are kept cooler than the surrounding photosphere and appear dark. At the top of the loop, however, the plasma is heated and so there is an increased buoyant force and the loop tends to stretch against gravity in an upward direction. If the field is really strong, the extension of the loop into the chromosphere leads to a loop prominence, and if the chromospheric winds and currents twist the field lines to encourage reconnection then the loop may break off explosively in a solar flare. Field lines on the outer edges of the spot pair do not reconnect because they generally are bundles of flux tubes,

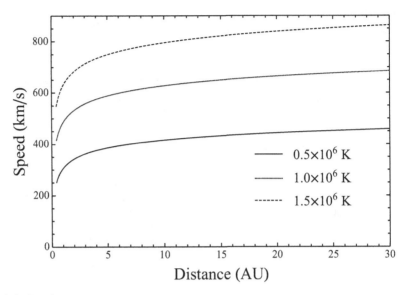

Fig. 7.10 Solar wind velocity for various temperatures.

but still stretch upward into the corona and form a coronal loop or coronal ray pattern that is often centrifugally and magnetically supported against gravity.

The solar wind has its origin in x-ray dark areas called coronal holes. Apparently the solar wind starts out in the lower coronal (or perhaps high chromosphere) regions where the temperature is too low and the magnetic field is mainly "open" and not in a loop so that the plasma is accelerated along flux tubes to the observed supersonic speeds without major reconnection. In the corona, the motion is radial for both the plasma and the magnetic lines. But beyond that, solar rotation provides an azimuthal displacement relative to the lower solar corona from whence the radially outward flowing plasma originated. Thus to a stationary observer, successively emitted parcels will follow a spiral path in space. Because these parcels are presumed to be riding along the same flux tube, the frozen-in magnetic field lines are spirals as well.

Near to its point of origin, the magnetic field has an inverse square distance relationship so that

$$B_r(r) = B_o \left(\frac{r_o}{r}\right)^2. \tag{7.116}$$

This is the field that occurs assuming the speed in the radial direction is $u(r)$ as derived previously. The solar rotation provides an azimuthal speed that increases with distance, and relative to the rotating sun it will provide an apparent negative speed that depends on solar latitude in two ways. The first is through the differential rotation rate $\omega(\theta)$, while the second is due to the latitudinal shortening of the radius of the rotational path. Thus

$$u_\phi(r, \theta) = -\omega(\theta)r \cos\theta. \tag{7.117}$$

Assuming the magnetic field strengths in the two orthogonal directions are in the same ratio, we obtain the following useful relationship

$$\frac{B_\phi}{B_r} = \frac{u_\phi}{u} = -\frac{\omega r \cos\theta}{u}. \tag{7.118}$$

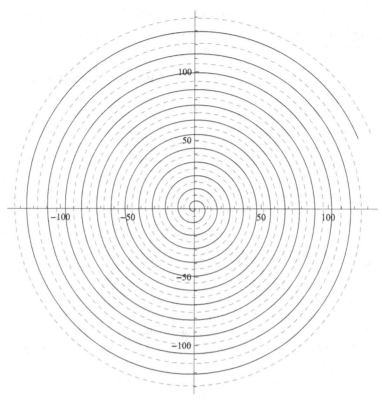

Fig. 7.11 Magnetic field lines of the solar corona.

Resolving the velocities as differentials in time,

$$\frac{r\,d\phi\cos\theta}{dr} = -\frac{\omega r\cos\theta}{u}, \tag{7.119}$$

or

$$\frac{dr}{d\phi} = -\frac{ur\cos\theta}{\omega r\cos\theta}. \tag{7.120}$$

Integrating with the differential rotation parameters,

$$\phi = \phi_o - \left(A + B\sin^2\theta\right)\int\frac{dr}{u(r)}. \tag{7.121}$$

According to Allen's compilation (Cox, 2000), the coronal features in degrees per day are $A = 13.46$, $B = -2.99$. From this we can determine the magnetic field components

$$B_\phi = -B_r\frac{\omega(\theta)r\cos\theta}{u(r)} = -B_o\left(\frac{r_o}{r}\right)^2\frac{\left(A + B\sin^2\theta\right)r\cos\theta}{u(r)}. \tag{7.122}$$

The canonical field tilt at 1 AU is 45°. Given a realistic solar rotation value, this can be matched to our solution given a $u(r)$ for a temperature of about 850 000 K.

In figure (7.11) we show two magnetic field lines starting and ending at the solar corona. Such configurations are said to represent the field lines from different magnetic sectors of the sun. We have chosen two line sets that are $180°$ apart for clarity in the graph. In actual practice, the magnetic polarity of adjacent sectors switches quite rapidly with solar longitude. The Archimedes spiral character of the field (and hence the MHD flow) is quite evident.

7.6.4 Termination of the solar wind

When Hundhausen wrote his chapter on the solar wind he could only speculate about how the termination of the solar wind would occur. Even then he was able to predict only that the termination is most likely to be in the form of a shock wave called the termination shock. Hundhausen offered speculations of where such a shock front would be, though he did not know the precise geometry of the interaction. Thanks to the Voyager 1 and 2 spacecraft, and more recently the IBEX spacecraft, we now know the truth about how the solar wind interacts with the interstellar medium. Here we only outline some of the early results concerning the heliospheric terminal shock:[17]

1. The inner heliosphere is a nearly closed "spherical" cavity, not the "comet-tail" open shape characteristic of planetary heliospheres.
2. Such a shape was predicted by Parker many years ago in the case where the external gas magnetic pressure exceeded the solar wind gas pressure. Because of the success of the "open" structure geometry, little attention was given to the spherical structure and it was a great surprise when such a shape was inferred from IBEX observations.
3. The required local interstellar magnetic field needed to maintain the spherical shape to the inner heliosphere is at least twice the original local interstellar B-field estimate. At present it is thought the value could be higher than $5 - 6\mu G$.
4. While the Voyager observations indicate that the TS is located at $\sim$90 AU, the most intense region where the neutrals interact is estimated at $\sim$150 AU from the sun.
5. The inability of most models to account for all the dominant features seen is striking.
6. The short-term variability of the particle fluxes and other features was not expected.

Exercises

7.1   Given a magnetic field strength of 5×10^{-5} T, compute the Larmor radius and cyclotron frequency
     1. For an electron with an energy of 10 KeV and a pitch angle of $0°$
     2. For a proton with a speed of 300 km/s and a pitch angle of $45°$

7.2   Using the dipole magnetic field equation in **7-1ChargePaths**, compute and graph the 100 keV electron gyro frequency around
     1. A white dwarf (10^9 gauss)
     2. A magnetic neutron star (10^{12} gauss)
     3. A magnetar (10^{14} gauss).
     Compare this with the 10 MeV gyro frequency.

7.3   The Arecibo radio telescope in Puerto Rico can be used in radar mode to transmit to planets and back.

---

[17] Taken from McComas et al., 2009; Fuselier et al., 2009; Funsten et al., 2009; Schwadron et al., 2009; Möbius et al., 2009 and Krimigis et al., 2009.

1. What should the transmitting polarization be if Earth's magnetic field has a component pointing directly into the 305-meter dish? What should the receiving polarization be?

2. The Arecibo telescope can also be used to perform ionospheric heater experiments. These are experiments where you want to maximize the energy absorbed by the electrons. What polarization should be transmitted to do that?

7.4 In **7-2EMWaves**, a method for calculating the ordinary and extraordinary polarization ellipses, is given with and without absorption. The comment is made that the shape of the ordinary ray is controlled by adjusting the wave frequency relative to the electron density. Play "what if" with this situation and produce a series of polarization diagrams as a function of the wave frequency to critical frequency ratio.

7.5 For a particular pulsar the signal at 1000 MHz arrives 3 seconds later than the signal at 2000 MHz.

1. What is the dispersion constant?

2. If the electron density is taken to be 0.03 cm^{-3}, what is the distance to the pulsar?

7.6 A pulsar yields a rotation measure RM $= -60$ rad m^{-2} and a dispersion measure DM $= 20$ pc cm^{-3}. Calculate the mean magnetic field along the line of sight.

7.7 For a certain pulsar the RM is observed to change by a factor of 10 and the DM is observed to change by a factor of 5 between when the pulsar is observed near the sun and opposite the sun. Assume the effect is entirely due to the solar wind inside Earth's orbit. What is the enhancement of the electron content and the average magnetic field due to the inner part of the solar wind?

7.8 The Parker equation (7.114) was derived assuming an isothermal equation of state. Derive the equation if one assumes an adiabatic equation of state:

$$p\rho^{-\gamma} = A.$$

7.9 In **7-6Boltzmann** (Section 4c) the characteristic transformations for the constant magnetic field situation are given. Use the method for producing a characteristic function when two transformations are given as outlined in Section 2a to produce a set of boundary conditions for a NDSolve check of the functions given by DSolve. Put those conditions into the equivalent NDSolve and plot the results.

7.10 In **7-9MHD** it was commented that it is faster to compute a double angle formula than it is to square the equivalent single angle function. Look up some multiple angle formulae in a standard math handbook and see which takes longer in *Mathematica*. (For example, $\sin 3x = \sin x - 4\sin^2 x$.)

7.11 In notebook **7-1ChargePaths**, case (c) considers the motion of charged protons and electrons in a dipole field. In that notebook, the calculation of proton and electron drifts in a generic dipole field was considered with arbitrary scaling. That model can also be used with proper scaling to model actual trajectories around real objects. The strongest stellar magnetic fields exist in white dwarfs and neutron stars, but the real record holders are neutron stars called magnetars.[18] Here are representative surface B fields for these objects:

1. White dwarfs: 10^3 to 10^9 gauss

2. Pulsars: 10^{12} gauss

3. Magnetars: 10^{14} gauss

Create and compare a proton-electron drift model for each type of object.

---

[18] A complete list of magnetars is maintained online by the McGill University Pulsar Group (http://www.physics.mcgill.ca/pulsar/magnetar/main.html).

7.12 In **7-6Boltzmann**, Section 3a, `DSolve` had difficulty producing characteristic functions for a sinusoidally varying source for the BTE, but a form was suggested as being possible. Confirm these results by changing the two functions into a single characteristic and use it to find the boundary conditions for an `NDSolve` analysis of the problem. Plot the results with the original function.

7.13 In **9-7MHD** a zero temperature Monte Carlo, force-balanced stability analysis is carried out for a sunspot. This was done by letting $\rho = 0$. Restore a pressure term by making the density nonzero and repeating the Monte Carlo analysis.

8

Galaxies

In this chapter we explore several aspects of dynamics within the Milky Way galaxy, including spiral density waves, the effects of dark matter, and the region near the central black hole. We also examine the effects of galactic rotation on stellar clusters by returning to the Orion Trapezium cluster studied in Chapter 6. Finally we look at the role of gas and dust within the galaxy on astrochemistry.

8.1 The existence of dark matter

Because of dust obscuration along the galactic equator, radial velocity studies of bright O and B stars in the 1930s and 1940s were able to discern only three or possibly four spiral arm structures in the part of the galaxy nearest the sun. At the time it was assumed that the derived circular orbit velocities fit Kepler's laws, and from that a "reasonable" mass for the galaxy was obtained. That assumption was accepted without much question. After World War II, with the prediction and subsequent discovery of the atomic hydrogen ground state emission called the 21 cm radiation, a new tool was available for measuring the distribution of matter in the Milky Way and other galaxies.

The proton and electron in an ordinary hydrogen atom are both fermions with a spin of 1/2. As such they both have magnetic moments that can be either parallel or antiparallel. The energy of the bound electron is slightly higher when parallel rather than antiparallel. This creates an energy difference within the ground state, an effect known as hyperfine splitting. When the electron flips a photon is emitted with a frequency of 1420.4 MHz, which corresponds to a wavelength of 21.1 cm. Unlike visible light, these 21 cm radio waves can penetrate cold ($\sim$10 K) dust clouds, which have predominantly molecular rather than atomic hydrogen, that are found within spiral arms. The atomic hydrogen that produces the 21 cm radiation are found in abundance in the "warmer" ($\sim$100 K) space between the cold molecular clouds. When the Dutch radio astronomer Hendrik C. van de Hulst predicted and subsequently found this radiation, the full extent of the galaxy could be observed and characterized.

Unlike most textbook accounts of the dark matter distribution, in this chapter we extend the analysis of 21 cm radial velocities for emission features that originate beyond the sun's orbit. We will do this by extrapolating the orbital velocity versus distance rotation law beyond 8.5 kpcs, and then solve for the distance using the observed velocity. The dominant structure in the profiles are two velocity systems. One low-velocity system is located, as expected, at 10–12 kpcs, which is closer to the sun's orbit. The high-velocity system is located at 13–19 kpcs. The combination of these two data sets shows very good agreement, so we are able to perform an effective but not well known method of discovering the dark matter density in the outer Milky Way.

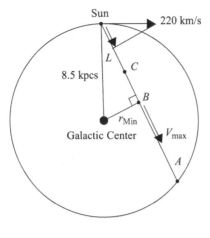

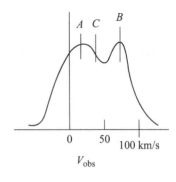

$$V_{obs} = V_{max} - 220 \text{ km/s } \sin(L)$$

$$r_{Min} = 8.5 \text{ kpcs } \sin(L) \qquad V_{max} = V_{obs} + 220 \text{ km/s } \sin(L)$$

Fig. 8.1 Measurement of 21 cm radiation in the Milky Way.

8.1.1 Dynamics of the Milky Way within the sun's orbit

Observationally the spectra of 21 cm radiation shows a systematic Doppler effect as one observed different locations along the galactic equator. This is evidence of differential rotation as a function of distance from the center of our galaxy. We begin with the observations published by van de Hulst et al. (1954), converted to circular velocities. To this we add observations gathered using a small radio telescope by a group of SUNY Geneseo undergraduate students as a laboratory experiment. As we will see, the combination of these two data sets is in good agreement. From this we can make a very classical analysis for the matter within the Sun's orbit at 8.5 kpcs.

The configuration for a 21 cm spectrum measurement can be seen in figure (8.1). The radio antenna is located at the sun, and the line of sight makes an angle with line to galactic center. The longitude angle L is between the galactic center and the line of sight upon which points A, B, and C lie. At the right is shown the emission intensity in the vicinity of 21 cm (1420 MHz), converted to Doppler velocity in km/s.

Because the line of sight intersects a continuum of distances, the observed Doppler shift is not a single value; however the peak with the greatest redshift is that which originates at the closest point to galactic center. To carry out our analysis, we must measure the velocity of the spectrum peak along multiple longitudes in order to determine the galactic rotation curve. This is not an easy task at some longitudes, where the maximum velocity peak is extremely blended with the other cloud positions.[1]

The measured speeds $V_{obs}(L)$ are those relative to the sun. Because the sun has an orbital speed $V_\odot$, we must compensate for its speed at each longitude; thus

$$V_{max}(L) = V_{obs} + V_\odot \sin L. \qquad (8.1)$$

This function of longitude is then converted to a function of radius by noting

$$R = 8.5 \sin L \text{ kpcs.} \qquad (8.2)$$

---

[1] For images of the original observations, see **8-1MW21cm**.

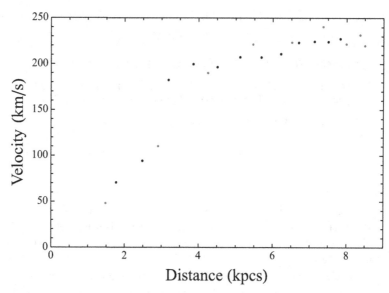

Fig. 8.2   Orbital velocity vs. radius for the Milky Way.

The observations of both van de Hulst and the Geneseo students are not numerous, so they can be entered in *Mathematica* by hand:

```
velobs1 = {10., 35., 80., 80., 55., 50., 15., 15., 0};
...;
vHlist = Table[{longH[[jj]], velH[[jj]]}, {jj, 1, 12}];
```

These are then converted to velocities relative to galactic center:

```
rG = Thread[8.5 × Sin[longG °]]; velRG = 220. × Sin[longG °];
rH = Thread[8.5 × Sin[longH °]]; velRH = 220. × Sin[longH °];

vel1 = velobs1 + velRG; vel2 = velH + velRH;

long = Join[longG, longH]; r = Join[rG, rH]; vR = Join[vel1, vel2];
```

Here we take the orbital velocity of the sun to be 220 km/s, which is the accepted value. The resulting velocity curve can be seen in figure (8.2), with the Geneseo data in gray and van de Hulst in black. It is clear the two data sets are in reasonable agreement.

We can now use these data to derive a rotation curve for the galaxy as a function of distance. This will be needed to extrapolate beyond the solar distance of 8.5 kpcs. For simplicity we will assume this function is a power law; therefore we may take the logs of radius and velocity to make a linear fit:

```
xypoint1 = Table[{Log[10, r[[ii]]], Log[10, vR[[ii]]]}, {ii, 1, 21}];
```

There is a clear discontinuity of the distribution at about 3 kpcs. Because $\log 3 \sim 0.5$, we will parse points where $\log r < 0.5$ and fit the function to the remaining points:

```
xypoint2 = Table[{0, 0}, {ii, 1, 17}]; jj = 0;

Do[If [xypoint1[[ii, 1]] > 0.5, jj = jj + 1; xypoint2[[jj, 1]] = xypoint1[[ii, 1]];
    xypoint2[[jj, 2]] = xypoint1[[ii, 2]];], {ii, 1, 21}]

lmf5 = LinearModelFit[xypoint2, {1, rL}, rL]
```

The result is a linear fit of $0.230\text{rL} + 2.152$, which yields a velocity curve function:

$$V(r) = 142 r^{0.23} \text{ km/s}. \tag{8.3}$$

This velocity function predicts a speed for the sun of 232 km/s, which is slightly higher than the accepted value. This is a systematic error of about 5%, which is reasonable given the observational errors. The discrepancy occurs due to the sun's peculiar motion relative to the background stars of some 30 km/s toward the star Vega.

In celestial mechanics, power laws frequently result from theory with exponents that are distinct fractions. Kepler's third law, for example, is a power law of 3/2 or 2/3 depending on its form. Because the fitted power law is very near $0.25 = 1/4$, it is tempting to set this fraction as the power law and fit the velocity curve to the accepted solar speed. Thus,

```
amp1 = Solve[220 == amp 8.5^0.25, amp][[1, 1, 2]]
```

This yields a velocity curve of

$$V(r) = 129 r^{1/4} \text{ km/s}. \tag{8.4}$$

Because we do not know in fact that there is such a theoretical form we will use the fitted formula in calculations for the mass outside 8.5 kiloparsecs.

8.1.2 Mass as a function of distance from galactic center

With the velocity function determined, we can now derive galactic mass as a function of radius. If galactic mass is axially symmetric then the mass enclosed within a given radius is equivalent to placing that mass at the center of the galaxy. Thus

$$\frac{mv^2}{r} = \frac{GM(r)m}{r^2}. \tag{8.5}$$

Thus

$$M(r) = \frac{v^2 r}{G}. \tag{8.6}$$

Because our velocity function is in km/s with radius in kiloparsecs, we express the gravitational constant in similar units,

$$G = 4.302 \times 10^{-6} \text{ kpc} M_\odot^{-1} \text{ (km/s)}^2. \tag{8.7}$$

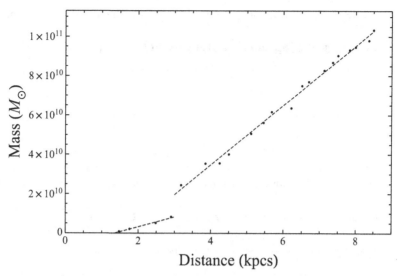

Distance (kpcs)

Fig. 8.3   Mass as a function of radius for the Milky Way.

The enclosed mass (in solar masses) as a function of radius is then

```
amp1 = Solve[220 == amp 8.5^0.25, amp][[1, 1, 2]]; rA = Sort[Join[rG, rH]];
massA = Sort[Join[mass1, mass2]];
xypoint0 = Table[{rA[[ii]], massA[[ii]]}, {ii, 1, 21}];
```

which can be seen in figure (8.3). The data show not only a break in the mass distribution at 3 kpcs, but also an anomalous increase in density just outside the break. In the last decade the Milky Way has had its classification changed from an ordinary spiral to a "barred" spiral galaxy. This bump in the mass distribution is evidence for a ring that is a feature of barred galaxies. In the Milky Way, this region is noted for the abnormally large number of molecular hydrogen clouds (also, presumably, containing dust). A similar dusty ring exists in our neighbor the Andromeda galaxy.

We are interested in the mass density as a function of radius. If we consider the volume density,

$$M = \int \rho \, dV, \tag{8.8}$$

or for an area density,

$$M = \int \sigma \, dA, \tag{8.9}$$

the integral can be converted to a differential equation by taking the derivative of both sides of the equation. To take a derivative we must first fit the mass data to a function. Given the split at 3 kpcs, we must fit $r > 3$ kpcs and $r < 3$ kpcs separately:

```
xypoint1 = Table[{rA[[ii]], massA[[ii]]}, {ii, 5, 21}];
xypoint2 = Table[{rA[[ii]], massA[[ii]]}, {ii, 1, 4}];
lmfm = LinearModelFit[xypoint1, {1, (rL - 3)}, rL]; eqnm = Normal[lmfm];
lmfm2 = LinearModelFit[xypoint2, {1, (3 - rL)}, rL]; eqnm2 = Normal[lmfm2];
```

If we assume a spherical mass distribution, only the radial derivative is needed, for example:

sol3S = Solve[∂_{rL} eqnm == ρ[rL] 4 π rL^2, ρ[rL]]

This yields

$$\rho(r) = \frac{C}{r^2}, \tag{8.10}$$

where $C = 1.204 \times 10^9 M_\odot$ per cubic kiloparsec for $r > 3$ kpcs, and $C = 3.963 \times 10^8 M_\odot$ for $r < 3$. For the disk model it is assumed that the vertical mass distribution is either Gaussian or exponential along the z-axis, with a half-width of about 300 parsecs. Thus, for the exponential case

sol3D = Solve[∂_{rL} eqnm == ρ[rL, z] 2 π rL dz, ρ[rL, z]]

$$\sigma DEg3[rL\_] = \frac{2.42227}{rL} \int_0^\infty e^{-z/300} \, dz$$

which yeilds:

$$\sigma(r) = \frac{C}{r}, \tag{8.11}$$

where $C = 727 M_\odot$ per square parsecs for $r > 3$, and 1453 for $r < 3$. The Gaussian model produces results that are 10% to 15% lower. The mass values derived from the orbital velocities is that due to both visible matter and dark matter. This can then be compared to the mass distribution of visible matter observed directly.

8.1.3 A luminous matter model for the Milky Way

It is generally assumed that our galaxy has the same brightness distribution of visible matter as other similar spiral galaxies. From this assumption we can calculate the enclosed visible mass as a function of radius. To obtain the central density, it is assumed that the enclosed mass derived by the rotation speed within 1 kiloparsec (the central bulge) is composed entirely of visible matter. Thus at $r = 1$ the observed orbital velocity is set equal to the 21 cm velocity in the computation of density. The list of mass values are then based on the computations of the galactic longitude of each observed hydrogen cloud[2]:

r0 = 3.70; r1 = 1.48; h = 0.5; vel0 = 80;

$$\rho 0 = \frac{2.3 \times 10^5 \times vel0^2 \times r1}{r0^2 \times h \times 2 \times \pi \times \int_0^{r1/r0} e^{-x} x \, dx};$$

...;

lummasslist = Table[{rA[[ii]], mass[[ii]]}, {ii, 1, num1}];

---

[2] The order of Mathematica code presented here differs from that in the notebooks.

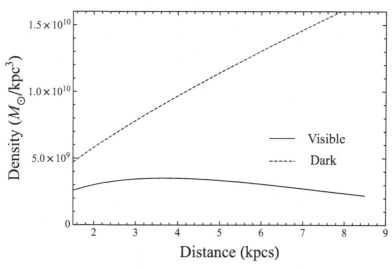

Fig. 8.4 Visible and dark matter density functions.

The distribution of dark matter can then be found by taking the difference of the total mass derived from the observed velocity function and the visible mass:

```
xypoint1 = Table[{rA[[i]], (massA[[i]] - mass[[i]])}, {i, 1, 21}];
```

Both the dark matter and visible matter distributions are data tables. To find the density function we must first find an interpolated function for each distribution. For visible matter we can just use `Interpolation[]`. The dark matter distribution must be 0 in the galactic center, so we must do a least squares power law fit instead. The volume density for each function is then found by taking the radial derivative:

```
lum = Interpolation[lummasslist]; slope = lum'
nlmf1 = NonlinearModelFit[xypoint1, a xp^b, {a, b}, xp]
dark = Normal[nlmf1]; slope1 = ∂_xp dark;
```

The result, seen in figure (8.4), shows that the visible matter density is not nearly sufficient to account for the velocity rotation curve, hence the presence of dark matter.

8.1.4 Dynamics of the Milky Way beyond the sun's orbit

The analysis for mass inside the orbit of the sun given above is relatively simple and straightforward. Maps of galactic regions beyond the sun's orbit exist, but are often presented with no explanation of how that structure is found. Here we explore some ways the rotation of the galaxy and the presence of dark matter beyond the sun's orbit can be deduced.

The geometry for a hydrogen cloud beyond the sun's orbit can be seen in figure (8.5). The distance of the sun from galactic center is 8.5 kpcs, the distance from the sun to the cloud is x, and the distance of the cloud from galactic center is r. The longitude angle L is again measured from galactic center. The

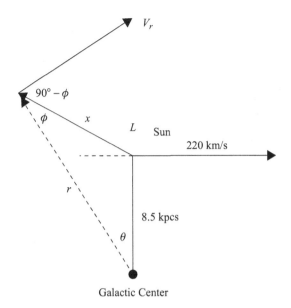

Fig. 8.5 Measurement of 21 cm beyond the sun's orbit.

angles ϕ and θ are related through the law of sines; thus if $R_\odot = 8.5$ kpcs is the solar distance, we have

$$\sin\phi = \frac{R_\odot \sin(360° - L)}{r} = -\frac{R_\odot \sin L}{r}. \tag{8.12}$$

The quantities r and V_r are unknowns. These can be determined through the dynamics of the problem. Consider now the Doppler effect along a line of sight where $270° > L > 180°$. This implies that $\phi < 90°$. From figure (8.5) we have along the line of sight that, due to the clockwise rotation of the galaxy, the V_r component is negative relative to the line of sight. The component velocity of the sun is also in the same direction (and negative). The difference between the two components is the observed radial velocity; thus from the sun the object will look as if it has a positive radial velocity,

$$V_{obs} = -V_r \cos(90° - \phi) - V_\odot \sin L, \tag{8.13}$$

where $V_\odot = 220$ km/s is the velocity of the sun. From the law of sines this can be expressed as

$$V_{obs} = \left(\frac{R_\odot V_r}{r} - V_\odot\right)\sin L. \tag{8.14}$$

If we suppose the previously derived rotational velocity relation, equation (8.3), holds outside the sun's orbit, we can add the component of the sun's motion to the observed maximum velocity just as before. Because we do not know the constraint equation that the right angle at the line of sight provides, we will put the assumed model directly into the equation and solve for the distance required to satisfy the dynamics. This in turn will lead directly to a mass–distance relationship from which the dark matter

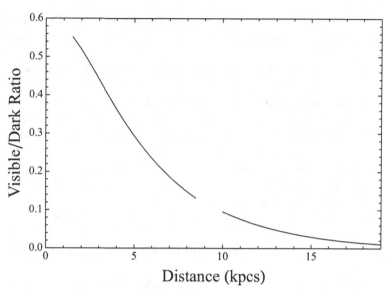

Fig. 8.6   Ratio of visible matter to dark matter in the Milky Way.

can again be directly estimated. The distances will again be derived from the spectra of van de Hulst et al.[3]:

```
longH2 = {182, 187, 192, 197, 202, 207, 212, 217, 222, 227, 232, 237, 242, 247, 252};
vser1 = {10, 10, 10, 20, 25, 40, 45, 50, 60, 70, 60, 65, 60, 80, 80};
vser2 = {10, 10, 10, 10, 10, 15, 10, 25, 15, 10, 10, 25, 25, 25, 25};
```

From these measurements the velocity as a function radius can be derived:

$$\texttt{vobs1 = Simplify}\left[\texttt{vR1 stdr}\ \frac{\texttt{Sin[el2]}}{\texttt{rr}} - \texttt{vSun Sin[el2]}\right]$$

```
. . .
vRdist = Join[vRdist1, vRdist2];
```

This distribution can be interpolated to a velocity as a function of radius, from which the mass distribution functions can be derived as before. The result can be seen in figure 8.6.

If we define the edge of the Milky Way as the distance where luminous matter is not significant, then the radius of the Milky Way is about 19 kiloparsecs with an enclosed mass of $3.5 \times 10^{11} M_\odot$. By that distance dark matter is essentially the only mass present. It used to be stated that the sun was located some 2/3 of the way out from the galactic center, but that was derived mainly with respect to the luminous mass. Actually the sun is more centrally located, with a ratio closer to 1/2.

[3] For an image of these spectra, see **8-1MW21cm**.

8.2 Revisiting the Trapezium cluster

One main theme of this textbook is illustrating computationally a number of analytical results produced by acknowledged master astrophysicists. Chandresekhar's work on stellar dynamics in general and dynamic friction in particular is an excellent case in point. Modern work on stellar clusters solves the dynamical friction problem by simply adding more and more objects into a computerized N-body situation and takes care of the dynamical friction computationally without adding it explicitly. Because adding objects is a major contributor to increasing the run time in an N-body problem, we do not have that luxury here. We can, however, look at the effect of adding details to the 4-body Trapezium problem. In Chapter 6 we examined the motion of the Trapezium cluster in the Orion Nebula. As we noted, our derivation did not include galactic effects on stellar motion. Because we now have a description of the mass distribution within the Milky Way, we can include the galactic effects on the cluster.[4]

8.2.1 The Milky Way gravitational potential

Because we have the velocity curve as a function of r, the radial force per unit mass is given by

$$f(r) = \frac{v^2}{r}, \tag{8.15}$$

and the potential is found by

$$U(r) = \int_r^\infty f(x)\,dx, \tag{8.16}$$

where the potential must vanish as $r \to \infty$.

In Section 8.1, the linear fit of our velocity, equation (8.3), seemed to fit a power law of $1/4$. We will therefore make an exponential fit:

$$\frac{v^2}{r} = a e^{bx^{1/4}}. \tag{8.17}$$

Thus `veldist` as the velocity vs. radius table calculated earlier,

```
lhsdata = Table[{veldist[[ii, 1]], (veldist[[ii, 2]]^2 / (10^3 veldist[[ii, 1]]))},
    {ii, 1, num1}];

nlm5 = NonlinearModelFit[lhsdata, a e^-b √x , {a, b}, x]; veleq = Normal[nlm5];
```

which gives

$$\texttt{veleq} = 43\, e^{-1.13 x^{1/4}}. \tag{8.18}$$

The radial potential is then calculated as

```
first = ∫_x1^∞ veleq dx; potential1[r0_] := first[[1]] /. x1 → r0;
```

<hr>

[4] We look at only the stellar motion aspects here; however, these calculations are also done in the Chapter 8 notebooks. The N-body notebooks for Chapter 8 follow the same notation and format as those for Chapter 6; therefore the output of the Chapter 8 notebooks can be run without modification in the Chapter 6 notebooks on Lyapunov exponents and Fourier analysis.

For the z-axis (out of the galactic plane) potential, we refer to the data given in Allen (Cox, 2000, p. 574), where the vertical potential is given at the position of the sun in $(km/s)^2/pc$.

```
zdata = {{0, 0}, {.3, 1.60}, {.5, 1.92}, {.7, 2.07}, {1, 2.22}, {1.5, 2.38},
    {2.0, 2.5}, {2.5, 2.60}, {3.0, 2.67}}
```

Again, an exponential function seems to provide the best fit; thus

```
nlmz1 = NonlinearModelFit[zdata, a (1 - e^(-b z)), {a, b}, z]
zpotential0 = Normal[nlmz1]
```

which yields

$$U(R_\odot, z) = 2.51\left(1 - e^{-2.87z}\right). \tag{8.19}$$

These two potentials can then be combined to give the general galactic gravitational potential. To be suitable for computation near the center of mass of the Orion Trapezium cluster the radial distance must be converted to (x, y) coordinates. Also, because the radius is given in kiloparsecs it must be converted to AU for the Cartesian form.

```
rratio = zpotential0[[1]] / potential1[8.5]

mwpotential[r0_, z0_, x_, y_, z_] = (first[[1]] (1 + rratio zpotential0[[2]])) /.
```

$$\left\{x1 \to \sqrt{\left(r0 + \frac{10^{-3} x}{206\,265.}\right)^2 + \left(\frac{10^{-3} y}{206\,265.}\right)^2}, z \to Abs\left[z0 + \frac{10^{-3}}{206\,256} z\right]\right\}$$

Here the quantities r0 and z0 represent the values at the Trapezium center of mass. They are calculated by converting the distance and position to galactic coordinates.[5]

8.2.2 Gravitational potential for the Orion Nebular Cluster

In Chapter 6 we incorporated the gravitational potential of the five Orion molecular clouds into the Trapezium integration using the existing coordinate system based on the standard RA and Dec heliocentric system. Here because of the dominance of the galactic plane we must use offset galactic coordinates:

```
cloudlist = {{"trap", 209, -19.4, 470, 0}, {"OriA1", 207, -20, 521, 12 300},
    {"OriA2", 210, -19.3, 465, 69 500}, {"OriA3", 213, -19, 412, 23 300},
    {"OriB1/2", 206.5, -16, 422, 65 300}, {"OriB3", 205, -11.5, 383, 18 000}};
```

---

[5] See **3aRatoGal200** and **8-2MWpotential** for details.

This is then transformed to an *xyz* coordinate system centered on the Orion Trapezium. In the RA-Dec system the $+x$ distance is toward the star, which is also that for the galactic coordinates:

```
xAUc = Table[206 265 (cloudlist[[ii + 1, 4]]) - cloudlist[[1, 4]]) , {ii, 1, 5}]
. . .

phi = ArcCos[cosphi] / Degree
```

The angles are then converted to linear distances in AU for the *y–z* directions. The *x*-direction is an outward extension of the heliocentric direction vector to the cluster after passing the cluster center, the *z*-direction is in the galactic latitude plane, and the *y*-direction is in the direction of increasing galactic longitude to match the convention of Chandrasekhar:

```
difB = Table[lat[[1]] - lat[[ii + 1]], {ii, 1, 5}]; cosΔB = Cos[difB];

Δlong = ArcCos[cosphi / cosΔB]; zAUc = 470 Sin[difB] 206 265;
```

Here we adopt a peculiar convention for the distances along the longitudinal direction. The astronomical convention of RA and Dec are measured in a "leftward manner," which makes $+y$ distances plot to the left. But as noted in Chapter 6, the *Mathematica* plot routines never allow this, so we put minus signs in as before. With this inversion it makes the positive *y*-axis point in the direction of the actual galactic rotation, which is opposite to the direction of galactic longitude:

```
yAUc = -470 Sin[Δlong] 206 265; yAUcg1 = yAUc;
```

The *y* coordinates are already parallel to the galactic plane and pointing in a direction of increasing galactic longitude, but *x* and *z* are not. To get *x* and *z* parallel to the galactic plane they must be rotated back through the galactic latitude of the Trapezium center of mass. We must also find r_o and z_o for the Trapezium. If we assume a geocentric or heliocentric distance of 470 parsecs and $B = -19.4°$, then

```
xAUcg1 = zAUc Sin[19.4 °] + xAUc Cos[19.4 °]; zAUcg1 = zAUc Cos[19.4 °] - xAUc Sin[19.4 °];
z0 = .47 Sin[-19.4 °];

r0 = √( (.47 Cos[-19.4 °])² + (8.5)² - 2 × .47 Cos[-19.4 °] 8.5 Cos[209 °]
```

Because the *x* and *y* axes are oriented with respect to the galactic center direction at the Orion Trapezium we need a final rotation angle of 61° counterclockwise to get the cloud positions in the same coordinate system as the overall galactic potential. Thus,

```
xAUcg = -yAUcg1 Sin[61 °] + xAUcg1 Cos[61 °]; yAUcg = yAUcg1 Cos[61 °] + xAUcg1 Sin[61 °];
zAUcg = zAUcg1;
. . . ;
cloudlist2 = Table[{idcloud[[ii]], masscL[[ii]],
    xAUcg[[ii]], -yAUcg[[ii]], zAUcg[[ii]]}, {ii, 1, 5}];
```

The potential function for the molecular clouds is then

$$
\texttt{cloudVG} = \sum_{jj=1}^{5} \sum_{k=1}^{nB} -\texttt{gc mass[[k]] masscL[[jj]]} \Big/ \sqrt{\big((\texttt{xAUcg[[jj]]} - \texttt{qx[[k]]})^2 +}
$$

$$
(\texttt{yAUcg[[jj]]} - \texttt{qy[[k]]})^2 + (\texttt{zAUcg[[jj]]} - \texttt{qy[[k]]})^2 \big);
$$

where qx, qy, and qz are the star coordinates in the galactic coordinate system.

In Chapter 6 we initially ignored the mass within the Orion cluster by assuming the cluster was spherical for computational efficiency. Of course the Orion Nebular Cluster (ONC) is not spherical (Hillenbrand and Hartmann, 1998) and once the galactic and molecular cloud potentials are included the Trapezium stars are no longer confined to the center of the ONC. Thus we must consider including some sort of cluster model in the calculations.

Because the elliptical contours for the ONC occur mainly in the outer bounds, we will avoid the computational difficulty of a multipole mass expansion by assuming spherical symmetry once again. In particular, we will adopt a "canonical" or average central density model (King, 1966) whose parameters are given in Hillenbrand and Hartmann. King models are obtained from a steady-state solution of the Fokker–Planck equation, a particular form of the Boltzmann equation in which the collision term is calculated on the basis of small-angle collisions. It is well suited for describing Coulomb (and hence also gravitational) interactions.[6] The Boltzmann collisionless equation, a collisionless relative of the Fokker–Planck equation called the Vlasov equation, is discussed in **7-6Boltzmann**. If one has a collision model available then one can change a Vlasov equation into a Fokker–Planck equation.

The King models, which have shown to be very good representations of globular clusters as well as dense open clusters, are those having an "isothermal" core where the mass density is essentially constant. For the ONC, from Hillenbrand and Hartmann, we adopt 0.2 parsecs (41 000 AU) for the approximate core radius, and 3×10^4 solar masses per cubic parsecs as the nominal mass density. We assume the center of mass of the Trapezium is the center of mass for the cluster.

8.2.3 A new Lagrangian for Trapezium

In the early 1940s, Chandrasekhar took on the task of trying to explain why spiral galaxies had their particular shape. His attack on the problem was typical of his style: go back to the beginning and start with fundamentals. Unfortunately he never achieved this goal because he could never find a mechanism that would produce spiral patterns in preference to the known circular orbits. But in the process of investigating the spiral problem he managed to develop a generalized analytical galaxy model based on potential theory that included a model for star clusters. Along with this he perfected his virial approach to cluster parameters and theoretically discovered dynamical friction.[7] Two obstacles kept him from applying his theories to real objects, both of which we can now overcome. First, he did not have a reliable model of the Milky Way mass distribution, and hence he had no knowledge of the potential produced by that model. Second, he had no understanding of the existence of density waves (Shu, 1992) that could explain spiral structures. In spite of these difficulties however, Chandrasekhar's observations for stars and star clusters on nearly circular orbits remains a beautiful and valid theory we will now adapt to the Trapezium model.

---

[6] See Sturrock, 1996, p.147 for example.
[7] See the *Mathematica* notebooks for Chapter 6, as well as **8-2MWpotential**.

To apply his theory directly requires that we work in the same cylindrical galactic coordinates that the galactic potential operates (as derived in Section 8.1). In addition because the centers of mass (including the molecular clouds) are constrained to move in circular galactic orbits (and are accelerated in the process) we must transform to a rotating frame. That means that only equations of motion derived from the Lagrangian as described by Chandrasekhar and appropriately modified will produce meaningful solutions. There is no valid Hamiltonian representation that can be used for computation.

We begin with the initial conditions calculated in Chapter 6, which provides the position and momentum of the Trapezium stars relative to their center of mass. The molecular cloud parameters are also imported, but with the y-axis inverted:

```
clouds = Import["cloudgalc" <> ToString[filenumber] <> ".csv"]
. . .

cloudvectors =
Table[{N[xAUc[[ii]] / 1000], N[-yAUc[[ii]] / 1000], N[zAUc[[ii]] / 1000]}, {ii, 1, 5}]
```

To this we add the galactic potential calculated in Section 8.1, which we enter directly to save computation time:

```
mwpotential[r2_, z2_, x3_, y3_, z3_] = 43.012774824987375`
    e^{-1.127895301120314` ((r2+4.848132257047972`*^-9 x3)²+2.3504386381829065`*^-17 y3²)^{1/8}}
. . .

    ((r2 + 4.848132257047972`*^-9 x3)² + 2.3504386381829065`*^-17 y3²)^{3/8});
```

We also need the radial position $r0 = 8.89007$ and the z position $z0 = -0.135987$ of the Trapezium center of mass in kiloparsecs.

The gravitational potential we derived has velocity units of km/s that must be converted to AU/century to match the Trapezium data. For computational purposes we again enter the conversion formula by hand:

```
mwpotentialAUA[r2_, z2_, x1_, y1_, z1_] = -vc1[[1]]² 43.012774824987375`
    e^{-1.127895301120314` ((r2+4.848132257047972`*^-9 x1)²+2.3504386381829065`*^-17 y1²)^{1/8}}
. . .

    ((r2 + 4.848132257047972`*^-9 x1)² + 2.3504386381829065`*^-17 y1²)^{3/8})
```

Computationally there is no reason to use the "exact" function in the equations of motion when a Taylor series will do just as well. Thus the zero-order term becomes

```
potential0 = mwpotentialAUA[r0, z0, 0, 0, 0]
```

The coefficients for x and z are found from

```
termr = FullSimplify[∂_{r1} mwpotentialAUA[r1, z0, 0, 0, 0]];
termz = FullSimplify[∂_{z1} mwpotentialAUA[r0, z1, 0, 0, 0]];
```

and setting z1 to 0. The approximate potential is then

$$\texttt{mW\Phi[xx\_, zz\_] := potential0 + termr0x} \frac{\texttt{xx}}{206\,265 \times 1000} + \texttt{termz0z} \frac{\texttt{zz}}{206\,265 \times 1000};$$

When multiplied by stellar mass this simple function becomes a kinetic energy, which is a computationally superior expression.

The cluster star-star interaction of Trapezium is then added.[8] Because the function generates a change in momentum it can be added directly to the force.

$$\texttt{mvDF[mass1\_, vxx\_, xxx\_, yyy\_, zzz\_] := Module}\Big[\ldots$$
$$- \texttt{vxx} \frac{\texttt{vc1[[1]] mass1}^2\ \texttt{2.3 factor[rau]}}{3 \times 1.24451 \times 1.9 \left(0.6 \times 10^4\right)} \Big]$$

As noted earlier, the rotation of the center of mass about galactic center must be taken into account. The easiest way to get this number is to use the empirical circular velocity from the 21 cm observations directly:

$$\omega\texttt{c1[rr\_]} = \frac{128.845\ \texttt{rr}^{0.25}\ \texttt{vc1[[1]]}}{\texttt{rr}\ 1000 \times 206\,265}; \ \omega\texttt{c0} = \omega\texttt{c1[r0]};$$

It should be emphasized that the sense of rotation is left-handed relative to the north galactic pole, so this value would be negative in a right-handed system.

We can now apply the galactic potential as formulated earlier as well as the potential of the cluster stars among themselves including the molecular clouds. Because initial tests of this tidal case indicated that some or all of the Trapezium stars may have come from outside the Orion cluster itself we have included a model potential for the ONC as well. The Lagrangian given by Chandrasekhar is for one star in a coordinate system fixed on the galactic center. Notice that we have changed the radial distance r0 to r1 and z0 to z1 so *Mathematica* will not do the substitution immediately. In this case, outside the potential itself, r1 is to be in AU, not kiloparsecs. The terms containing the ω_c's are from the kinetic energy of the rotating frame only. The molecular cloud potentials are included in the Ω term.

$$\texttt{eL1 = mstar} \left(\frac{1}{2} \left(\texttt{xdot}^2 + \texttt{ydot}^2 + \texttt{zdot}^2\right) - \omega\texttt{c xdot y} + \omega\texttt{c ydot (r1 + x)} + \frac{1}{2} \omega\texttt{c}^2 \left((\texttt{r1 + x})^2 + \texttt{y}^2\right)\right) -$$
$$\texttt{mstar mwpotentialAUA[x, y, z]} - \Omega\texttt{[x, y, z]};$$

Here ωc is the angular velocity of the cluster around the galactic center and Ω is the interstellar potential. The terms involving ωc are the added Coriolis terms. Notice also that we must multiply the galactic potential by stellar mass, as it was not included in its derivation. However, the y coordinate above is defined in a right-hand coordinate system that is the same as that used by Chandrasekhar. In that

---

[8] See **8-2MWpotential** for a detailed derivation.

coordinate system the $\omega c > 0$ is a right-hand angle. But galactic rotation is left handed. In this we substitute a negative rotation; thus

$$
\begin{aligned}
\texttt{eL2 = mstar} &\left(\frac{1}{2}\left(\texttt{xdot}^2 + \texttt{ydot}^2 + \texttt{zdot}^2\right) + \omega c\ \texttt{xdot}\ \texttt{y} - \omega c\ \texttt{ydot}\ (\texttt{r1} + \texttt{x}) + \frac{1}{2}\,\omega c^2\left((\texttt{r1}+\texttt{x})^2 + \texttt{y}^2\right)\right) - \\
&\texttt{mstar mwpotentialAUA[x, y, z]} - \Omega\texttt{[x, y, z]};
\end{aligned}
$$

We can use the same form of Ω as was used in the Hamiltonian to find the derivatives needed in the equations of motion. If, however, we correct the distances to the molecular clouds to take the rotation into account we must be sure to get the direction to time correct by keeping or changing the sign yet again, as NDSolve allows only positive times.

8.2.4 Calculating the N-body problem

To calculate the motion of the Trapezium cluster we must first calculate the complete Lagrangian for the N-body problem including terms for the Chandrasekhar rotation and the potential using the same variables as before. The signs of the y coordinate and velocity would be negated to follow the RA-Dec convention used earlier, but the galactic rotation direction is backwards (left-handed) compared with that assumed by Chandrasekhar so the signs are unchanged here:

$$
\begin{aligned}
\texttt{eL} = &\left(\sum_{\texttt{ii=1}}^{\texttt{nB}} \texttt{mass[[ii]]}\left(\frac{1}{2}\left(\texttt{qx}_{\texttt{ii}}\texttt{'[t]}^2 + \texttt{qy}_{\texttt{ii}}\texttt{'[t]}^2 + \texttt{qz}_{\texttt{ii}}\texttt{'[t]}^2\right) - \right.\right.\\
&\qquad \cdots \\
&\left.\left. + (\texttt{zAUc[[jj]]} - \texttt{qz}_{\texttt{k}}\texttt{[t]})^2\right)\right)\right) \texttt{ /. \{r1} \rightarrow \texttt{r0, z1} \rightarrow \texttt{z0\};}
\end{aligned}
$$

From this we determine the Lagrange equations:

$$
\frac{\mathrm{d}}{\mathrm{d}t}\left(\frac{\partial L}{\partial \dot{q}}\right) = \frac{\partial L}{\partial q}, \tag{8.20}
$$

For the RHS of the equation we have:

```
firstderx = Table[∂qx_kk[t] eL, {kk, 1, nB}];
firstdery = Table[∂qx_kk[t] eL, {kk, 1, nB}];
firstderz = Table[∂qz_kk[t] eL, {kk, 1, nB}];
```

and for the LHS:

```
secondx = D[Table[∂qx_kk'[t] eL, {kk, 1, nB}], t];
secondy = D[Table[∂qy_kk'[t] eL, {kk, 1, nB}], t];
secondz = D[Table[∂qz_kk'[t] eL, {kk, 1, nB}], t];
```

These are then combined into the equations of motion, which are also combined with the initial conditions to derive a solution:

```
eqqx = Table[secondx[[jj]] == firstderx[[jj]] +
    mvDF[mass[[jj]], qx_jj'[t], qx_jj[t], qy_jj[t], qz_jj[t]], {jj, 1, nB}];
eqqy = Table[secondy[[jj]] == firstdery[[jj]] +
    mvDF[mass[[jj]], qy_jj'[t], qx_jj[t], qy_jj[t], qz_jj[t]], {jj, 1, nB}];
eqqz = Table[secondz[[jj]] == firstderz[[jj]] +
    mvDF[mass[[jj]], qz_jj'[t], qx_jj[t], qy_jj[t], qz_jj[t]], {jj, 1, nB}];
eqs = Flatten[Join[eqqx, eqqy, eqqz]];
sol1 = NDSolve[{eqs, ics1}, vars, time,
    Method → {"Projection", Method → "ExplicitRungeKutta",
    "Invariants" → tplusVB, MaxIterations → 10 000 000}]
```

For calculating the motion backward in time we must reverse the signs in the Lagrangian as well as reversing the signs of the velocity. The presence of the second derivative and the rotational velocity factor means we cannot simply reverse the initial velocities as is done in the Hamiltonian method. We therefore start with the Lagrangian:

$$
\text{eL1} = \left(\sum_{ii=1}^{nB} \text{mass}[[ii]] \left(\frac{1}{2} \left(qx_{ii}'[t]^2 + qy_{ii}'[t]^2 + qz_{ii}'[t]^2 \right) - \right. \right.
$$

$$
\cdots
$$

$$
\left. \left. + \left(zAUc[[jj]] - qz_k[t] \right)^2 \right) \right) \right) /. \{r1 \to r0, z1 \to z0\};
$$

and follow the same procedure as before.

The result can be seen in figure (8.7). It is clear that Chandrasekhar's theory is sound even if the mathematics and derivations are a bit obscure to modern thought processes. From a deeper analysis of these computations[9] we find the following:

1. The alignments toward the molecular clouds are no longer obvious.
2. The dynamical friction term modifies the association of the stars considerably. In particular, the previous tight binding of C and D is reduced so the two are only loosely bound now. This is reflected in smaller Lyapunov coefficients as well as decreasing total energy and virial. Only component C has a negative Lyapunov exponent and therefore will not escape the cluster.[10]
3. Due to the two-dimensional nature of the galactic potential, orbit stretching in the x–z plane is expected, and found in the results. It is this type of motion that is described by the Hénon–Heiles potential.

---

[9] See **8-3TheCLFunction**.

[10] Component C is now known to be a multiple system; however the largest mass is still $40 M_\odot$ (Kraus et al., 2009) and therefore capable of a supernova explosion, perhaps within the 10^6 year calculation period calculated here. In that case the other components will be free to leave the ONC entirely.

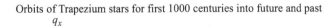

Orbits of Trapezium stars for first 1000 centuries into future and past

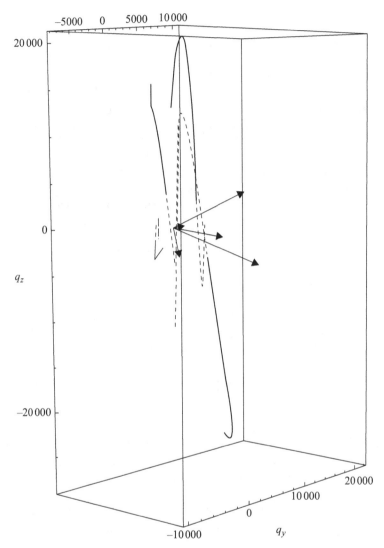

Fig. 8.7 Motion of Trapezium including galactic gravitational potential.

4. Because the galactic rotation speed increases with an increase in the x distance[11] differential rotation is responsible for model results stretching the orbits in the $+y$ direction.
5. As suspected the galactic tidal effects gradually rip clusters apart, particularly open clusters. Globular clusters, because of their much higher masses and their much larger distances from the galactic disk and center, are less subject to such tidal disruption.

[11] See the galactic rotation curve of Section 8.1.

8.3 The Hénon–Heiles equations

The Hénon–Heiles equations arise from an approximation of the gravitational potential of the Milky Way in the vicinity of the sun, and describe the motion of a mass in that potential. It turns out the equations are chaotic for a number of parameter sets. The result of this chaos is the possibility that stars can be ejected from clusters or galaxies. The Hénon–Heiles problem is so famous it is included in *Mathematica*'s own set of examples to use with `NDSolve`. Because the invariant given is energy, the SPRK method works very well.

8.3.1 Background and history

As pointed out previously, if we limit our discussion to the information given in books that offer more than a passing interest in Hamiltonian systems, our choice becomes quite narrow. If we restrict our interest to Hénon–Heiles the number dwindles to three: Goldstein, Poole, and Safko (2002); Hilborn (1994); and Rasband (1990). There is also one author (Drazin, 1992) who assigns problems of interest for students.

We have seen that the actual galactic potential can be well fit to exponentials, and they lead to what look like epicyclic motions of stars in our part of the galaxy. Rasband (p. 169) shows that the Hénon–Heiles Hamiltonian can be obtained from the Hamiltonian of the Toda lattice. The Toda lattice arises in the mathematical theory of a monatomic crystal through a lattice connected by nonlinear springs. The Toda lattice and Hénon–Heiles are both examples of nonintegrable Hamiltonians, that is Hamiltonians for which numerical integration is required. While the Hamiltonian may be a constant of motion, others are not so immediately evident, if they exist at all. Rasband points out that the Toda problem does have a second constant of motion in addition to the Hamiltonian, and that in approximation the exponential can be written to become the Hénon–Heiles expression itself. It is this approximation that captured the imagination of chaos scientists.

8.3.2 The Hénon–Heiles Hamiltonian

The Hénon–Heiles equations describe the motion of individual stars around a galactic center where motion is restricted to the xy-plane. The Hamiltonian is given by

$$H = \frac{p_x^2}{2m} + \frac{p_y^2}{2m} + \frac{1}{2}k\left(x^2 + y^2\right) + \lambda\left(x^2 y - \frac{1}{3}y^3\right). \tag{8.21}$$

This can be simplified by letting $k = \lambda = 1$ and noting that $p_x = m\dot{x}$ and $p_y = m\dot{y}$. The Hamiltonian then reduces to a dimensionless normalized form

$$E = \frac{1}{2}\left(\dot{x}^2 + \dot{y}^2\right) + \frac{1}{2}\left(x^2 + y^2\right) + \left(x^2 y - \frac{1}{3}y^3\right). \tag{8.22}$$

This yields the equations of motion

$$\ddot{x} = -2xy - x, \qquad \ddot{y} = y^2 - x^2 - y. \tag{8.23}$$

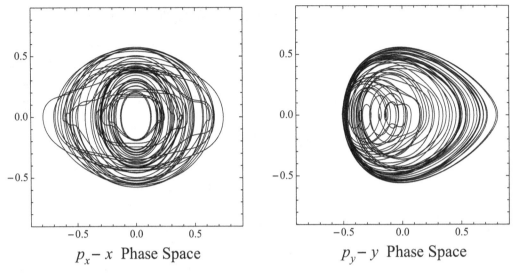

$p_x - x$ Phase Space $p_y - y$ Phase Space

Fig. 8.8 Phase spaces $p_x - x$ and $p_y - y$ for the Hénon–Heiles equation.

It is this form of the Hénon–Heiles equation that is included in *Mathematica*. The Hénon–Heiles equation is typically presented in terms of Poincaré sections.[12] These cannot be generated in EquationTrekker because it cannot handle a system of two 2nd order equations. So in *Mathematica* we use NDSolve and then show the x–y trajectories along with the x phase plane and the y phase planes without the time strobing that characterizes a Poincaré section, as seen in figure (8.8).

8.3.3 Hénon–Heiles in *Mathematica*

The Hénon–Heiles equation must be loaded into mathematica via GetNDSolvedProblem[]. It has a default set of initial conditions that can also be loaded:

```
system = GetNDSolveProblem["HenonHeiles"];

initial = system["InitialConditions"]; time = system["TimeData"];
```

Once loaded it can by solved by the various methods of NDSolve[]:

```
sol = NDSolve[system, time,
  Method → {"FixedStep", Method → {"ImplicitRungeKutta",
      "DifferenceOrder" → 10, "ImplicitSolver" → {"Newton",
        AccuracyGoal → MachinePrecision, PrecisionGoal → MachinePrecision,
        "IterationSafetyFactor" → 1}}}, StartingStepSize → 1 / 10]
```

Because of the "hidden" structure in the internal calls of the problem, one must be careful when entering new initial conditions. This is usually best done by editing values of the default initial conditions.

---

[12] See, for example, Goldstein, Poole, and Safko (2002), p. 496; Hilborn (1994), p. 343; and Rasband (1990), p. 168.

As in all chaos situations, the problem is very sensitive to initial conditions. If the initial conditions exceed a satisfactory value, the `NDSolve` method will fail.

8.4 The central black hole of the Milky Way

The center of the Milky Way lies in the direction of the constellation of Sagittarius approximately 8.5 kiloparsecs from the sun. This information is very much taken for granted in modern times, but the establishment of that fact was not easy. Prior to World War II, the distance to the galactic center was obtained indirectly to be between 8–10 kiloparsecs. This was first done by measuring the positions of globular clusters surrounding our galaxy. Globular clusters are scattered in a roughly spherical halo around the Milky Way and other galaxies. From the sun, the globular cluster system is asymmetrical with a concentration in the hemisphere where the galactic center lies. The distance to the galactic center was originally determined by taking the centroid of the projections of the globular clusters on the galactic equatorial plane and assuming galactic center coincided with the center of the globular cluster halo.

Optical studies in the plane of the galaxy did not fare as well because dust limited the view severely, with a limiting distance of about 1–2 kiloparsecs. The visible light from the galactic center is cut down by a factor of $10^{13.6}$. In the infrared and radio regions, however, the extinction is not nearly so severe. We have seen how 21 cm radio waves can be used to build a model of the disk of the Milky Way, including parts that cannot be observed in the visible spectrum.

Just as radio astronomy has made dramatic strides since WWII, so has infrared astronomy. Observations at infrared wavelengths penetrate the galactic dust reasonably well. When large infrared (IR) telescopes are used in tandem as interferometers, the IR diffraction (being larger than for optical wavelengths) of the individual telescopes is overcome as well. Thus it has been possible for the last 20 years to monitor the vicinity of the galactic center and discover a wide variety of individual radio and IR objects all within less than a one parsec radius (24″ arc) of the center.

In a number of other spiral galaxies whose central regions can be viewed even at optical wavelengths, there is much accumulated evidence of one or more very massive (even supermassive) compact objects. In some galaxies the central object is self-luminous and given the name quasar. As we will see in Chapter 9, the physical dimensions of such objects are quite small. So small in fact that they often show light variations that allow estimates of their sizes to be obtained. In other instances, the central objects are mostly dark, which brings to mind the most dramatic objects thought to exist in general relativity, black holes.

8.4.1 Calculation of black hole mass

In Chapter 1 we laid out two procedures for obtaining a visual binary orbit with the main aim of obtaining the mass information of the orbiting objects via Kepler's third law. In elementary texts that process is taken for granted with little indication of the practical difficulties involved. For a valid error analysis the task is even more complex. We presented this rather rude shock because it is instructive for advanced undergraduate students to see some of the real issues behind accomplishing an "elementary" textbook task with precision. Except for the fact that the computations are done with an electronic computer, the approach used in Chapter 6 is not new. In this chapter we examine a modern application of the "ancient" arts of astrometry and orbit analysis. In particular its application to the center of the Milky Way itself.

In the very center of our galaxy is an object that is able to shine (in the infrared) through all the interstellar dust, and appears to have some luminosity in the radio range as well. This source is called Sgr A*. In the vicinity of this central source there are many other objects including gas and stars. It is the stars in the vicinity of this compact object that has attracted the attention of many, including the team of Gillessen et al. (2009). The work of Gillesen et al. is a masterful accomplishment, and should be considered one of the classic papers of modern times. It can be read and appreciated by undergraduate students, and is a beautiful interface between three areas of modern astrophysics: astrometry, celestial mechanics, and general relativity. Readers are encouraged to study the original paper, as we will here give only a brief outline of the work to validate the information we are most interested in, the stellar orbital elements and related data.

The paper by Gillessen et al. uses the following process in analyzing the central cluster, known as the S-cluster:

1. Identification and location of the objects in the S-cluster, labeled from S1 to S112 found in the vicinity of Sgr A*; determination of the presumed location of the galactic massive black hole (MBH); obtain high-quality IR spectra of these objects.
2. Transform to a common system by performing high accuracy astrometry of the S objects, including
   a. Fainter objects relative to brighter ones
   b. S-stars relative to brighter reference stars
   c. Reference stars relative to SiO maser stars whose positions are determined using the VLA very accurately compared with Sgr A* positions
   d. Location of stars on pixels
3. Fit the data with a model of the potential and gather in that way reliable orbit parameters as well as valuable information about the potential itself and the object responsible for it.
4. Describe the database and instrumentation available to use in this study.
5. Statistical issues explicitly discussed:
   a. Intrumentation issues (the SHARP, NACO, SINFONI systems and others)
   b. Errors in calibration between systems when making astrometric meausurements
   c. Final coordinate system
   d. Relating S-stars to reference stars
   e. Estimates of astrometric errors, image distortions, and transformation errors
   f. Differential effects in field of view and source confusion (from interferometry)
   g. Gravitational lensing
   h. Spectroscopic data and radial velocity errors
6. Orbital fitting
   a. Generalized orbit program that considers an "arbitrary" potential function
   b. High dimensional chi-square minimization process
   c. Multiple mass components
   d. Measure positions of Sgr A* at maximum light
   e. Consideration of four relativistic effects (light retardation, relativistic Doppler effect, gravitational redshift, Schwarzschild correction to the Newtonian potential)
   f. Entry of independent information on selected parameters (known as priors)
   g. Orbit fitting done on the periastron distance rather than semi-major axis
   h. Designed to be used for parallel computation

Our main goal is to make use of the Gillesen et al. elements to calculate the S-star positions relative to the BH mass center. However, it is instructive to use Kepler's third law to infer the mass of the central object. A data table for the S-stars is found in GCStars.cvs:

starlist = Import["GCStarsR.csv"];

The first column is the designation of the star, and the subsequent data columns are semi-major axis a (in seconds of arc), eccentricity ϵ, inclination i, longitude of ascending node Ω, argument of periapsis ω, the periapsis date p_t, and the period T in years. The first six columns are necessary to completely define a stellar orbit.

A basic estimate of the central mass can be found from Kepler's third law,

$$\frac{a^3}{T^2} = \frac{GM}{4\pi^2}. \tag{8.24}$$

When a is expressed in AU and T in years, their ratio yields the central mass in solar masses. Because the semi-major axis is given in seconds of arc, it must be converted to AU by multiplying by the distance to galactic center. The central mass can then be found by taking the average of the orbital ratios. Thus

massBH = Table[(206 265 starlist[[ii, 2]] 8330 / 206 265)³ / starlist[[ii, 8]]², {ii, 2, 29}];

massMBHGC = Mean[massBH]; StandardDeviation[massBH];

This yields a value of

$$M_{GC} = (4.3338 \pm 0.0158) \times 10^6 M_\odot, \tag{8.25}$$

which agrees very well with the value given by Gillessen et al.[13]

8.4.2 Stellar motion near galactic center

To plot the orbits of S-stars we must find the coordinate positions in their orbital plane, and then transform those relative to the plane of the sky as appropriate for a binary star orbit. This is almost, but not quite, the inverse of the binary orbit determination procedure. This involves an equation known as Kepler's equation, which should not be confused with Kepler's laws.

The Kepler equation is a transcendental equation that relates the eccentric anomaly E of an elliptical orbit and its mean anomaly M,

$$M = E - \epsilon \sin E, \tag{8.26}$$

where ϵ is the orbital eccentricity. The mean anomaly can be expressed in terms of the orbital period T and the periapsis time p_t; thus

$$\frac{2\pi (t - p_t)}{T} = E - \epsilon \sin E, \tag{8.27}$$

which gives the eccentric anomaly as a function of time t. In the orbital plane one can define a Cartesian x–y plane with its origin in the center of the orbit (not galactic center). For elliptical motion, the positions

---

[13] Here we give our value beyond the customary rounding rules as this is often characteristic of computer output. Following the rounding rules our value would be 4.33 ± 0.02, although some authors would give 4.334 ± 0.016 because the error begins with a 1.

are then given by[14]

$$x = a\,(\cos E - \epsilon),\qquad y = a\sqrt{1 - \epsilon^2}\,\sin E, \tag{8.28}$$

which indirectly gives x and y as functions of time. With $r = \sqrt{x^2 + y^2}$, the velocities are given by

$$\dot{x} = -\frac{2\pi}{T}\frac{a^2}{r}\sin E,\qquad \dot{y} = \frac{2\pi}{T}\frac{a^2}{r}\sqrt{1 - \epsilon^2}\,\cos E. \tag{8.29}$$

From these we can then plot the orbits in their orbital planes.

In the solar system, the definitions of the coordinate systems are generally determined by the sense of motion of Earth in its orbit, but with visual binaries the orbits are defined by their appearances on the plane of the sky. Apparent motions are often defined in a polar coordinate system with the position of the primary star at the origin. Motion of the secondary star (if observed) is defined as moving from north through east, and is considered retrograde if it moves in the opposite sense.

Green (1985) recognized an ambiguity problem between visual binary and spectroscopic binary definitions. He thus draws his orbit diagram with the correct aspect for the angles that are consistent with the astrometric definition of the observed angles, but revises his definitions of the Cartesian system centered on the primary star. The method of orbit determination most consistent with Green's coordinate formulation is the Thiele–Innes method, but we were unable to use that method for our binary orbit example in Chapter 6 because the observations needed to cover more than one orbit period. Such data overlap is needed for precise determination of the center of the orbit ellipse, something that our example data failed to do. However, we can still generate the galactic center orbit paths using Green's formulae. For the coordinates in this frame, $+x$ is along the Right Ascension, and $+y$ is northward along the Declination. The $+z$ axis is toward the primary star, and points away from us so that it will have the same convention as the radial velocity. Because Gillessen et al. had spectroscopic information available, we can have confidence that they located the ascending nodes (following Green's convention) properly so that the z-axis forms a right-hand system in the sky centered at Sgr A*. Thus we can plot

$$
\begin{aligned}
x &= r\,(\cos(v + \omega)\sin\Omega + \sin(v + \omega)\cos\Omega\cos i)\\
y &= r\,(\cos(v + \omega)\cos\Omega - \sin(v + \omega)\sin\Omega\cos i)\\
z &= r\sin(v + \omega)\sin i
\end{aligned}
\tag{8.30}
$$

as functions of time.

The computational procedure of generating a table of orbital positions as a function of time is known as producing an ephemeris, and the resulting tables are called ephemeridies. In the case of solar system objects it is considered most efficient to separate time-dependent factors from the time-independent ones. We use *Mathematica* to expand Green's expressions and use the same approach here:

```
orbitplane[incl_, Ω1_, ω1_] := Module[. . .{sX, cX, sY, cY, sZ, cZ}]
```

$$\texttt{keplereq[ts\_, tP\_, tmP\_, eccl\_]} := \texttt{Module}\left[\ldots 2\,\texttt{ArcTan}\left[\sqrt{\frac{1 + \texttt{eccl}}{1 - \texttt{eccl}}}\ \texttt{Tan[eE1 / 2]}\right]\right]$$

---

[14] Danby, 1988.

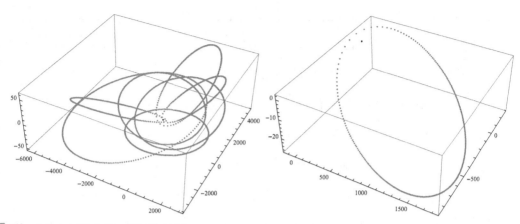

Fig. 8.9   Example S-star orbits (left) and S2 orbit (right).

```
plotorb[kk_, pts_] := Module[...Table[
    {8330 rRadius[nu[t], asec, ecl] (Sin[nu[t]] coeff[[1]] + Cos[nu[t]] coeff[[2]]),
     8330 rRadius[nu[t], asec, ecl] (Sin[nu[t]] coeff[[3]] + Cos[nu[t]] coeff[[4]]),
     8330 rRadius[nu[t], asec, ecl]
       (Sin[nu[t]] coeff[[5]] + Cos[nu[t]] coeff[[6]])}, {t, 0, pyrs, Δt}]]
```

It is possible on a fast computer to graph all of the orbits together, as seen in figure 8.9, but this does not improve one's understanding of the types of objects to be found at the center of the Milky Way. One object, S2, is of particular interest due to its close approach to the central black hole. It is sometimes referred to as the "event horizon grazer."

8.4.3 Relativistic orbits near the central black hole

The orbits we have derived assume standard Newtonian gravity and do not take into account any relativistic effects. This is generally valid for "distantly" orbiting stars, but may not hold for close grazing orbits such as that of S2. We now examine S2 in this relativistic context.

In Section 5.4 we derived the Schwarzschild metric for a relativistic mass, given as

$$ds^2 = \left(1 - \frac{2m}{r}\right)dt^2 - \left(1 - \frac{2m}{r}\right)^{-1}dr^2 - r^2 d\theta^2 - r^2 \sin^2\theta\, d\phi^2. \tag{8.31}$$

Because our S2 orbit is given in Cartesian coordinates, it is useful to express the metric in isotropic form. By isotropic, it is meant that the metric has spatial coordinates such that one might set up (x, y, z) coordinates. This form can be found by expressing the proper time radius r in terms of the Cartesian radius r_{iso}.

$$r = r_{iso}\left(1 + \frac{m}{2r_{iso}}\right)^2. \tag{8.32}$$

The metric then becomes

$$ds^2 = \left(\frac{1 - m/2r_{iso}}{1 + m/2r_{iso}}\right)^2 dt^2 - \frac{1}{c^2}\left(1 - \frac{m}{2r_{iso}}\right)^4 dr_{iso}^2 - r^2\, d\theta^2 - r_{iso}^2 \sin^2\theta\, d\phi^2, \tag{8.33}$$

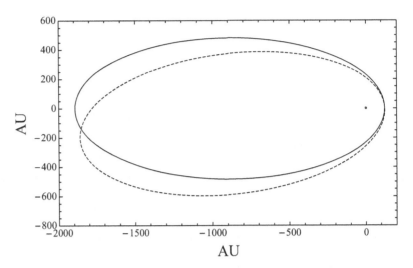

Orbital precession of S2 after 200 orbits.

which can be expressed in Cartesian coordinates as[15]

$$ds^2 = \left(\frac{1 - m/2r_{iso}}{1 + m/2r_{iso}}\right)^2 dt^2 - \frac{1}{c^2}\left(1 - \frac{m}{2r_{iso}}\right)^4 \left({dx_{iso}}^2 + {dy_{iso}}^2 + {dz_{iso}}^2\right). \tag{8.34}$$

This form of the metric is useful when examining astrometric observations under general relativity.

Instead of deriving the orbital equation as a second-order differential equation of r and ϕ, Green (1985) and others derive an orbit equation that is of first order,

$$\left(\frac{dr}{d\tau}\right)^2 = -\frac{GM f(r)}{ar^3}, \tag{8.35}$$

where τ is the proper time. It is shown by Green that the equation $f(r)$ is cubic, and hence influences the limiting values of r.

$$f(r) = r^3 - 2ar^2 + a\left(1 - \epsilon^2\right)r - 2ma^2\left(1 - \epsilon^2\right). \tag{8.36}$$

The three roots of the equation are the periapsis, apoapsis, and the Schwarzschild semi-radius m. These roots are of interest here because they show how to convert between the GR corrected boundary values and the Newtonian ones. Although there is a correction to the semi-major axis, the correction to the eccentricity is more important. Green shows that to an approximation of m/a in the standard Schwarzschild coordinate system

$$a_{GR} = a - m \qquad \epsilon_{GR} - \epsilon\left[1 + \frac{m}{a\epsilon^2}\left(1 + \epsilon^2\right)\right]. \tag{8.37}$$

Although the above approximate GR values apply to the standard coordinate system, the advance of the periastron is not uniform in that coordinate system. Thus the value one gets depends on how long the observations have been made. It is because of this nonuniformity that Green recommends and uses

[15] Green (1985).

the isotropic system. He shows that the classical expressions must be modified further in the isotropic system

$$a_{GRiso} = a - 2m, \tag{8.38}$$

$$\epsilon_{GRiso} = \epsilon \left[1 + \frac{m}{a\epsilon^2} \left(1 + \epsilon^2 \right) \right] \left(1 + \frac{m}{a - m} \right). \tag{8.39}$$

Then in one orbital period (periastron to periastron) there is a constant change of the argument of the periastron, given by

$$\Delta\omega = \frac{6\pi m}{a_{GRiso} \left(1 - \epsilon_{GRiso}^2 \right)}. \tag{8.40}$$

When considering the application of GR to observations, it must always be remembered that numerical integrations (even those of extremely high precision) are still approximations, even if the differential equations they approximate are reasonably sophisticated models. In the case of the ordinary Schwarzschild metric, the coordinates are assumed to be "congruent" to Newtonian coordinates. For the isotropic metric, transformations to "flatness" are built into the theory. Even in the case of S2 (figure 8.10), the difference is noted only in the advance of the periastron. This is because although S2 comes closest to galactic center at 112 AU, which is much larger than the Schwarzschild radius (0.0856 AU), so even Newtonian physics is not an unreasonable description of the orbits.

In the event that objects that do pass rather close to the galactic center are discovered an entirely different approach seems to be needed for the orbit analysis. Additionally, the Schwarzschild geometry is for a static black hole. If the galactic center is spinning, then the Kerr metric is appropriate. Chandrasekhar (1983) showed the Kerr metric to be separable. Should objects that pass the galactic central MBH much closer than S2 be found, Angelil, Saha, and Merrit (2010) and Angelil and Saha (2010) have shown how equations developed from the perturbative Hamiltonian can be used to analyze the resulting orbits.

8.5 Spiral density waves

Chandrasekhar seemed to abandon interest in stellar dynamics when he found he could not readily explain the spiral arm patterns seen in many spiral galaxies as the simple streaming of stars. Of course in 1942, he was unaware of the amazing amount of interstellar hydrogen gas in both atomic and molecular form throughout the galaxy, as discovered by radio means.

The concept of the galactic material density underwent an even more profound alteration when dark matter could be shown to be predominant in the outer parts of the Milky Way. In Section 8.2 we showed that galactic tides from the rest of the galactic mass exerted a significant influence on the Orion Trapezium cluster, inducing what appears to be epicyclic motion in the star paths round a point that itself describes a circular orbit around the galactic center. This type of motion is encountered again in the study of the Hénon–Heiles potential. In the Hénon–Heiles case, the choice of initial condition influences the regularity of the motion from periodic to quasi-periodic, to chaotic. Furthermore we note that in the case of the solar system, the epicycles of Ptolemy were able to generate elliptical paths for the planets. It is a bit ironic that epicycles arise once again in modern times, but now in connection with an explanation of the widely observed co-moving galactic spiral patterns. This is a "steady-state" wave solution which evaded Chandrasekhar's prodigious intellect.

The quantitative methods and ideas that eventually led to the solution of the galactic spiral structure in a convincing manner are traced to Shu in the second of his texts on the Physics of Astrophysics (1992). Shu starts with viscous accretion disks, progresses through fluid instabilities, viscous shear flow, and turbulence, and lands squarely on spiral density waves. Here we consider only the parts of Shu's exposition that has a direct bearing on the density wave explanation for spiral structures. The interested reader is referred to the appropriate parts of Shu and his references for more background and details.

8.5.1 Basic principles

We begin by considering the Milky Way as an infinitesimally thick fluid disk. Disk problems of any type are typically expressed in cylindrical coordinates (r, ϕ, z), and for an infinitesimal disk the volume density ρ becomes a product of the surface mass density σ with the Dirac δ function,

$$\rho(r, \phi, z, t) = \sigma(r, \phi, t)\, \delta(z). \tag{8.41}$$

The integrated pressure in the vertical direction is

$$p = \int_{\infty}^{\infty} a_o^2 \rho \, \mathrm{d}z, \tag{8.42}$$

where a_o is the "thermal" speed of the fluid at $z = 0$, or the 1D dispersion of velocities known as the dispersion velocity. Furthermore we suppose that

$$\Delta p = -\sigma \, \Delta h, \qquad h = \int \frac{\mathrm{d}p}{\sigma}. \tag{8.43}$$

If we let $u_r(r, \phi, t)$ be the r-component of the fluid velocity, while $j_z(r, \phi, t)$ is the specific angular momentum about the z-axis, then integrating in the z-direction we obtain three dynamical equations, derived from the Navier–Stokes equations for fluid flow,

$$\frac{\partial \sigma}{\partial t} + \frac{1}{r}\frac{\partial}{\partial r}(r\sigma u_r) + \frac{1}{r^2}\frac{\partial}{\partial \phi}(\sigma j_z) = 0, \tag{8.44}$$

$$\sigma\left(\frac{\partial u_r}{\partial t} + u_r \frac{\partial u_r}{\partial r} + \frac{j_z}{r^2}\frac{\partial u_r}{\partial \phi} - \frac{j_z^2}{r^3}\right) = -\frac{\partial p}{\partial r} - \sigma\left(\frac{\partial V}{\partial r}\right)_{z=0}, \tag{8.45}$$

$$\sigma\left(\frac{\partial j_z}{\partial t} + u_r \frac{\partial j_z}{\partial r} + \frac{j_z}{r^2}\frac{\partial j_z}{\partial \phi}\right) = -\frac{\partial p}{\partial \phi} - \sigma\left(\frac{\partial V}{\partial \phi}\right)_{z=0}. \tag{8.46}$$

A fourth equation is the Poisson equation, which relates the gravitational potential to the mass distribution,

$$\frac{1}{r}\frac{\partial}{\partial r}\left(r\frac{\partial V}{\partial r}\right) + \frac{1}{r^2}\frac{\partial^2 V}{\partial \phi^2} + \frac{\partial^2 V}{\partial z^2} = 4\pi G\left(\sigma \delta(z) + \rho\right). \tag{8.47}$$

These equations are quite complex, so we explore only a linearized form.

As Shu indicates, if one starts with an axisymmetric disk in an equilibrium flow state then the equations reduce considerably. With this assumption there is now a relationship between the differential rotation about the z-axis with equilibrium angular speed $\Omega(r)$, surface density σ_o, and sound speed a_o. This simplifies the Poisson equation considerably:

$$r\Omega^2(r) = \frac{a_o^2}{\sigma_o}\frac{d\sigma_o}{dr} + \left(\frac{\partial V_o}{\partial r}\right)_{z=0}, \tag{8.48}$$

$$\frac{1}{r}\frac{\partial}{\partial r}\left(r\frac{\partial V_o}{\partial r}\right) + \frac{\partial^2 V_o}{\partial z^2} = 4\pi G\left(\sigma_o\delta(z)\rho_{\text{ext}}\right). \tag{8.49}$$

Given this equilibrium state, we can then do the usual substitutions of linear expansions (one constant plus one small variable term) of the main variables (σ, u_r, j_z, and V) back into the three dynamical equations plus the simplified Poisson equation to obtain a set of wave dispersion relations (Shu, p. 138). The constant terms in the linear expansion drop out upon taking the derivatives, leaving only equations containing the perturbation derivatives. Solutions are then found by assuming the perturbation terms are sinusoidal (wave-like) in form and back substituting. The equilibrium terms are subtracted to arrive at the linearized perturbation equations. It is then assumed the perturbations are "epicyclic," that is, the vorticity is not 0. This is in good accord with our previous results using tidal dynamics equations.

One approach is to use Fourier decomposed solutions to obtain the dispersion equation. In Fourier decomposition it is assumed that the argument of the exponential term is the usual "wave propagation" one, $\omega t - k\sqrt{\mathbf{r}\cdot\mathbf{r}}$, but we could likewise use the backward propagating wave $\omega t + k\sqrt{\mathbf{r}\cdot\mathbf{r}}$. Here $k = 2\pi/\lambda$ represents the wave number and the phase angle $m\phi$ is given by $\sqrt{\mathbf{r}\cdot\mathbf{r}}/\lambda$.

The full Fourier decomposition in matrix form is given by Shu as

$$\begin{pmatrix} \sigma_p \\ u_p \\ j_{zp} \\ V_p \end{pmatrix} = \Re\left[\begin{pmatrix} S(r) \\ U(r) \\ J_z(r) \\ V(r,z) \end{pmatrix}e^{i(\omega t - m\phi)}\right], \tag{8.50}$$

where p indicates "perturbation." Here m is an index indicating how many "arms" the given situation will have.

When substituted into the perturbation equations above we find

$$i(\omega - m\Omega)S + \frac{1}{r}\frac{d}{dr}(r\sigma_o U) - im\sigma_o\frac{J_z}{r^2} = 0, \tag{8.51}$$

$$i(\omega - m\Omega) - 2\Omega\frac{J_z}{r} = \frac{d}{dr}\left(a_o^2\frac{S}{\sigma_o}\right) - \left(\frac{\partial V}{\partial r}\right)_{z=0}, \tag{8.52}$$

$$\frac{1}{r}\frac{\partial}{\partial r}\left(r\frac{\partial V}{\partial r}\right) - \frac{m^2}{r^2}V + \frac{\partial^2 V}{\partial z^2} = 4\pi GS(r)\delta(z). \tag{8.53}$$

Here $(\omega - m\Omega)$ represents the instantaneous rotational frequency difference from the equilibrium rotation. The boundary conditions can be applied as

$$U = J_k = 0 \text{ at } r = 0, \qquad S, V \to 0 \text{ as } \sqrt{r^2 + z^2} \to \infty. \tag{8.54}$$

8.5.2 The WKB and WKBJ method

At this point the equations are still quite complex, but if we solve the third equation in a reasonable fashion the remaining two become total differential equations. These describe locally propagating

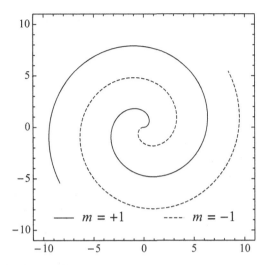

Fig. 8.11   Spiral density maxima for $m = 1$ and $m = -1$.

waves, but we desire global descriptions, so we will explore solutions for which ω is now a rotational eigenvalue of the perturbation equations. As pointed out by Shu, approximate global solutions can be found through the application of the WKB (or WKBJ) method. This is similar to solving the 1D quantum wave equation with the same assumptions.

Students who have had a quantum mechanics course may recall something called the Wentzel–Kramers–Brillouin (WKB) method that is sometimes used to find approximate solutions to the Schrödinger equation. In the spiral density wave problem, Shu (1992) uses a method he calls the WKBJ method without clarifying that the two are indeed the same thing.[16] The applicability of the method hinges critically on having a slightly inhomogeneous situation such that the spatial variation is on the scale of the wavelength of the waves and slowly varying compared to the frequency of the waves.

The method of Shu simply augments the phase expression of the Fourier decomposition $\omega t - m\phi$ by adding a "rapidly changing" phase function $\Phi(r)$ such that

$$\left| \frac{1}{r} \frac{d\Phi}{dr} \right| >> 1. \tag{8.55}$$

This becomes the so-called "tightly wound" assumption. The variation of the surface density under WKB thus has the form

$$S(r) = S_o(r) e^{i\Phi(r)}, \tag{8.56}$$

where S_o is much more slowly varying than $\Phi(r)$. The full solution is then

$$\sigma_p(r, \phi, t) = \Re \left[S_o(r) e^{i(\omega t - m\phi + \Phi(r))} \right]. \tag{8.57}$$

In figure (8.11) the locus of maximum (or minimum) surface density is plotted, and they follow a familiar spiral pattern. This pattern is stationary. Although the disk is locally differentially spinning a

---

[16] A reference to *Wikipedia* (2012) under "WKB Approximation" clarifies the historical context, and adds that the J stands for Sir Harold Jeffreys who actually discovered the method several years (1923) before WKB (1926), though WKB were unaware of that.

ω, the patterns of maximum density spin at ω/m. This means the $\pm m$ solutions superpose to create a stationary pattern, just as two opposite waves in a string create a standing wave.[17]

8.5.3 Epicyclic frequency, radial wave number, and the asymptotic dispersion relationship

Shu defines the epicyclic frequency as

$$\kappa^2 = \frac{1}{r^3}\frac{d}{dr}\left[r^4\Omega^2\right],$$ (8.58)

and the negative of the radial wave number as

$$k = \frac{d\Phi(r)}{dr},$$ (8.59)

from which he derives a matrix dispersion relationship. From equation (8.50) we have

$$\begin{pmatrix} \sigma_p \\ u_p \\ j_{zp} \end{pmatrix} = D \begin{pmatrix} S \\ U \\ J_z \end{pmatrix},$$ (8.60)

where

$$D = \begin{pmatrix} i(\omega - m\Omega) & ik\sigma_o & 0 \\ ia_o^2 k/\sigma_o & i(\omega - m\Omega) & -2\Omega/r \\ 0 & r\kappa^2/2\Omega & i(\omega - m\Omega) \end{pmatrix},$$ (8.61)

is the coefficient matrix. The solution is found simply by multiplying by the inverse matrix,

$$\begin{pmatrix} S \\ U \\ J_z \end{pmatrix} = D^{-1}D \begin{pmatrix} \sigma_p \\ u_p \\ j_{zp} \end{pmatrix}.$$ (8.62)

Although this can be tedious to calculate manually, it is easily done in *Mathematica*:

```
matrixdisp = {{i (ω - m Ω), i k σ0, 0},
    {i k a0² / σ0, i (ω - m Ω), -2 Ω / r}, {0, (κ² / (2 Ω)) r, i (ω - m Ω)}};
vector1 = {ss, uu, jj}; vector2 = {0, -i k vv, 0};
vector3 = FullSimplify[Inverse[matrixdisp] . vector2]
```

This solves the matrix equation for S, U, and J, although we only need S to find the dispersion relation as shown by Shu. Dividing the resulting solution by i gives expressions identical to those of Shu (pp. 143–144). If the expression for S is compared with the asymptotic solution of Poisson's equation we arrive at the dispersion relationship for spiral density waves.

Shu solves the Poisson problem using an approximation that obtains S as a function of the derivative of the potential V in the out-of-plane direction, then expands the potential in the complex plane that includes the z direction. With

$$k = \frac{d\Phi(r)}{dr},$$ (8.63)

---

[17] An animation of this effect can be seen in **8-9SpiralDwaves**.

then

$$S(r) \approx -\frac{|k|V(r, z = 0)}{2\pi G}. \tag{8.64}$$

when this is equated with matrix solution of S, the desired dispersion relationship is obtained,

$$(\omega - m\Omega)^2 = \kappa^2 + k^2 a_o^2 - 2\pi G|k|\sigma_o. \tag{8.65}$$

The eigenvalue frequency (angular speed wavelength) is then

$$\omega = m\Omega \pm \sqrt{\kappa^2 + k^2 a_o^2 - 2\pi G|k|\sigma_o}. \tag{8.66}$$

8.5.4 A spiral density wave equation

Most readers will have some acquaintance with wave equations in electricity and magnetism, so there should be no surprise that spiral density waves satisfy a wave equation as well. Shu shows that such equations in lowest order would be in two forms,

$$- a_o^2 \frac{d^2 S}{dr^2} + 2\pi G\sigma_o \frac{dS}{dr} + \left(\kappa^2 - (\omega - m\Omega)^2\right) S = 0, \tag{8.67}$$

$$- a_o^2 \frac{d^2 S}{dr^2} - 2\pi G\sigma_o \frac{dS}{dr} + \left(\kappa^2 - (\omega - m\Omega)^2\right) S = 0. \tag{8.68}$$

The former is the plus form for leading spiral waves, while the latter is the minus form for trailing waves.

Shu continues with a nice discussion of how an external potential might serve to "pump" the formation of density waves in the early history of the galaxy, which is a bit beyond the scope of this text. What we have done is to show why the density wave theory is a serious model of spiral galaxies, the Milky Way in particular. Certainly the fact that the wave equation solutions are not symmetric with respect to a change of the sign of the gravitational forcing term seems to indicate that the spiral structure will have a progressive nature in response to an external potential. But the details of the theory also indicate why Chandrasekhar's theory involving mainly dynamical friction never yielded a satisfactory solution to having "permanent" spiral structures. What was needed was a structure that permitted dynamical "self-driving" of the system, i.e., wave amplification as well as wave damping.

8.6 The interplay of gas and dust

One of the great joys of astrophysics is the fact that nature is capable of producing an enormous variety of phenomena and circumstances that first seem inexplicable, and then with more study and additional observations begin to make sense. Interstellar molecules is a case in point. Early observations in the optical region found lines of CH, CH+, and CN in the spectra of hot stars. Theoretical work indicated that only such molecular fragments could exist between stars because the ultraviolet starlight was energetic enough to swiftly break down heavier (mainly hydrocarbon) molecules into these by-products. This view was verified by the action of solar UV on comet molecules where the same fragments were seen. In the case of ions that one detected in comet atmospheres, it turned out that bombardment by solar wind protons produced those by ion exchange. In the interstellar situation cosmic rays did the same thing. It all fit. There was also dust on the same lines of sight as the diatomic molecules, and ions and light scattered was not only subject to extinction but was shown to be polarized as well. A connection between the dust and the molecules was not considered likely because of the low temperatures and low densities.

As more and heavier molecules were discovered by radio techniques (Burke and Graham-Smith, 2002; Herbst 1995; and Herbst and Dishoeck, 2009), a pattern emerged. All the discoveries were coming from regions called molecular clouds, and within those clouds was an immense amount of dust. Surrounding these clouds were huge halos of CO. The CO molecule is the most stable of the simple diatomic molecules and is always the last to dissociate. The mechanism was now clear. In the interior of the cloud the temperature was very low ($\sim$10 K) but kept above the 3 K cosmic background by cosmic ray heating. The gas within the dust was shielded from the UV by the dust, and so collisions with both gas and dust could build up molecules even at 10 K that the external UV could not break down. Of course even at 10 K, molecules would drift out of the cloud, and exposure to additional UV photons would break the molecules down leaving only dust, atoms, and CO behind.

As we have pointed out several times in this chapter, dust is the main factor in limiting an optical line of sight in or near the equatorial plane of the Milky Way. Similar situations exist for other spiral galaxies where gas and dust are mixed together. In globular clusters and elliptical galaxies dust is not common. Of course that means that photons of all varieties, including x-ray and UV, stream between the stars, making interstellar conditions harsher in dustless galaxies than in dusty ones. But as we have indicated, the optical extinction between the sun and the Milky Way center is 30 magnitudes, and that seems to have been sufficient to have shielded us from a flood of harsh radiation at the times when the galactic center was in a more active phase.

8.6.1 Examples of chemical calculations

Although the ideas behind the breakdown of complex molecules formed in dense dust clouds are fairly clear, the details behind the formation of complex, mainly organic, molecules are less evident. It turns out that the statistical mechanical principles needed to explain low temperature, nonequilibrium chemistry in dense dust clouds have already been briefly discussed in earlier chapters featuring the Boltzmann transport equation. The difference now is that since the collisions drive the chemistry we can no longer use "collisionless" versions of the theory. Even here we will not model a realistic chemical situation. That is too complicated, as it would require many thousands of reactions including ion–molecule, atom–molecule, and atom–atom reactions as well as dust interactions. Our models will be simple and serve as illustrations of what *Mathematica* has available for solving much larger models.

Conditions within interstellar molecule-forming regions are reviewed in detail in Herbst (1995) and Herbst and Dishoeck (2009). Representative values for some of the parameters can be found in Allen (2000). The constraints can be summarized as:

1. By far the most abundant molecule is H_2, but because it is homopolar it is not the most easily observed.
2. The average number density of H_2 is $10^4/cm^3$ with excursions as high as $10^5/cm^3$ in small regions near protostars.
3. Low average kinetic temperatures are $\sim$10 K, but can rise to 50 K in some regions.
4. Heavy molecules observed are mostly composed of H, C, N, and O, as might be expected from atomic abundances.

5. To build a complete model of molecule formation requires 4000+ reaction equations.
6. Basic chemistry is based on gas phase reactions, but many species require dust interactions at some point.

If temperatures and densities were higher and more uniform throughout dense dust clouds, then we could use the Law of Mass Action as demonstrated in the appendix, but cosmic ray heating is too random and the heat "conductivity" (dominated by IR radiation transport in most instances) is too low to even expect an equilibrium configuration in which mass action operates. But even so, molecule formation and maintenance is still collision driven, either by gas–gas collisions or gas–dust collisions.

The zero-order collision term is usually not sufficiently accurate in plasmas. Instead the binary collision terms that arise as part of the Chapman–Enskog formalism[18] need to be considered. McLennan (1989) starts with the Navier–Stokes equations and develops the complete first-order situation in detail, and mentions that some problems required the use of second-order approximations called the Burnett equations. The Burnett equations not only describe binary collisions, but also compensate for tertiary ones that often occur in moderately dense gases. We assume that a first-order treatment without tertiary events is all that is required, and follow in outline form the transport equations derived by Laurendeau (2010).

8.6.2 Molecular cloud conditions and kinetic theory

For setting up our chemical model integrations we use the known constraints summarized earlier. In determining each part of a reaction total rate needed to set up the differential equations we need to estimate the fraction of colliding particles that branch into each reaction mode. But even before that we need to know the time scale of the collisions, whether molecule–molecule, molecule–atom, atom–atom, etc. at 10 K. We must also set the velocity distribution so that the range of particle speeds can be estimated. Likewise the mean free path (from simple theory) for each species interaction can be estimated.

But there are problems in setting up even the conditions for H and H_2 because of conflicting abundance data in the open literature. For example, in Allen (2000) we find the hydrogen atomic density in molecular clouds is assumed to be twice that of H_2, and hence $n(H) = 4 \times 10^3 R/cm^3$, where R (in parsecs) is the average radius of the cloud. That must mean $n(H_2) = 2 \times 10^3 R/cm^3$. But even as that chapter's author complains about the scatter in the values, we find in the adjoining table the rule appears to contradict many of the other entries. In addition, there are no atomic species depletions known for within the molecular clouds, so we must assume the young star atmosphere abundances in the same compilations.

A similar challenge occurs in estimating the amount of dust surface area per cm^3. How do we estimate it? What is the mean free path of the gas with respect to the dust? Unfortunately we must engage in a bit of creative numerology to get estimates before being able to go further. Allen lists a gas to dust ratio of ~ 200. We interpret this as a mass ratio. So if the hydrogen atoms plus hydrogen molecules total some 6×10^3 protons/cm^3 in mass (for a cloud of 1 parsec radius) we have about 30 times a proton mass in "dust" over the same volume. That could be divided up in a number of ways, but Si is often mentioned as the main "dust" atom. Silicon is chosen because its next oxidation state can build solids that are no

---

[18] See McLennan (1989) and Laurendeau (2010).

ices. We will take the dust to be represented by one Si atom per cubic centimeter, tied up in a form that would not be counted as a gas such as CO, SiO, NO, etc.[19]

If we can estimate the mass of a typical interstellar dust particle, we can figure out over what volume of space the Si atoms that make it up have to be gathered from. Of course that means observationally Si and the other dust elements would be depleted over the same volume. Such depletion is observed (Allen, 2000, p. 535). If we assume the largest dust particles are about 1 micron in radius, as per interstellar meteoroids detected at Earth (Meisel, Janches, and Mathews, 2002), then we would have one such "large" particle within a spherical volume of 120 cubic meters with an absorbing area of 1.2×10^{-7} cm^2. According to Herbst (1995), the average dust particle is only 0.1 micron, so by our scheme there is only one average dust particle within a sphere 10 times smaller with an area 100 times smaller.

In standard kinetic theory, mutual collision frequency (in MKS) of gas and dust particles is given as

$$C_f = 10^6 \rho \frac{3}{4R_v^3} (R_m + R_d)^2 v_r, \tag{8.69}$$

where ρ is the density of the gas in cubic centimeters, R_v is the radius of the volume occupied by one dust particle, R_m is the molecular radius, R_d is the dust radius, and v_r is the relative velocity, usually taken to be the thermal velocity of the smaller object. Thus the relative velocity between gas and dust is

$$v_{gd} = \sqrt{\frac{8kT}{\pi m}}, \tag{8.70}$$

where T is the thermal temperature of the gas and m is the mass of a gas atom or molecule. The relative velocity between different gases is

$$v_{gg} = \sqrt{\frac{8kT}{\pi \left(\frac{m_1 m_2}{m_1 + m_2}\right)}}, \tag{8.71}$$

where m_1 and m_2 are the two atomic/molecular masses. Calculating the collision frequencies, we find a typical collision frequency between H$_2$ molecule and a 1 micron dust particle is about 14 days. The collision frequency between C$_2$ molecule and a dust particle is about 40 years. Conversely, the collision times between H and H$_2$ or C and C$_2$ is on the order of a thousandth of a second.

Gas phase reactions, in particular ion–molecule reactions, have been the main stay of interstellar molecule investigations (Herbst, 1995) for a decade or more, and these calculations show why. So then why bother with examining the much slower colliding dust reactions? As we will see, at low temperatures the dust is able to gather up a wide variety of molecular types before reacting them together. But the real value of the dust is to break the chemical bonds of the molecules on its surface, resulting in lower activation energies than possible with gas phase reactions only, and reassembling heavier ensembles by catalytic action. Such mechanisms seem to be required to produce the plethora of heavy organics observed (Herbst and Dishoeck, 2009).

---

[19] If one assumes the solar abundance, then Si accounts for only 1/10 of that density, so our figures may be a bit off. In fact using the solar carbon abundance, we get 1.1 atoms per cubic centimeter, so graphite is also a possibility. On the other hand, most of the carbon in dense dust clouds appears to be tied up in molecules, and the IR Si signature seen in solar system dust is also seen in interstellar clouds, so we will stick with Si as the main dust element.

8.6.3 Gas phase reactions in interstellar molecular clouds

Before examining dust reactions we must first study gas-phase mechanisms. As shown by Laurendeau (pp. 301–311) and discussed in Trusler and Wakeham (2003), the microscopic equivalents of the traditional transport coefficients (momentum = viscosity, energy = thermal conductivity, and mass = diffusion) can be derived from a simple collision theory (assuming hard spheres and no potentials) via the Boltzmann velocity distribution for binary collisions.

If a certain physical property has a gradient in one direction (say the z-direction) dp/dz, then the flux in the z-direction becomes for a pure gas,

$$f_p(z) = -\frac{1}{3}n\Lambda\bar{v}\frac{dp}{dz}, \tag{8.72}$$

where n is the number density, Λ the mean free path, and $\bar{v}$ the mean velocity. The dynamic viscosity becomes

$$\eta = \frac{1}{3}nm\Lambda\bar{v}, \tag{8.73}$$

where m is the mass of the particle. The thermal conductivity becomes

$$\lambda = \frac{1}{3}\left(\frac{nC_V}{N_A}\right)\Lambda\bar{v}, \tag{8.74}$$

where C_V is the specific heat and N_A is Avogadro's number, and the self-diffusion coefficient becomes

$$D = \frac{1}{3}\Lambda\bar{v}. \tag{8.75}$$

The Λ and $\bar{v}$ both can be traced back to kinetic theory and through the volume collision rate, which is what we need for our theory of chemical reactions. If the collisions are for two different molecules, then the collision rate becomes

$$\xi_{12} = \pi n_1 n_2 (\rho_1 + \rho_2)^2 \bar{v}_{12}, \tag{8.76}$$

where ρ_1 and ρ_2 are the approximate rigid sphere molecular radii in the same units as the volume of the n's. For a reduced mass μ_{12} between two molecules in the collision,

$$\bar{v}_{12} = \sqrt{\frac{8kT}{\pi\mu_{12}}}. \tag{8.77}$$

There are two mean free paths, one for each mass.

$$\Lambda_{1\to2} = \sqrt{\left(\frac{m_2}{m_1+m_2}\right)}\frac{1}{\pi n_2 (\rho_1 + \rho_2)^2}, \tag{8.78}$$

$$\Lambda_{2\to1} = \sqrt{\left(\frac{m_1}{m_1+m_2}\right)}\frac{1}{\pi n_1 (\rho_1 + \rho_2)^2}. \tag{8.79}$$

The $(\rho_1 + \rho_2)$ particle radii are for approximate rigid spheres (ars) assuming no molecular potentials. Laurendeau indicates that an effective value of this needs to be computed if there is a significant force during the collision. Without going into detail it turns out that corrected values for the viscosity conductivity, and diffusion can be calculated from the so-called collision integrals $\Omega(l, s)$ as functions of temperature. Thus

$$\eta_{eff} = \frac{32\Omega(2, 2)}{15\pi} \eta_{ars}, \tag{8.80}$$

$$\lambda_{eff} = \frac{64\Omega(2, 2)}{75\pi} \lambda_{ars}, \tag{8.81}$$

$$D_{eff} = \frac{16\Omega(1, 1)}{9\pi} D_{ars}. \tag{8.82}$$

We can make use of this correction procedure to correct the $(\rho_1 + \rho_2)$ appearing in the collision rate equation above; hence

$$\sigma_{sq} = \frac{15\pi}{32\Omega(2, 2)} (\rho_1 + \rho_2)^2, \tag{8.83}$$

where the ρ values are the approximate rigid sphere radii for the species.

Laurendeau provides tables (Appendix P) of the normalized $\Omega(2, 2)$ and $\Omega(1, 1)$ integrals based on the Lennard–Jones potential as a function of a reduced "temperature":

$$T^* = \frac{kT}{\epsilon}, \tag{8.84}$$

where ϵ is an energy parameter specific to a particular gas. For H_2, Laurendeau lists $\epsilon/k = 38$ K, so at 10 K $T^* = 0.263$. But such a value is not within the values of T^* given in the table. The prevailing temperatures in molecular clouds are too low (by nearly a factor of 3), so we will extrapolate the values to that temperature.[20]

The Lennard–Jones 6–12 potential is suggested by Laurendeau as the simplest realistic intermolecular potential to have been extensively tested for a variety of common monatomic, diatomic, and polyatomic gases.[21] With ϵ and σ as constants for a particular model, the L–J potential is

$$\phi = 4\epsilon \left(\left(\frac{\sigma}{r}\right)^{12} - \left(\frac{\sigma}{r}\right)^{6} \right), \tag{8.85}$$

From this, the unnormalized collision integral is defined by Laurendeau as

$$\Omega(l, s) = \sqrt{\frac{2\pi kT}{\mu}} \int_0^\infty \int_0^\infty e^{-\alpha^2} e^{2s+3} \left(1 - \cos^l \theta\right) b \, db \, d\alpha, \tag{8.86}$$

where

$$\alpha = \sqrt{\frac{\mu}{2kT}} v_f \sqrt{\frac{8kT}{\pi\mu}} = \frac{2v_f}{\sqrt{\pi}} \qquad b = \sigma_f \sigma_{LJ}, \tag{8.87}$$

---

[20] For a justification of the validity of such an extrapolation, see **8-11GasNDust**.

[21] Those familiar with diatomic models will understand why this potential is 6–12 while most textbooks quote the diatomic electrostatic force as 7–14. Of course to really do the problem correctly one should attempt the quantum mechanical scattering problem on a supercomputer, but this is way beyond our scope here.

with σ_f is the fraction of the L–J radial parameter for the particular molecule and v_f is the mean velocity at that temperature. For hydrogen molecules at 10 K, this can be calculated as

```
Ω1sH2[ss_, ll_] := Module[{}, Quiet[NIntegrate[Re[
```

$$\sqrt{\frac{2\,\pi\,k\,temp}{\mu}}\ e^{2\ ss+3}\ e^{-\left(\frac{2\,vf1}{\sqrt{\pi}}\right)^2}\ (1 - Cos[eq\theta\ /.\ \{x \to \sigma f1,\ y \to vf1\}]^{11})\ 2\ \sigma f1\ \sigma LJ],$$

```
    {σf1, 0.0001, 2.}, {vf1, 0.0001, 8.}]]];
num1 = Ω1sH2[1, 1]; num2 = Ω1sH2[2, 2];
```

But these need to be normalized by the integrals for rigid spheres. The only thing that needs replacing is the formula for the scattering angle. For hydrogen molecules as rigid spheres[22] the angle is given by

```
eqθRS = If[σf1 < 1 , π - 2 ArcSin[σf1], π];
```

thus

```
Ω1sH2RS[ss_, ll_] := Module[{}, Quiet[NIntegrate[Re[
```

$$\sqrt{\frac{2\,\pi\,k\,temp}{\mu}}\ e^{2\ ss+3}\ e^{-\left(\frac{2\,vf1}{\sqrt{\pi}}\right)^2}\ (1 - Cos[eq\theta RS]^{11})\ 2\ \sigma f1\ \sigma LJ],$$

```
    {σf1, 0.0001, 2.}, {vf1, 0.0001, 8.}]]];
denom1 = Ω1sH2RS[1, 1]; denom2 = Ω1sH2RS[2, 2];
norm1 = num1 / denom1; norm2 = num2 / denom2;
```

This gives normalized values of $\Omega(1, 1) = 0.28$ and $\Omega(2, 2) = 2.09$.

8.6.4 Calculating rate coefficients

To evaluate gas phase reactions at low temperatures (whether equilibrium reactions or not) the calculation or measurement of the reaction rates is essential. As we have shown, the L–J correction is about a factor of 2 over the hard sphere results even at 10 K. Laurendeau shows that the empirical Arrhenius law is a reasonable description of chemical kinetics with Maxwellian velocity distributions when the temperature is low and few internal molecular levels are excited during collisions. We have shown the corrected total collision rate with mean velocity converted to temperature is

$$\xi_{12} = 2n_1 n_2 \frac{15\pi}{32\Omega(2, 2)} (\rho_1 + \rho_2)^2 \sqrt{\frac{2\pi kT}{\mu}}. \tag{8.88}$$

The numerical correction factor is ~ 0.7, so that the hard sphere approximation is not bad and the respective radii as listed in most tables is probably adequate for our purposes.

[22] See Allen (2000), or Laurendeau (p. 324).

The type of reaction we are going to describe is the so-called bimolecular reaction where $A + B \rightarrow C + D$ where there is one molecule on each side (A and C for example) and on each side is an atom (B and D for example). Because H and H_2 are so abundant the reactions involving the two of them must be common, but we are not interested in that one. Much more interesting are reactions involving either C, N, or O atoms. The specific reaction rate for each type of collision (in particles per volume) is given at 10 K in usual chemical concentrations by

$$\text{rate}_{12} = 1.4 \, (\rho_1 + \rho_2)^2 \sqrt{\frac{2\pi kT}{\mu}} e^{\epsilon_{AB}/kT} [A][B].\tag{8.89}$$

The exponential term arises by summing over Maxwellians from some reaction threshold energy ϵ_{AB} to infinity. The exponential result is derived by Laurendeau (pp. 320–323) by considering what fraction of collisions have a certain relative velocity distribution along their line of closest approach. This approach assumes that we can look up the values of the ρ's and ϵ's somewhere and proceed. Certainly atomic and molecular radii are tabulated in a number of places, but the ϵ's have to be obtained from tables of molecular bond strength (usually energy/mol) and then converted to energy/reaction. The function before the exponential, derived from the classical case, is not usually correct in real functions because of quantum effects during collision. Called the pre-exponential factor, our kinetic theory derivation assumes that no energy is available from the molecular or atomic levels in the reaction. But this is rarely the case, even at low temperatures.

Notice that the reaction shown is for the forward reaction. There is an equivalent reaction going in the other direction, where $[C][D]$ and ϵ_{CD} are involved. If we denote the rates in each direction by the chemical notation

$$\text{rate}_f = k_f[A][B], \qquad \text{rate}_b = k_b[C][D],\tag{8.90}$$

then the equilibrium constant becomes simply

$$K_{eq} = \frac{[C][D]}{[A][B]} = \frac{k_f}{k_b}.\tag{8.91}$$

If there are any (quantum mechanical) differences between the two reaction thresholds and/or in the pre-exponential terms that drive the reaction in one way or the other, that determines the value of the equilibrium constants for the reaction. Such differences often involve the formation of chemical complexes that serve as unique intermediaries. In the interstellar case it is often the formation of complexes on the surface of dust particles that are involved in addition to the gas phase reactions. Calculations involving quantum collision theory are necessary to produce the correct nonequilibrium k_f and k_b expressions, but they are extremely complicated and so we do not pursue them here.

When it became evident that the neutral–neutral reactions (particularly with those involving abundant molecular hydrogen) as we described earlier would not be sufficient to explain the many complex organic molecules observed, attention turned to other mechanisms. When the rotation temperatures for the nonhomopolar diatomic species came out to be in excess of that of the microwave background (10 K–12 K rather than 3 K) it was realized that cosmic rays were providing the extra energy through ionization. This meant that a fraction of both atoms and molecules would be ionized. Another indication of ionization by cosmic rays was the fact that easily detectable x-rays have been detected coming from the same molecular clouds, no doubt because of the relatively high mass densities that are found within them. Thus not only is the kinetic temperature of the gas elevated, but the supply of ions is orders of magnitude higher than would be expected in a photon-shielded volume of low temperature.

Ion–molecule reactions behave differently from those of neutral–neutrals. As pointed out by Herbst (1995), such reactions are in fact much simpler. First of all the exothermic processes have little activation energy, and the behavior is determined by whether the neutral species is polar or not:

1. If the neutral is nonpolar (such as CH) then the charged induced dipole is the potential, which becomes isotropic at long range. This leads to the temperature independent Langevin rate coefficient

$$k_L = 2\pi e \sqrt{\frac{\alpha}{\mu}} \approx 10^{-9} \text{ cm}^3/\text{s}, \tag{8.92}$$

where α is the polarizability in cm^3, e is the electronic charge in ESU, and μ is the reduced mass in grams. The polarizability is related to the refractive index n through

$$\alpha = \frac{n-1}{2\pi N/V}, \tag{8.93}$$

where N is the number of particles, V the volume, and $n-1$ is specified at STP. Most $n-1$ tabulations[23] are of the form

$$n - 1 = A + \frac{AB}{\lambda^2}, \tag{8.94}$$

from which the parameters are obtained.
2. If the species has a permanent dipole moment[24] the theories indicate that the rate is temperature dependent to the $-1/2$ power, not the $1/2$ power. This enhances the k_L at low temperatures up to 10^{-7} at 10 K.
3. If H_2 or a similar homopolar molecule is involved, then quantum tunneling and complex formation complicates the calculations considerably. There is also an enhancement of the rates at low temperature.
4. Radiative association may also be involved in ion–neutral reactions through complex formation.
5. Neutralization of ionized products probably occurs through recombination with free electrons with or without dissociation.

Other details are described by Herbst, but we do not cover those here. The true situation is extremely complicated chemistry, much of which is uncertain. Progress in recent years has been considerable, but not yet complete, and is described in Herbst and Dishoeck (2009).

We can now set up a "simple" model of the creation of diatomic molecules (molecular ions) following a cosmic ray event within a molecular cloud. We will need to calculate the ion–molecule rates beginning with the Langevin formula:

kL = 2 π 4.8 × 10⁻¹⁰ √(α1 / μ1) ;

---

[23] Allen (2000), for example.
[24] See, for example, Nelson, Lide, and Maryott (1967) for a list of such substances.

Our initial list of elements and ions will be those of H, C, N, and O. We can then calculate the rates for ion–molecule collisions leading to triatomic species. We begin by creating all the atomic ionic and molecular masses along with their reduced masses for each stage:

```
mass1 = {1.0079, 12.001, 14.007, 15.999}; mass2 = {1.0079, 12.001, 14.007, 15.999};

mass3 = Table[0., {ii, 1, 16}];

Do[mass3[[ii + 4 (jj - 1)]] = mass1[[ii]] + mass2[[jj]];, {ii, 1, 4}, {jj, 1, 4}]
mass4 = Table[0., {kk, 1, 4 × 4 × 4}];

Do[mass4[[ii + 16 (jj - 1)]] = mass3[[ii]] + mass2[[jj]];, {ii, 1, 16}, {jj, 1, 4}]
mu3 = Table[0., {ii, 1, 16}];
```

$$Do\left[mu3[[ii + 4 (jj - 1)]] = \frac{(mass1[[ii]]\ mass2[[jj]])}{mass3[[ii + 4 (jj - 1)]]};, \{ii, 1, 4\}, \{jj, 1, 4\}\right]$$

```
mu4 = Table[0., {ii, 1, 64}];
```

$$Do\left[mu4[[ii + 16 (jj - 1)]] = \frac{(mass3[[ii]]\ mass2[[jj]])}{mass4[[ii + 4 (jj - 1)]]};, \{ii, 1, 16\}, \{jj, 1, 4\}\right]$$

Now we tackle the polarizability of the neutral atom or molecule. What is desired is an empirical formula that will allow the prediction of polarizability for atoms or molecules for which there are no tabular data. Allen (p. 100) gives values for the atomic polarizability for a number of cosmically abundant atoms including our most abundant four.

```
alpha1 = {6.76 × 10⁻²⁵, 17.6 × 10⁻²⁵, 11.0 × 10⁻²⁵, 8.03 × 10⁻²⁵}; id1 = {"H", "C", "N", "O"};
```

To expand the number of α tabular values we must develop a suitable correlation equation. This requires a bit of trial and error[25] but one finds

$$\alpha(m, q) = 4.584 \times 10^{-25} m^{0.386} + e^{55.1q} (q - 0.53), \tag{8.95}$$

where m is the mass in AMU and q the charge in electron units. For light elements and molecules it is reasonably accurate to set $q = m/2$ and be spared the agony of devising an algorithm that calculates the molecular electronic structure. We will define the resulting function as $pol\alpha$.

Next we need to create identifiers for the molecules created at each step starting with the creation of diatomic molecules:

```
idAi = Table["", {ii, 1, 16}]; idA = Table["", {ii, 1, 16}];

Do[idA[[ii + 4 (jj - 1)]] = StringJoin["(", id1[[ii]], id1[[jj]], " ⁺⁻)"];,
  {ii, 1, 4}, {jj, 1, 4}]

Do[idAi[[ii + 4 (jj - 1)]] = StringJoin["(",
    StringJoin[id1[[ii]], "←"], id1[[jj]], ")"];, {ii, 1, 4}, {jj, 1, 4}]
```

[25] See **8-11GasNDust**.

Here `idA` are the products and `idAi` are the reactions. For the creation of triatomic molecules:

```
idB = Table["", {ii, 1, 64}]; idBi = Table["", {ii, 1, 64}];

Do[idB[[ii + 16 (jj - 1)]] = StringJoin["(", idA[[ii]], id1[[jj]], " +-"];,
  {ii, 1, 16}, {jj, 1, 4}]

Do[idBi[[ii + 16 (jj - 1)]] = StringJoin["(", idA[[ii]], "←", id1[[jj]], " +)"];,
  {ii, 1, 16}, {jj, 1, 4}]
```

We can then calculate the creation rates of the diatomic and triatomic species:

```
kL1 = Table[0., {ii, 1, 16}]; kL2 = Table[0., {ii, 1, 64}];
```
$$Do\left[kL1[[ii]] = 2\,\pi\,4.8 \times 10^{-10} \sqrt{pol\alpha[mass3[[ii]]] / mu3[[ii]]}, \{ii, 1, 16\}\right];$$
$$Do\left[kL2[[ii]] = 2\,\pi\,4.8 \times 10^{-10} \sqrt{pol\alpha[mass4[[ii]]] / mu4[[ii]]}, \{ii, 1, 64\}\right];$$

We are now ready to calculate the gains and losses moving to a steady state. The calculations start with the initial elements/molecules as ions/neutrals as listed in table (8.1) as taken from Allen (p. 29). The ratio of ions to neutrals is assumed to be 10^{-6} for simplicity, following Herbst (1995).

```
diatomic = {h2'[t] == kL1[[1]] hp[t] h[t], c2'[t] == kL1[[6]] cp[t] c[t], ...
    h[0] == 100 000, cp[0] == 36 × 10⁻⁶, c[0] == 36, ...};
```

$h[0] == 100\,000$, $cp[0] == 36 \times 10^{-6}$, $c[0] == 36$

```
vars = {hp[t], cp[t], np[t], op[t], h[t], c[t], n[t], o[t],
    h2[t], c2[t], n2[t], o2[t], ch[t], nh[t], oh[t], cn[t], co[t], no[t]};
```

We then solve the set of equations via `NDSolve[]` choosing a time of 10^{15} seconds as roughly equal to the lifetime of a molecular cloud.

$$diasol = NDSolve\left[diatomic, vars, \{t, 0, 10^{15}\}\right]$$

The results are seen in table (8.1). Although we make no claim about the quantitative accuracy of these results, the relative ranking of the molecular abundances shown are in accord with expectation and even observations outside molecular clouds. That is, H_2 will build up the fastest, with CH next followed by NH and OH. Even this simplistic approach yields reasonable results.

8.6.5 Surface interactions with dust

Looking out into space is also looking back into time. Galaxy observations at large distances reveal not how they are now, but how they were when the light left for earth. Thus it was surprising how fast dust formed after the first wave of star formation and subsequent surge of supernovae provided heavy elements. As a result in the giant molecular clouds there is always an association of gas and dust. Recent research in surface physics has shown how rapidly solid surfaces interact with gases to adsorb them. We do mean adsorb rather than absorb because the surface potential holds the molecules to the surface rather than letting them go deeper. While on the surface, molecules may roam around and interact with each other and that tends to make a thin layer first. Once the surface density of molecules is high enough, then surface chemistry takes place much faster than would happen in the gas phase alone. This is why many catalysts used in gaseous environments are solids (Resch and Koel, 2003).

Table 8.1 Calculated Diatomic Reactions

Species	Initial	Final
H^+	0.1	0.0962489
C^+	0.000036	0.0000347657
N^+	0.000011	0.0000106156
O^+	9×10^{-6}	8.67961×10^{-6}
H	100000	100000
C	36	36
N	11	11
O	9	9
H_2	0	0.00374914
C_2	0	1.95947×10^{-10}
N_2	0	1.74396×10^{-11}
O_2	0	1.12034×10^{-11}
CH	0	2.46593×10^{-6}
NH	0	7.68191×10^{-7}
OH	0	6.40323×10^{-7}
CN	0	1.17156×10^{-10}
CO	0	9.43775×10^{-11}
NO	0	2.79987×10^{-11}

The processes involved in building up large organic molecules on interstellar dust grains is outlined in Rensch and Koel. The important surface processes are

1. Adsorption of incoming molecules
2. Surface diffusion
3. Surface chemical reactions (only those not resulting in solid bonding to the surface)
4. Desorption

Interaction of gas molecules with the surface is actually not much different from the interaction of molecule to molecule collisions considered above. The Lennard–Jones potential is also applied to yield[26]

$$E_{LJ} = 4\pi\epsilon N_s \sigma^3 \left(\frac{1}{45} \left(\frac{\sigma}{z}\right)^9 - \left(\frac{\sigma}{z}\right)^3 \right), \tag{8.96}$$

[26] Rensch and Koel, p. 769.

where ϵ is the L–J energy parameter as calculated above, σ is the hard sphere diameter of the atoms in the solid, and N_s is the number density of the atoms in the solid. This expression can be used to estimate the holding (or desorption) potential due to van der Waals forces. Chemical bonds will result in a much higher activation energy than this formula gives, but that is very dependent on the composition of the dust, which cannot be taken into account very well. The range of chemical bond energies for various combinations of CNO and hydrogen are in the range of a few eV, while the van der Waals potentials are about $1/10$ of those.

The standard classical treatment of surface adsorption/desorption processes does not function well at 10 K. For example, classically one of the first indications that something is wrong is the use of the classical adsorption/desorption potential to determine the mean residence time of a molecule. This concept is needed to determine how much of the incoming flux "sticks" to the surface. The mean residence time for a molecule to stay on the surface when the gas temperature is T_g is given by the inverse of the classical classical energy level occupancy probability

$$\tau = \tau_o e^{(11604.5\Delta\epsilon)/T_g}, \tag{8.97}$$

where $\Delta\epsilon$ is the desorption potential in eV, and τ_o is a constant $\sim 10^{-13}$. If this is applied to $T_g = 10$ K, one obtains a residence time $\sim 10^{349}$, which is effectively infinite.

Surface physics at 10 K requires quantum theory because at that temperature atoms no longer follow Maxwell–Boltzmann statistics, but rather Fermi–Dirac, Bose–Einstein, or both. This is the same quantum degeneracy we encountered in our discussion of electrons in white dwarfs, but it is now more difficult. To solve the interstellar heavy molecule situation rigorously we would need to consider quantum degeneracy in mixtures of "gaseous" particles in the presence of a quantum degenerate solid.

8.7 Galactic mergers

There is modern evidence that galaxies, particularly the giant elliptical ones, have grown at the expense of their previous neighbors, as we can often see multiple nuclei and in some cases the actual swallowing process in progress. In the early universe such processes were likely proceeding at a much higher rate than observed today. Mathematically this process can be modeled by so-called predator–prey problems. Such problems are usually studied in biology, often by solving what are known as integral equations. Integral equations are usually formulated as either Volterra or Fredholm type (Arfken and Weber, 2001; Hassani, 1999; Davis, 1966; Margenau and Murphy, 1956). However, *Mathematica* has no explicit formal structure for analytically solving such equations. *Mathematica* does not even list integral equations in their documentation.[27] But as is well known (Davis, 1966), integral equations can be converted into equivalent differential equations, many with invariants of some type. The user is expected to recognized and perform the transformation themselves. Then NDSolve can be used. An example of the solution of a set of differential equations resulting from an integral equation is in the NDSolve suite called Lotka Volterra equations. If solving integral equations in Mathematica is of interest, then that is a good starting point.

---

[27] But that does not mean that *Mathematica* can't deal with some types of integral equations in some of its regular functions. See the discussion in **9-5galaxy2D** for details.

Exercises

8.1 The Large Magellanic cloud is roughly 48 kpc from the center of the Milky Way, and an orbital speed of about 220 km/s. From these facts estimate the mass of the Milky Way galaxy. How does your estimate compare with the mass as calculated in the text?

8.2 In **8-1MW21cm** we concentrated on obtaining the density of matter beyond the sun's orbit. With the exception of the central black hole, the nature of the volume of the Milky Way with $r < 4$ kiloparsecs is largely ignored for both the 21 cm observations and the luminous matter. Try to fit the internal-most radial velocity data (even if piecewise) assuming 0 at the galactic center, and compare your result with the "magic" radial velocity distance relationship found for the outer regions.

8.3 In **8-1MW21cm** we derived the Milky Way luminous mass versus distance by integration, and then plotted the results in discrete points.

1. Fit the points with a least-squares polynomial in r and extrapolate back to $r = 0$.
2. Compare with the least-squares fit of the 21 cm derived mass for the central regions.
3. Comment on these results compared with the black hole mass derived in **8-10GCStars**.

8.4 An alternative model to dark matter is known as Modified Newtonian Dynamics (MOND).[28] This model proposes that the acceleration due to gravity is given by

$$F_g = m \frac{a^2}{a_o},$$

where a is the gravitational acceleration and a_o is a small constant.

1. From the observational data of the text, determine the value of a_o.
2. A more general MOND model assumes the gravitational force to be

$$F_g = \mu \left(\frac{a}{a_o} \right) ma,$$

where μ is a function such that $\mu \to 1$ for large a. Approximate μ as a polynomial series to third order and fit the data in **8-1MW21cm** to determine the coefficients.

8.5 The Hénon–Heiles potential can be expressed as

$$V(x, y) = \frac{1}{2} \left(x^2 + y^2 \right) + \left(x^2 y - \frac{1}{3} y^3 \right).$$

Create a plot (surface, contour, etc.) of this function. What is the condition on the (dimensionless, normalized) energy for the motion to be bounded?

8.6 Study the behavior of the Hénon–Heiles problem

1. Using the SPRK method by autocorrelation techniques for the two coordinates separately
2. Using the Hamiltonian and Lyapunov methods (**6-9Lyapunov**)

8.7 Given the mass of the central black hole of the Milky Way to be $4.3 \times 10^6 M_\odot$, calculate the Schwarzschild radius of the black hole.

---

[28] This model is generally viewed to be invalidated by observations of the Bullet Cluster. See *Wikipedia* (2012) http://en.wikipedia.org/wiki/Bullet_Cluster.

8.8 The Roche limit for an orbiting mass can be approximated by

$$d = 2.44R \left(\frac{\rho_M}{\rho_m} \right)^{1/3},$$

where R is the radius of the central mass, ρ_M is the average density of the central mass, and ρ_m is the average density of the orbiting mass. Estimate the Roche limit of the star S2. You may assume the "radius" of the central black hole is its Schwarzschild radius, and S2 is a $15M_\odot$ main sequence (polytropic) star. How does this compare to the star's distance of closest approach?

8.9 Using the notebook 11**MassAction** (Appendix/thermonotebooks), study one of the interstellar hydride molecules (other than CH) to obtain an estimate of its abundance relative to H_2 under equilibrium conditions. Compare your result with the ionic calculations.

8.10 Repeat the Trapezium N-body analysis from Chapters 6 and 8 using a different x-distance file set (found in the Chapter 6 files).

8.11 Repeat the notebook 8-3 to 8-7 runs using the default NDSolve[] and compare with the "Projection" method NDSolve[].

8.12 Using the methods of the 8-4 and 8-5 notebooks do a more complete chaos/stability analysis of the Hénon–Heiles results found in Section 8.3.

9 Cosmic structures

In the first part of the twentieth century, the study of the cosmos (hence cosmology), meaning the study of the universe, was mainly theory. Much of the observational material was beyond the grasp of the telescopes of the time, photography was the only detection medium, and spectroscopy was very primitive. Astrophysics had to thrive by observational studies of the sun, the major planets, and bright stars. Nebulae and clusters had been cataloged by Messier, William Herschel, Caroline Herschel, and John Herschel by the thousands, but few astronomers of the time ever thought they would have the technology to truly study such things in detail.

Lack of observational data has historically been the achilles heel of cosmology. As recently as 1990 Kolb and Turner complained that astrophysics in general and cosmology in particular needed more observations. The present situation is in marked contrast to this view, such that there are now so much data publicly available that it strains our computational ability to analyze it. So many people are currently involved with data reduction and analysis that in the last decadal survey there was a call for the formation of a new area of astronomy and astrophysics called *astroinformatics* (Borne, 2009).

We start this chapter with several topics that are active parts of this frontier subfield of astronomy and astrophysics, as much of the data in the large databases are galactic and extragalactic. As we shall see, our examination will be limited not by the available data, but by our ability to analyze these data on a personal computer. We begin with an examination of the Hubble law, including evidence of universal acceleration. We then examine galaxies, including radio galaxies and quasars. Finally we look at cosmic structure and evolution.

9.1 The Hubble law

The systematic redshift of galaxies was first discovered by V. M. Slipher in 1912, and later investigated by Edwin Hubble. Hubble gets more credit for his work than Slipher because Hubble was able to get distance indications through his work on Cepheid variables, a technique in widespread use even today. The spectral redshifts, combined with the Cepheid distances, led to the ubiquitous graph in every elementary astronomy book, known as the Hubble Law. On the theoretical side, in 1922 A. Friedman used the equations of general relativity to show that expansion was one of the possible motions of "the universe" and cited the observations as indicating the GR approach was valid. In a quirk of fate, this theoretical result was missed by Albert Einstein in what folklore says he called "his greatest mistake" (Feynman, 1995, p. 149). What Einstein did was to search for the "static" solutions that he (and most everyone else) thought were correct. He realized a "cosmological constant" Λ was needed to keep the universe static. After the expansion was discovered, Einstein later recanted the conclusions of that

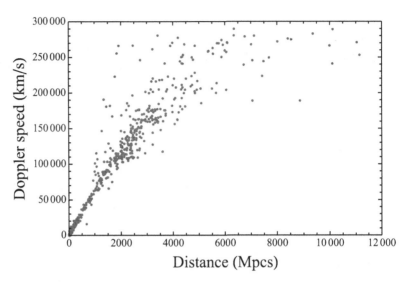

Fig. 9.1 Speed vs. distance for observed galaxies.

paper. His argument and those of others led to the opinion that $\Lambda = 0$ even as late as 1962–63.[1] A nonzero value was almost never discussed in polite company. But nature had the last laugh. By the dawn of the twenty-first century, not only had the expansion failed to go away, but $\Lambda \neq 0$ after all.

9.1.1 Galactocentric velocity versus distance

Today's astronomy and astrophysics students are so steeped in the various aspects of the structure of the universe that the Hubble relationship is often taken for granted. But it is the most fundamental aspect of the universe at large, and because we have the computational power to investigate statistical questions in some depth we will explore them here.

To explore the fit of observational data we have merged two very important lists of galaxy properties into a single edited list,[2] known as NED-1D and NED-4D. Because it is the distance determinations to these objects that is the most uncertain part of any study, we will use the estimated distance variances to weight our least-squares solutions, with $w = 1/\text{variance(mpc)}$. In a number of cases no such variances are given, so we simply use the average variance for each original list, which is 0.2 megaparsecs for the "nearby" (NED-1D) list and 0.4 megaparsecs for the "far" (NED-4D) list. A graph of the combined measurements can be seen in figure (9.1).

As figure (9.1) shows, the galactic speeds saturate at the speed of light, exactly as expected for relativistic motion. It is therefore useful to convert the speeds back to a measure of their redshift before attempting to fit the data. In Chapter 1 we derived an expression for the relativistic Doppler effect for a light emitting object,

$$v_{\text{obs}} = \frac{\sqrt{c - v}}{\sqrt{c + v}} v_e, \qquad (9.1)$$

---

[1] The dates Feynman gave the lectures that were to become his book on gravitation.
[2] NASA/IPAC Extragalactic Database under Level 5 (Madore, 2012).

where ν_{obs} is the observed frequency and ν_e the frequency emitted at the source. For optical astronomy, wavelength is preferred; thus

$$\lambda_{\text{obs}} = \frac{\sqrt{c + v}}{\sqrt{c - v}} \lambda_{\text{rest}}, \tag{9.2}$$

where λ_{rest} is the wavelength emitted at 0 relative velocity. This redshift is typically expressed in terms of a value z, which is the ratio of the wavelength difference to the rest wavelength,

$$z = \frac{\lambda_{\text{obs}} - \lambda_{\text{rest}}}{\lambda_{\text{rest}}} = \frac{\sqrt{c + v}}{\sqrt{c - v}} - 1. \tag{9.3}$$

Unlike observed speed, the z redshift has (in principle) no upper limit, which makes it more suitable for data fitting.

9.1.2 Fitting the NED data

The scatter of the galactic measurements is considerable. For this reason we compare different statistical hypotheses to see which provides the best fit of the data. We will consider four model fits:

1. Quadratic fit with zero constant term (lmw)
2. Quadratic fit with constant term (lmw1)
3. Unweighted quadratic fit without a constant term (lmw2)
4. Variance weighted quadratic fit without a constant term (lmw3)

We begin by importing the galactic data. The imported data has a header row, so we must remove it. The columns of the file are galaxy name, galaxy type, error of the distance D, D in megaparsecs, galactic longitude, galactic latitude, redshift in km/s, and the weight of the distance determination. From this we can extract a data list and a weight list.

```
galaxylist = Import["GalaxyList.csv"];

vgc = Table[galaxylist[[ii, 7]] + 220 N[
        Cos[(galaxylist[[ii, 5]] - 90) °] Cos[galaxylist[[ii, 6]] °]], {ii, 2, numline}];

galaxylistR = Table[{galaxycopy[[ii, 1]], galaxycopy[[ii, 2]],
        galaxycopy[[ii, 8]], galaxycopy[[ii, 4]], galaxycopy[[ii, 5]],
        galaxycopy[[ii, 6]], vgc[[ii - 1]]}, {ii, 2, numline}];

datalist = Table[{galaxylistR[[ii, 4]], vgc[[ii]]}, {ii, 1, numline - 1}];

wtlist = Table[galaxylistR[[ii, 3]], {ii, 1, numline - 1}];
```

Because the speeds are in km/s, we must convert them to their equivalent z values:

```
zee = FullSimplify[ λobs - λrest
                    ───────────── ]
                        λrest

zeeobs[v1_] := zee /. {c → c2, v → v1};

datalistA = Table[{galaxylistR[[ii, 4]], zeeobs[vgc[[ii]]]}, {ii, 1, numline - 1}];
```

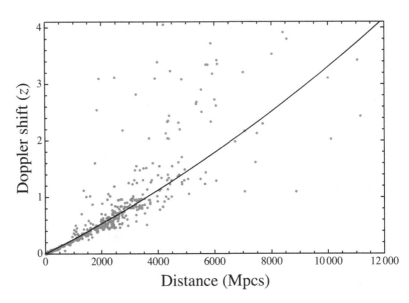

Fitting the redshift data.

The data fits[3] can then be made:

```
lmw3 = LinearModelFit[datalistA, {dist, dist²},
   dist, Weights -> wtlist², IncludeConstantBasis → False]
```

The results can be seen in figure (9.2). If we look at the correlation of each function we find the last function produces by far the best fit to the data, and is indicated by the black line.

It should be noted that we have used a somewhat conservative weighting scheme by taking the inverse of the standard distance errors. Sometimes the weights are taken to be quadratic and hence true variance weighting. Variance weighting enters naturally in least-squares fitting because the covariance matrix is used to find the coefficients and the errors. Thus we should examine its residuals as a function of distance to see if it contains systematic effects that might influence the values of the parameters.

```
fitres3 = lmw3["FitResiduals"];

resid3 = Table[{galaxylistR[[ii, 4]], fitres3[[ii]]}, {ii, 1, numline - 1}];

resZ = NonlinearModelFit[datalistD, a - b x, {a, b}, x]

ccoefZ = √(resZ["RSquared"])
```

The residuals of our fit as seen in figure (9.3) imply a bit of a linear trend; however, the linear correlation is only 0.58, which indicates it is not significant in the context of the available sample.

---

[3] In 9-1HubbleLaw several other fits are tried, but their correlative coefficients are less than 0.9.

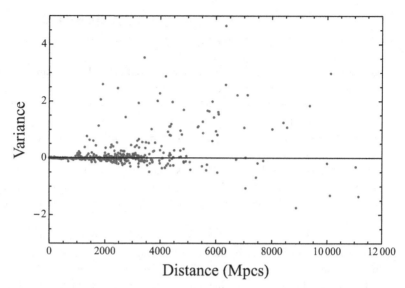

Fig. 9.3   Variance of our Hubble model fit.

Our best fit function can then be expressed as

$$z(D) = 0.00249D + 8.194 \times 10^{-9}D^2. \tag{9.4}$$

In interpreting this result, one should remember that the distance axis is really a "time into the past" axis, such that $t_{past} = D/c$ is the light travel time. This is why we can interpret the Hubble parameter (the slope of the redshift vs distance graph) as a "speed." This also means that the quadratic term in the distance is in reality an acceleration term, where the coefficient divided by 2 corresponds to a "Newtonian" value of acceleration.

When the NED (2009) data are fitted with a variance weighting, the value for the Hubble parameter is in agreement with the accepted values now appearing in textbooks and papers. This same data set produces an acceleration term that is accurate within 10%. We must consider this to be a confirmation of an expanding, accelerating universe, but the results do not offer a clue about what the nature of the accelerating force really is.

One concern with these data is the rather large values of z that occur with distances between 2000 and 5000 megaparsecs. At that distance one would expect $z \sim 1$, not 3–5 times that. In the final part of **9-1HubbleLaw**, we isolate the outliers and repeat the regression of residuals. A stronger anti-correlation with distance is obtained than when using the entire sample and thus strengthens the reality of the nonlinear term in the main Hubble flow considerably. Other velocity perturbations to the Hubble flow are discussed in Ruben and Coyne (1988). Although central black holes in some of the objects might provide a mechanism for such redshift anomalies, there is a known limit to how much internal redshift can be provided. Having established the main properties of the Hubble flow, we next move to Quasars, Radio Galaxies, and AGN as they appear to represent objects where such such redshift perturbations seem to be indicated.

9.2   Quasars and radio galaxies

Nowhere in astrophysics is the synergy between optical and radio astronomy more evident than in the relationship between radio galaxies and quasars. The radio galaxy side of the story is easy to trace. One of the strongest sources in the sky, Cygnus A was on most of the early radio maps. It can be detected easily with a small radio telescope not only because it is radio bright, but also because it is nearly a minute of arc across. Eventually the twin lobe radio source was found (in 1951) to be related to an 18th magnitude optical object whose $z = \Delta\lambda/\lambda = 0.06$ indicated it was approximately 300 megaparsecs away. At that distance, 1 minute of arc is 900 000 parsecs across. The optical Milky Way is roughly 30 000 parsecs across, so the radio lobes are 30× larger than our own galaxy. The optical counterpart of Cygnus A is a tiny thing compared to the Milky Way, and it is connected with one lobe by a thin filamentary "jet."[4]

Then in 1960, the source 3C295 was found to be associated with a similar optical source, but at $z = 0.45$. This was followed by the southern "nearby" galaxy NGC 5128, with its characteristic double lobe radio source 4 degrees across. Radio galaxies have been discovered beyond $z = 1$ and have a variety of lobe shapes, but always with the radio source resulting from synchrotron radiation and powered by the small central object. The cause of these radio galaxies was fairly mysterious at the time. But then general relativity theorists realized that thermonuclear blasts were not capable of producing that kind of energy, so it was suggested the cause was some sort of gravitational interaction with a black hole.

In 1963 a lunar occultation of 3C273 showed radio emission from a very small source that had a nearly stellar optical counterpart. This "star," which is 13th magnitude and visible in amateur optical telescopes of the time, was dubbed a quasi-stellar object (QSO) or quasar for short. Soon after, Maarten Schmidt published a paper claiming a redshift of 0.158 based on a single spectrum line. The spectrum line was OII at 3729 angstroms, and this line was one of the brightest emission lines in nearby "normal" galaxies, but Schmidt was really risking all with his paper. Of course it turned out he was correct. After 3C273, there followed 3C48 ($z = 0.37$), 3C147 ($z = 0.57$), and 3C9 ($z = 2.012$). The era of the quasar was opened.

9.2.1   Quasar/AGN optical luminosity

Research into quasars and their relationship to radio galaxies has continued for the past nearly half century to be a frontier subject with many discoveries and connections made. It turns out that quasars are more closely related to galaxies that have active nuclei (AGNs) as evidenced by bright emissions over a wide spectral range, x-rays to radio, coming from a relatively small part of the galactic nucleus. Although there was an apparent relationship in the early days between quasars and radio objects as illustrated earlier, there are now about twice as many radio-quiet quasars than radio-loud ones.

The list of radio sources we use here is the catalog published by Véron–Cetty and Véron (2010) and available through the VizieR database server. It is not known upon what measurements the absolute magnitudes are based, but it is assumed that corrections for intergalactic and galactic absorption have been included.[5]

---

[4] Burke and Graham-Smith (2002).
[5] See **9-2quasars** for the details of the imported data table.

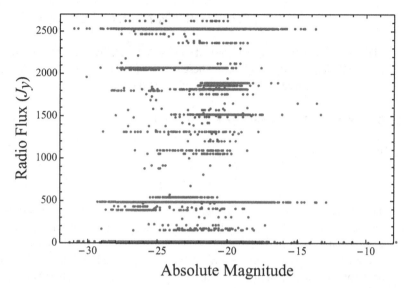

Fig. 9.4   Correlation between radio flux and absolute magnitude.

In figure (9.4) we have plotted the radio flux versus the absolute magnitude. This shows a very surprising lack of correlation between them. It suggests that the optical luminosities of quasars and AGNs are strongly decoupled from the radio luminosity. The reason for this is not clear because the power source for both is likely to be the same.

In figure (9.5) we compare the optical luminosities of radio-loud and radio-quiet objects. Although there are some subtle differences, the two sources are quite similar. Because the combined data has almost 170,000 points, it is worth comparing the statistical fits of loud versus quiet radio sources.

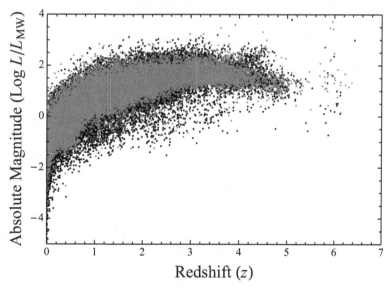

Fig. 9.5   Radio loud (gray) and radio quiet (black) absolute magnitudes compared.

9.2.2 Quantitative results on Quasars and AGN

The observational data can be found in two lists, `rQlist2B` for the radio-quiet observations and `rLlist2C` for the radio-loud ones. These data lists were used for figure (9.5), and give luminosity on a $\log_{10}$ scale compared to the luminosity of the Milky Way:

`qandAGN = Join[rQlist2B, rLlist2C];`

We then can fit the combined observations to a function of redshift z, in particular

$$L(z) = A + B\sqrt{z} + Cz + Dz^{3/2}. \tag{9.5}$$

This function is determined through a bit of trial and error, but produced the highest correlation coefficients to the data. The only physical justification for our chosen function is that $z \sim v/c$ is approximately proportional to the time into the past through the Hubble parameter, that is, $z \sim Ht$. The chosen variables gave a higher correlation than a regular power law, though it should be noted that there are "ripples" that appear in the data that are not reproduced or analyzed in these first-order correlations:

`lmfA = LinearModelFit`$\left[\texttt{qandAGN, }\left\{\sqrt{\texttt{zz}}\texttt{ , zz, zz}^{3/2}\right\}\texttt{, zz}\right]$

`eqfA = Normal[lmfA]`

`coeffA = lmfA["BestFitParameters"]; coefeA = lmfA["ParameterErrors"];`

$\sqrt{\texttt{lmfA["RSquared"]}}$

This gives values of $A = -1.9521 \pm 0.0087$, $B = 3.4266 \pm 0.0290$, $C = -0.8145 \pm 0.0284$, and $D = -0.0015 \pm 0.0085$, with a correlation of 0.873. The fits for radio-loud and radio-quiet AGNs vary slightly, as seen in figure (9.6).

At this point in a usual least-squares study of a data set, one normally examines the residuals for systematic effects that are not included in the original analysis. However, in this case it is extremely difficult to carry out an analysis in the usual way because the numbers are base 10 logarithms and not linear. About all we can say is that although the average quasar/AGN luminosity decreases with time (since the big bang) after reaching its maximum optical output as one would expect from a luminous object, the subsequent evolution gets less deterministic at the same time.

Because the Hubble parameter relates redshift z to time past, we have examined the equivalent of the temporal statistics of quasar and AGN luminosities. There are at least three timescales operating to produce the observed trends. There is one with an exponential increase proportional to the square root of time t, with two representing decays in t and $t^{3/2}$. Radio-loud objects show two roots where the object luminosity equals that of the Milky Way.[6] This is interpreted as an evolutionary effect where an object comes into "visibility" through an increasing luminosity in the period when $z > 6$. Radio-quiet objects come into "visibility" already at an elevated luminosity that increases with increasing z, and then decays from there as one goes to the present.

---

[6] See **9-2quasars** for the calculation of those roots.

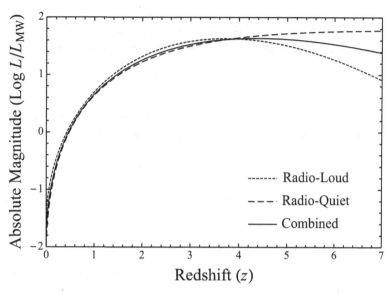

Fig. 9.6 Data fits for active galactic nuclei.

The weak radio-loud sources peak at $z \sim 6$, but at amplitudes less than the stronger radio-loud sources, which peak much later at $z \sim 3.5$. At their peaks, the optical energy represents a maximum of some 100–200 times the total estimated luminosity of the Milky Way, all coming from the nuclear regions that are known to be extremely small compared with a whole galaxy. Certainly such a luminosity cannot be from nucleosynthesis, as it would take a total conversion of several solar masses per second to fuel the radiation process. Thus some sort of spinning, collapsed object or set of collapsed objects seem to be required. But then the source of luminosity would have to be very tiny. We next consider evidence on that point.

9.2.3 Temporal fluctuations in the brightness of quasars

In the decade after the discovery of the first quasars, the claimed variability of quasars as gleaned from archival photographs was used as evidence that quasars and AGNs were basically small objects compared with normal galactic dimensions. But as pointed out by Terrell (1977), the cosmological distances at which the quasars are presumed to be (based on their redshifts) makes reconciliation of the observations and theory difficult. We will not get into this controversy theoretically, but show that some modern data on brightness fluctuations in the optical range certainly support limiting sizes on the order of a parsec or so.

Terrell demonstrates that the exact relativistic limit for an arbitrary dependence of surface brightness for a fixed sized object is

$$\left| \frac{dL}{dt} \right| \le \frac{2cL_{max}}{R(1+z)}, \tag{9.6}$$

where R is the radius of the spherical object upon which the fluctuations in luminosity L occur. Terrell's concerns about the applicability of the above formula arose because synchrotron radiation was being considered as providing L_{max}, and as such may show complications due to the so-called so-called superluminal effects seen in a number of radio sources. If the variations in luminosity are due to overlapping shot noise pulses then the variance of the luminosity is related to the luminosity through

$$\frac{\sigma(L)}{L_{max}} = \frac{1}{\sqrt{n_t t_e}}, \tag{9.7}$$

where n_t is the pulse frequency in the same units as c, and t_e is the effective pulse length.

Terrell quotes that for a relativistically expanding source

$$t_e = 5R\frac{1 + v_r/c}{3(v_r/c)(3 - v_r/c)c}, \tag{9.8}$$

where v_r is the expansion speed. In the limit of $v_r \to c$, $t_e = 5R/3c$. But whether the surface is static or expanding, t_e has the limit

$$t_e \leq \frac{\kappa_1}{c}R(1 + z), \tag{9.9}$$

where κ_1 is a number between 1 and 5/3. To use the formula for σ/L, we substitute the limit as an equality, thus

$$\frac{\sigma(L)}{L_{max}} = \frac{1}{\sqrt{n_t\kappa_1 R(1 + z)/c}}. \tag{9.10}$$

Observationally the left hand side of the equation is not provided, but instead the variance of the mean apparent magnitude is given. If we let

$$L_{max} = L_{mean} + \sigma(L), \tag{9.11}$$

and convert to magnitudes, we obtain the approximation

$$\frac{\sigma(L)}{L_{max}} = \left(2.512^{-\sigma(m)} - 2.515^{-2\sigma(m)}\right). \tag{9.12}$$

This produces the curious result numerically that the ratio squared has the approximate value of the magnitude variance. Solving for R we obtain

$$R = \frac{c}{n_t\kappa_1(1 + z)(\sigma/L)^2}. \tag{9.13}$$

For the photometric observations and analyses we will make use of data from Hook et al. (1994) as downloaded and edited from VizieR (2012). The file includes a header in the first line:

```
varq = Import["Hooketal.csv"]; numlines6 = Length[varq];

ratioL2 = Expand[(N[(⁵√100)]^-sig - N[(⁵√100)]^-2 sig)²]

. . .

parameter = ───────────── /. {z1 → zz1, sig → sig1};
            (1 + z1) ratioL2
```

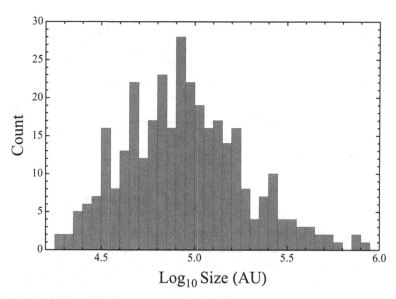

Fig. 9.7   Size of AGNs as a fully relativistic object.

The only number we do not know (or cannot estimate on our own) is the pulse frequency n_t, but as an average we adopt the 15 ± 5/yr (Terrell, p. 196) obtained from the power spectrum of 3C273.

The size of the emitting region can be estimated based on four separate models, two static and two fully relativistic.[7] Although the results are different in quantitative details, the fundamental properties are determined by the observed fluctuation amplitudes and the redshifts. More rapid "flickering" as expected produces smaller sizes.

As an example we consider a slow frequency for a fully relativistic ($v = c$) object. Here cauyr is the speed of light in AU/yr.

```
nt = 10 ; x1 = 5 / 3; radiusAURs =  cauyr   parameter;
                                    ──────
                                    nt x1
```

A histogram of the result can be seen in figure (9.7). The result confirms that the emitting region sizes for this sample of quasars and AGNs are certainly small considering the high luminosities observed. The statistical size distribution is log normal, with the most probable size being $\log_{10} \sim 5$, or about $100,000$ AU. While this is not "solar system" size, as was first estimated in the early 1970s, it is still small on a galactic scale. Terrell comments that although one has some difficulty reconciling the observed luminosities and the inferred sizes with the special relativity version of synchrotron emission, it is quite likely that a similar mechanism in general relativity occurring around a black hole will eventually explain the observations.

---

[7] For a full analysis of these models, see **9-2quasars**.

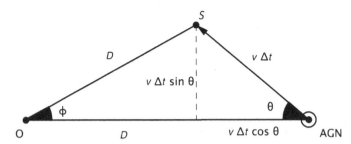

Fig. 9.8 Geometry of a superluminal jet.

9.2.4 Superluminal jets

There are cases where active galactic nebula will give off intense radio emission in the form of a jet of material moving at nearly the speed of light. Regions of these jets can be tracked at high resolution, and in some cases appear to be moving at superluminal speeds. These are known as superluminal jets.

Clearly the jet material is not traveling faster than light. Rather it is an optical illusion due to the relativistic speed of the material and small angle of deviation from our line of sight. The geometry of a superluminal jet can be seen in figure (9.8).

Consider jet material that leaves the AGN at a time t_o with a speed v at an angle θ from our line of sight, and reaching the point S at a time t. If we let t'_o be the arrival time of the radio emission of the material at t_o and t' be the arrival time from the material at t, then

$$t'_o = t_o + \frac{1}{c}\left(D + v\Delta t \cos\theta\right), \quad t' = t + \frac{D}{c}. \tag{9.14}$$

The time difference between these two signals is

$$t' - t'_o = \Delta t \left(1 - \beta \cos\theta\right), \tag{9.15}$$

where $\beta = v/c$. Because ϕ is small, $D\cos\phi \sim D\phi$, and thus the apparent transverse speed of the material is

$$v_T = \frac{\phi D}{t' - t'_o} = \frac{v \sin\theta}{1 - \beta \cos\theta}. \tag{9.16}$$

This (apparent) transverse speed is not bounded by the speed of light, which can be demonstrated by calculating the angle θ_{max} at which the motion has the greatest apparent speed

$$\left.\frac{dv_T}{d\theta}\right|_{\theta_{max}} = 0, \tag{9.17}$$

which gives the maximum transverse speed

$$v_{T max} = \frac{v}{\sqrt{1 - \beta^2}}, \tag{9.18}$$

which can clearly be superluminal for large $v < c$.

9.3 Synchrotron radiation

In the quasar, AGN, and radio galaxy papers, synchrotron radiation is often mentioned as being involved in active galaxy processes. Most undergraduate physics and astrophysics students have learned the qualitative aspects of synchrotron radiation in their first modern physics course, but learning quantitative details is usually postponed until after they have mastered electricity and magnetism.

A text reference tailored to the needs of astrophysics is provided by Shu (1991), where the equations of all the astrophysically important radiation processes are derived, including a fully relativistic theory of synchrotron processes. Shu starts with the derivation of the E and B fields emitted by a relativistically moving charge using the Lienard–Wiechert potentials. The most general result is

$$\mathbf{E} = \frac{q}{\left(R - \bar{\mathbf{R}} \cdot \bar{\mathbf{v}}/c\right)^3} \left\{ \left(1 - \frac{v^2}{c^2}\right) (\bar{\mathbf{R}} = Rv/c) + \frac{\bar{\mathbf{R}}}{c^2} \times \left[(\bar{\mathbf{R}} - Rv/c) \times \dot{\bar{\mathbf{v}}} \right] \right\}, \tag{9.19}$$

$$\mathbf{B} = \frac{\bar{\mathbf{R}}}{R} \times \mathbf{E}, \tag{9.20}$$

where the quantities $\mathbf{R} = \mathbf{x} - \mathbf{r}(\tau)$ and $\mathbf{v}$ refer to the retarded time τ such that $t = \tau - R(\tau)/c$, and then much effort is spent finding simplified versions for specific cases.

In most instances of astrophysical interest the observer is in the wave (or far) zone, so we can replace $\mathbf{R}$ by $x\mathbf{k}$ along some some unit vector $\mathbf{k}$ from the observer to the particle at an average distance x, which considerably simplifies the equations

$$\mathbf{E} = \frac{q}{c^2 x \left(1 - \beta \cos\theta\right)^3} \mathbf{k} \times (\mathbf{k} \times \dot{\mathbf{v}}), \tag{9.21}$$

$$\mathbf{B} = -\frac{q}{c^2 x \left(1 - \beta \cos\theta\right)^3} (\mathbf{k} \times \dot{\mathbf{v}}), \tag{9.22}$$

where $\beta = v/c$.

9.3.1 Electromagnetic field of accelerating charges

For a linear acceleration situation, the Poynting vector calculation gives the radiated power into a solid angle $d\Omega$ centered on $\mathbf{k}$ and located a distance x away as

$$\frac{dP}{d\Omega} = \frac{q^2}{4\pi c^3} \left(\frac{\dot{v}^2 \sin^2\theta}{(1 - \beta \cos\theta)^6} \right). \tag{9.23}$$

The cross section of an emitted radiation beam for various values of β can be seen in figure (9.9). The full shape is actually a figure of revolution in the x–z plane. This is also what would be seen by an observer approaching an oscillating charge.

Shu performs a similar derivation of the relativistic radiation pattern for the case where the velocity and acceleration are at arbitrary angles with respect to each other. It is assumed that v is in the z-direction, and $\dot{\mathbf{v}}$ is in the x–z plane. The $\mathbf{k}$ vector is pointing in the $(x, y, z) = (\sin\theta \cos\phi, \sin\theta \sin\phi, \cos\theta)$

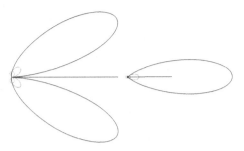

Fig. 9.9   Radiation of a charge accelerating parallel (left) and perpendicular (right) to velocity.

direction, and the acceleration makes an angle ψ with the velocity:

$$\frac{\mathrm{d}P}{\mathrm{d}\Omega} = \frac{q^2}{4\pi c^3} \left(\frac{1}{(1 - \beta \cos\theta)^4} + \frac{2(\sin\theta \cos\phi \sin\psi + \cos\theta \cos\psi)\beta \cos\psi}{(1 - \beta \cos\theta)^5} \right.$$
$$\left. - \frac{(1 - \beta^2)(\sin\theta \cos\phi \sin\psi + \cos\theta \cos\psi)^2}{(1 - \beta \cos\theta)^6} \right) \tag{9.24}$$

The cross section for $\psi = 90°$ is seen in figure (9.9). In the linear case the radiation is split into two beams, neither of which can be made to coincide. This splitting is why linear accelerators are not efficient producers of synchrotron radiation. In the case where acceleration is perpendicular to velocity the radiation beam is directed in the direction of the velocity vector. The circular motion achieved in a unidirectional magnetic field described by the familiar cyclotron formula is the easiest way to configure a current of charged particles for maximum radiation. It is assumed that the same configuration is characteristic of celestial sources, so Shu proceeds with that model in a simplified manner first before launching into a detailed model. The detailed model is very intricate, but is masterfully derived by Shu, and we refer those who need to learn the analytic theoretical procedure to consult that reference.[8]

9.3.2 Calculating synchrotron radiation

The detailed theory is based on evaluating the complicated integrals with the method of stationary phase, and is needed to characterize the spectrum and polarization of the optically thin synchrotron radiation with accuracy. Although the functions are complicated, the final results can be expressed in functions that *Mathematica* provides. The method of stationary phase is a technique for approximating integrals with rapidly varying sinusoidal parts, though this technique is rarely found under this name in modern treatments. Sometimes this method is listed as the method of steepest descent, but then it gets confused with the gradient method of root or singularity finding that goes by the same name in numerical analysis texts. Either way, this technique goes back to Kelvin and Stokes.

There are several parameters that enter the final results that need to be defined. The theory includes the case of helical rather than just circular motion around the magnetic field. Thus α is the pitch angle of the helix, such that it is equal to $90°$ for circular motion and $0°$ for motion parallel to the magnetic field. The other parameter is the cyclotron frequency ν_B, where

$$\omega_B = 2\pi \nu_B = \frac{qB\sqrt{1 - \beta^2}}{m_o c}. \tag{9.25}$$

[8] For computational examples at varying angles and speeds, see **9-4Synchro**.

Finally there is the characteristic frequency ν_c which is defined in terms of ν_B and α as

$$\nu_c = \frac{3eB_o \sin\alpha}{4\pi m_o c(1-\beta^2)} \tag{9.26}$$

To obtain the frequency dependent emission power components of the polarized synchrotron radiation we need to compute the so-called synchrotron integrals. Shu shows that

$$\begin{bmatrix} P_\nu^{em}(\mathbf{x}) \\ P_\nu^{em}(\mathbf{y}) \end{bmatrix} = \left(\frac{\sqrt{x}}{2}\right)\frac{e^3 B_o \sin\alpha}{m_o c^2}\begin{bmatrix} F(\nu/\nu_c) - G(\nu/\nu_c) \\ F(\nu/\nu_c) + G(\nu/\nu_c) \end{bmatrix}, \tag{9.27}$$

where $F(x)$ and $G(x)$ are defined in terms of the modified Bessel function of the second kind as

$$F(x) = x\int_x^\infty K(5/3, y)dy, \qquad G(x) = xK(2/3, x). \tag{9.28}$$

If we eliminate the magnetic field component in the coefficient to put it in terms of the critical frequency (treated as an observable), then for an individual electron the evaluation integrals become

```
emPx[v_, vc_, β1_] :=
```
$$\mathbf{N}\left[\frac{2\,\pi\,\mathbf{qq^2}}{\sqrt{3}\,\mathbf{c1}}\,\left(1-\beta1^2\right)\,\mathbf{v}\,\left(\int_{\mathbf{v/vc}}^\infty \mathbf{BesselK[5/3, r1]\,dr1 - BesselK[2/3, v/vc]}\right)\right]$$

```
emPy[v_, vc_, β1_] :=
```
$$\mathbf{N}\left[\frac{2\,\pi\,\mathbf{qq^2}}{\sqrt{3}\,\mathbf{c1}}\,\left(1-\beta1^2\right)\,\mathbf{v}\,\left(\int_{\mathbf{v/vc}}^\infty \mathbf{BesselK[5/3, r1]\,dr1 + BesselK[2/3, v/vc]}\right)\right]$$

For an ensemble of electrons, a usual approximation to the velocity distribution law is a power law relationship, primarily because of its ease of integration. Thus the volume emissivity equations with an electron energy spectrum of $n_o\gamma^{-p}$, where n_o is the electron number in the emitting volume, yields

$$\begin{bmatrix} j_\nu^{em}(\mathbf{x}) \\ j_\nu^{em}(\mathbf{y}) \end{bmatrix} = \left(\frac{\sqrt{3}}{2}\right)n_o\frac{e^3 B_\perp}{m_o c^2}\int_1^\infty \gamma^{-p}\begin{bmatrix} F(x) - G(x) \\ F(x) + G(x) \end{bmatrix}d\gamma, \tag{9.29}$$

where

$$\gamma = \sqrt{\frac{2\nu/(3\nu_L)}{x}}, \tag{9.30}$$

and

$$\nu_L = \frac{eB_\perp}{2\pi m_o c}, \tag{9.31}$$

is the Larmor precession frequency, and $B_\perp = B_o\sin\alpha$. To evaluate the integrals in *Mathematica* we change the variable from ν to x,

$$\mathbf{gamma} = \sqrt{\frac{2\,\mathbf{v}/(3\,\mathbf{vL})}{\mathbf{x}}}\;; \quad \mathbf{dgdx = PowerExpand[\partial_x\,gamma]};$$

$$\mathbf{ff[x\_]} = \mathbf{x}\int_\mathbf{x}^\infty \mathbf{BesselK[5/3, r1]\,dr1;} \quad \mathbf{gg[x\_]} = \mathbf{x\,BesselK[2/3, x]};$$

Because $F(x)$ contains an integration we must make sure the result is extracted from the conditional expression first using `ff[x][[1]]`. It should also be noted that neither $F(x)$ nor $G(x)$ converge properly. Solving these analytically is a task that leads to consideration of the method of stationary phase mentioned above.

The divergence problems of the integrals originate back in their definitions. We did not have to consider them initially because for a single electron we could stay close to the critical frequency. But in the power law generated spectrum we end up some distance away from ν_c, and those are in the unstable regions. However, *Mathematica* is able to provide exact solutions providing data is within certain boundaries.[9]

```
jF[v1_, vL1_, p1_] := N[jF1[[1]] /. {p → p1, v → v1, vL → vL1, x0 → 16}]

jG[v1_, vL1_, p1_] := N[jG1[[1]] /. {p → p1, v → v1, vL → vL1, x0 → 16}]
```

The exact polarization fraction can then be given by

$$P = \left| \frac{F - G}{F + G} \right|, \tag{9.32}$$

while Shu gives an approximate formula as

$$P_{Shu} = \frac{p + 1}{p + 7/3}. \tag{9.33}$$

A comparison of the two models can be seen in figure (9.10). Cosmic ray spectra give $p \sim 2.5$, and at that value the two are in reasonable agreement.

9.3.3 Synchrotron radiation and AGNs

Although our synchrotron model can be used in a variety of astrophysical situations, we will focus on the quasar/AGN case as a model for the prodigious energy outflow from these objects. The energy source is presumed to be a collapsed object at the center of a roughly 1 parsec radius volume where the greatest activity occurs. In making our estimates of the total number of electrons, we will choose the maximum physical limit rather than the nominal estimate. In the literature the temperature of 10^{12} K is mentioned as an upper limit. The electron density limit is the quantum degeneracy limit for that temperature.

```
temp = QuantityMagnitude[UnitConvert[Quantity[ 10^12, "Kelvins"]]];

       (2 π me kk temp)^(3/2)
nQ = ─────────────────────────
              hh^3
```

This gives $n_Q = 2.415 \times 10^{39}$. The volume of one parsec in cubic meters is approximately 10^{50}, so $n_o \simeq 10^{90}$. To calculate the power generated by a quasar or AGN we need to determine the Larmor frequency. If we assume the magnetic field is 0.05 microgauss averaged over the core, with a maximum of 1 microgauss, then $\nu_L = 9.34 \times 10^{-9}$ Hz.

---

[9] Here we choose $x = 16$ as a compromise bandwidth of plasma.

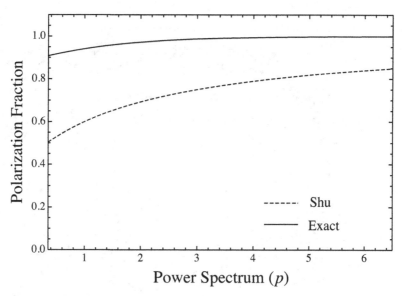

Fig. 9.10 Polarization fraction for a power spectrum source.

The Stokes parameters for polarization are given by

$$I(v) = \sqrt{\left(j_v^{em}(\mathbf{x})\right)^2 + \left(j_v^{em}(\mathbf{y})\right)^2}, \tag{9.34}$$

$$Q(v) = \sqrt{\left(j_v^{em}(\mathbf{x})\right)^2 - \left(j_v^{em}(\mathbf{y})\right)^2}, \tag{9.35}$$

and the total power can be estimated by

$$P = \int I(v)\, dv, \tag{9.36}$$

over a reasonable range of frequencies. Thus for a range of IR to optical frequencies

```
pjemx[v_, vL_, n0_, p_] := (√3/2) n0 (qq²/c1) vL Abs[jF[v, vL, p] - jG[v, vL, p]];
```

. . .

```
∫_{10¹²}^{10¹⁴} stokesIp[v5, vLagn, 1. × 10⁹⁰] dv5
```

which gives a radiated power on the order of 5×10^{35} watts.

The Milky Way has a luminosity of about 10^{36} watts, similar to the value we have calculated. But as we have noted there are some objects that are $10\times$ to $100\times$ more intense. Perhaps the AGN sources are larger than the sizes obtained using Terrell and Olsen's method of estimation. For example, our own estimates of the 3C 273 variation was about 80 days. This gives a radius of about 4 parsecs for the

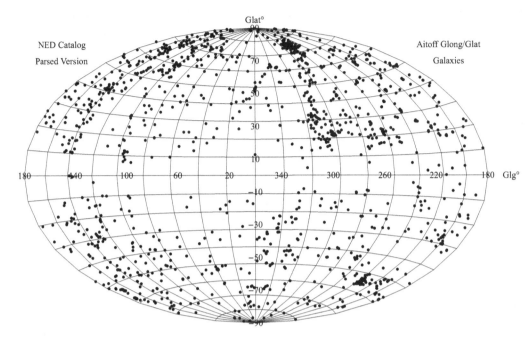

Fig. 9.11 Distribution of observed galaxies.

emitting region. In terms of the number of electrons this would increase n_o by a factor of 64, which would boost the energy output by a corresponding amount.[10]

9.4 Distribution of galaxies

While a major portion of the effort in galactic astrophysics is the determination of distance, there have been times when distances have not been available, so astrophysicists have resorted to doing statistical studies in 2D while waiting for enough distances to become available to do significant 3D work. In radio astronomy many years were spent collecting source counts all across the sky without really knowing what type of objects were responsible. In lieu of redshifts, source strength was considered a crude proxy for distance. Many of the statistical properties of the most distant objects in the universe were first found in radio studies.

Since we have used optical catalogs in the study of redshifts we will continue to use the same catalogs for exploring the 2D distribution of the same objects orthogonal to the distance. The objective is to characterize the distribution within volume slices at different distances, but first we select a suitable way to display such data, and then use it to examine the whole sky at all distance distributions to see what to expect in the slices.

[10] The effects of varying parameters in the calculation can be explored by running **9-4Synchro** with varying values.

Most people are familiar with the patterns of bright stars in the optical sky called constellations. In establishing those patterns the brightness of the stars in the constellations themselves are essential clues to the figures. If one draws a map of the stellar patterns without different sized dots, and include the faintest stars visible to the naked eye, the sky becomes a jumble of "random" dots. Certainly in such a situation one must resort to the same statistical methods we investigate here. But even to the naked eye the projected number density of the fainter stars is surprisingly uniform. If one needs to test this, just make a series of naked eye star counts through a short tube on a clear, dark night. A similar type of measurement which should be superior to just counting is to find the actual distribution of angular distances between objects as a function of those distances.

9.4.1 Plotting the galaxy distribution

Just as Fourier transforms have a discrete version called the discrete Fourier transform (DFT), the discrete version of the 2D spatial correlation function is known as the two-point correlation function. So why do we not handle this situation exactly as we did for the time autocorrelation using inversion of a power spectrum to get the autocorrelation? If the data were equally spaced, then it would be a simple matter to find the power spectrum for each quantity and then take the appropriate 2D inverse to get the 2D correlation. But our data are not collected at equally spaced intervals. Instead it is the actual angular distances between points that we are interested in, not some variable value at those unequally distributed points.[11]

A brute force method is to simply calculate all possible combinations of angular distances between all the objects in the sample, and then construct a histogram of the actual results. This is computationally extreme, particularly when the number of points is large. The brute force method quickly gets out of hand when calculating with large extragalactic catalogs. Thus much effort has been expended in finding alternative ways to make the equivalent measurements. These are still called two-point angular distance correlations, and using them instead of the brute force method is now standard, particularly with galaxies or galaxy clusters.[12]

The basis of these modern methods is to do a Monte Carlo analysis to compare the observations with computer randomized distributions of points. But no matter how the two point correlation function is obtained, to compare with cosmological models the angular distribution must be transformed into the 3D correlation function. That requires solving an integral equation originated by Limber (1952, 1954) involving an unknown distribution function. As discussed in more detail in **9-5galaxy2D**, *Mathematica* has only limited ability to solve integral equations directly so one usually reverts to solving the equivalent differential equation (Davis, 1962). But a specialized algorithm for just that type of numerical integral equation inversion has been widely used in galaxy counts (Baugh and Efstathiou, 1993) based on a technique originated by Lucy (1974).[13]

Because astrophysical problems often have data sets that cover the whole sky, most ways of displaying all the data at once are cumbersome because of distortions introduced by the various projection

---

[11] An example of the difficulty in finding the power spectrum when the observations are not equally spaced can be seen in **9-3Q3C273**.

[12] For example, see Crocce, Cabré, and Gaztañaga (2011); Sebok (1986); Haynes and Giovanelli (1988); Bahcall (1988); Dekel (1988); and Peebles (1980).

[13] For a nice compact summary of the modern concept of angular correlations, van de Weijgaert (2007) has one on his web site: http://www.astro.rug.nl/ weygaert/tim1publication/lss2007/computerII.pdf

schemes that are available. The Aitoff projection is a compromise system, and is particularly useful for astronomical displays. (The famous WMAP image of the cosmic microwave background is a Mollweide projection.) Unfortunately *Mathematica* does not provide a special function to produce such a map, so we must create one ourselves (figure 9.11).[14] The data for the galaxies are taken from the NED-D list compiled by I. Steer of the NASA/IPAC Extragalactic Database (NED) which is operated by the Jet Propulsion Laboratory, under contract with the National Aeronautics and Space Administration.[15]

9.4.2   Initial interpretation of the galaxy data

It is clear that the raw distribution of NED galaxies (at least the brightest ones) is not random or uniform. Many years of study of the apparent distribution as a function of distance (or redshift) show that there are clusters as well as "ridges" of density that run from cluster to cluster. Such ridges provide evidence for a "supercluster." The early ideas about superclusters treated them as roughly spherical ensembles of clusters. But the ridges have demolished that view. Instead there are huge networks of galaxy clusters that are interconnected by chains of galaxy enhancement regions separated by large "voids."

Some of the observed nonuniformities here could be due to various biases in the sample itself that have nothing to do with the intrinsic properties of the galaxies. First of all there is the "zone of avoidance" along the galactic equator. This will be in any catalog that uses optical data. It and other features that have symmetry in the galactic coordinate system are caused by dust absorption in the Milky Way (Limber 1952, 1954). The NED catalog is already biased to those objects bright enough to have accurate redshifts determined for the distance-redshift calibration, and the dust absorption adds to that bias in areas near the galactic equator as well as the "bands" running in galactic latitude at several longitudes. Another bias is that many sky surveys are not all-sky because of site or instrumental limitations. These examine only certain strips of the sky.

Although not completely balanced in terms of numbers, the points from the NED galaxy catalog visually shows no overtly artificial sharp boundaries that characterize "deep" surveys in the longitude/latitude plane. However, the NED catalog is not free from such survey biases because a number of linear features are evident in the z direction, as seen in figure (9.12). The number of points in each "line" of depth is probably not sufficient to use for studies of the 3D two-point correlation function, so we do not attempt that here.

It is interesting to note that in the 190° to 240° longitude area there seem to be several galaxies observed in the zone of avoidance on or close to the galactic equator. There are other places (longitude ~140° and longitude ~320°) where a few objects are also seen on or very near the galactic equator. These concentrations are most likely due to a thinning of the galactic dust in those directions. Such an interpretation is made more likely by the fact that the "holes" through which these objects seem to be seen are offset in the direction opposite to the galactic rotation by 20–30 degrees from known spiral arm dust concentrations (Orion and Cygnus, for example). The idea is that these "holes" are outward views along spiral arms. A less likely interpretation is that there is an abnormal overabundance of bright galaxies in those directions, but from these data alone we cannot tell if that idea has any validity. One would have to examine the interstellar reddening in the observed directions in some detail before deciding as was done first by Limber (1952, 1954).

---

[14] For details on generating an Aitoff projection in *Mathematica* see **9-5galaxy2DR**.

[15] NED-D is available online as http://ned.ipac.caltech.edu/Library/Distances/NED21.11.1-D-5.1.1-20111026.csv.

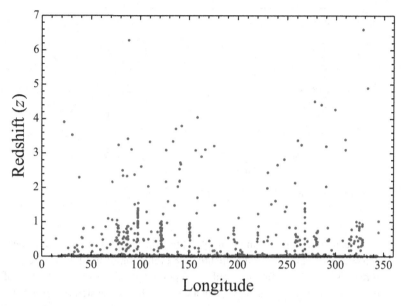

Fig. 9.12 Distances of observed galaxies.

9.4.3 Characterization of the two hemispheres

Figure (9.11) shows that for statistical reasons the first partition considered should be between galactic coordinate hemispheres. Thus we introduce a separation algorithm. For the Northern Hemisphere (northlist),

```
north = Table[{0., 0., 0.}, {ii, 1, galaxynum[[1]]}]; jn = 0;
Do[If[galaxylist[[jj, 3]] > 0, jn = jn + 1;
   north[[jn]] = {galaxylist[[jj, 2]], galaxylist[[jj, 3]], galaxylist[[jj, 4]]}],
   {jj, 1, galaxynum[[1]]}];

nnum = jn; northlist = Table[{0., 0., 0.}, {kk, 1, nnum}];

Do[northlist[[jj]] = {north[[jj, 1]], north[[jj, 2]], north[[jj, 3]]}, {jj, 1, nnum}]
```

A similar algorithm produces southlist.

The z histograms of the two hemispheres, figure (9.13), appear similar, particularly near the $z = 0$ origin. We therefore partition the data in the same manner. We choose as a definition of "local" all galaxies in the so-called local velocity anomaly to -780 km/s (Faber and Burstein, 1988), but we truncate at $+150$ km/s to exclude outer members of the Virgo cluster.

Having defined our "local" exclusion zone we construct two data sets for each hemisphere. Our choice of "near" and "far" puts all objects influenced by the great attractor (GA) in the same distribution group.[16]

---

[16] See Faber and Burstein (1988).

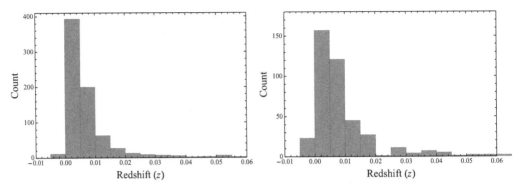

Distance histograms for Northern (left) and Southern (right) Hemispheres.

We can tell visually from figure (9.11) that the galaxies in each hemisphere are not uniformly spaced across the sky, but we need a quantitative evaluation of our intuitive feel. Our approach here is brute force, and will calculate the angular two point correlation via a histogram of the actual angular distribution of separation distances. Such an approach is suggested because of the rich set of *Mathematica* histogram options available not only for plotting, but also for collecting the histogram results for later analysis.

A main task in connection with the angular correlation function is the statistical characterization of certain parameters. There are theoretical reasons for making the simplifying assumption of an exponential decay in the angular distribution that takes the asymptotic form of a power law in angle, at least at small angles.[17]

We can test that out to large angles with our sample. The power law exponent of the angular correlation $\omega(\theta)$ is stated as $1 - \gamma$, where $-\gamma$ is the exponent of the 3D spatial correlation function $\xi(r)$.

$$\omega(\theta) = \left(\frac{\theta}{\theta_o}\right)^{1-\gamma}, \qquad \xi(r) = \left(\frac{r}{r_o}\right)^{-\gamma}. \tag{9.37}$$

Thus a determination of γ from the power law representation is our goal.

We start by separating near ($z < 0.0005$) galaxies in `northlist` from the far ($z > 0.0005$) galaxies.

```
nlocal = Table[{0., 0., 0.}, {ii, 1, nnum}]; jn = 0;
Do[If[northlist[[jj, 3]] <= 0.0005, jn = jn + 1; nlocal[[jn]] =
    {northlist[[jj, 1]], northlist[[jj, 2]], northlist[[jj, 3]]}], {jj, 1, nnum}];

nnum1 = jn; nlocallist = Table[{0., 0., 0.}, {kk, 1, nnum1}];
Do[nlocallist[[jj]] = {nlocal[[jj, 1]], nlocal[[jj, 2]], nlocal[[jj, 3]]},
    {jj, 1, nnum1}]
```

[17] Crocce, Cabré, and Gaztañaga, 2011; van de Weijgaert, 2007; Maller et al., 2005; Bahcall, 1998; Baugh and Efstathiou, 1993; Thieberger, Spiegel, and Smith, 1990; Fry, 1986; Sebok, 1986.

We then do a calculation of the angular separations, from which we generate a histogram of the separation angles:

```
radial = Table[Table[0., {ii, 1, nnum1}], {jj, 1, nnum1}];
```

. . .

```
neardata = HistogramList[radial2, 30]
```

Because of the connection between a power spectrum and a correlation function, we can compute and fit the power spectrum of the histogram.

```
psnear = Abs[Fourier[neardata[[2]]]]
pSnear = Table[{N[ii / 360], 2 Log[10, psnear[[ii]]]}, {ii, 1, 13}]

psfit = NonlinearModelFit[pSnear, a x^b, {a, b}, x]
```

We also determine the value of γ for the near data.

```
neardata1 = Table[{N[neardata[[1, ii]]] / °, N[neardata[[2, ii]]]}, {ii, 2, 20}];
nearfit1 = NonlinearModelFit[neardata1, b φ^{1-γ}, {b, γ}, φ]

nearcoeff = nearfit1["BestFitParameters"]; nearerr = nearfit1["ParameterErrors"];
```

This gives a value of $\gamma = 1.14 \pm 0.17$. The far Northern Hemisphere galaxies yield $\gamma = 0.83 \pm 0.10$, while for the near and far galaxies of the Southern Hemisphere we find $\gamma = 1.14 \pm 0.10$ and $\gamma = 0.98 \pm 0.11$ respectively.

9.4.4 Comparison to a Monte Carlo model

A common way of comparing the results obtained from actual data is to perform a Monte Carlo simulation of the actual data and see what differences occur between them. For this purpose the usual pseudorandom numbers may not be good enough, so we will test several.

```
randomset = Table[{RandomReal[{0, 360}], RandomReal[{0, 90}]}, {ii, 1, nnum1}];
```

. . .

```
Do[radialB[[kk]] = Re[radialA[[kk]]], {kk, 1, jq}]
```

From the random data we create a histogram and power spectrum, from which can derive near and far values for γ. For example, for the far distribution we have

```
randomdata = HistogramList[radialB, 30]
```

. . .

```
randomfitP = NonlinearModelFit[randomdataP, b φ^{1-γ}, {b, γ}, φ]
```

With the γ values determined from the Monte Carlo simulations, we can determine the probability of equivalence between the observed γ and the pseudo-randomly generated one. For this we use the

Student t-test for the comparison of means. The standard t-test routine in *Mathematica* produces the probability that the two values are samples of the same number from the same population.[18] Again for the Northern Hemisphere,

`resultsL = {{gamfarL, gamfarLerr}, {gamrandL, gamrandLerr}}`

$$t1 = \frac{\text{resultsL}[[1, 1]] - \text{resultsL}[[2, 1]]}{\sqrt{\text{resultsL}[[1, 2]]^2 + \text{resultsL}[[2, 2]]^2}}$$

`p1 = StudentTPValue[t1, 30, TwoSided → True]`

It should be noted that the resulting probabilities given by *Mathematica* are for agreement, not disagreement with the random model. Also, each run will give slightly different results due to the pseudo-random distribution.

It appears the *t*-test is unable to detect clustering against the strong random background in the data. There are, however, several qualitative trends:

1. Both north and south far samples at large angles (*R* stands for the right-hand side of the histogram) are in agreement with the γ given by the random simulation.
2. The far left hand side (*L* notation) for both north and south give similar probabilities of being similar to the model.
3. The far results for all angles are divergent, with the north more similar to the random model than the south.
4. The near samples for north and south both seem to disagree with the random model, with the south in greater disagreement than the north.

9.4.5 Is there a fractal connection?

It appears that the shape of the power spectra is much more sensitive to the subtle differences in the angular correlation than is the characterization of the power law parameters themselves. So why bother calculating γ at all? First of all, γ is more directly derivable from theoretical cosmological models and the mechanics of galaxy and galaxy cluster formation (Fry, 1987) than is the power spectrum.

It should also be noted that our values of γ are noninteger. Any time some relationship involving a dimension has a noninteger power law it is called a fractal. As first shown by Mandelbrot (1983) and elaborated upon observationally by Thieberger, Spiegel and Smith (1990) one can relate values of γ to various types of fractal dimension. In fact, because there seems to be no universal (cosmic) value of γ despite extensive searches, one is led to the supposition that galaxy clustering requires a fractal characterization where there exists a spectrum of γ values called multifractals.[19] But multifractals, though applicable, have a connection to geometric measure theory as well as Brownian motion, the cascade theory of turbulence, and random vector projections (Feller, 1966) and it would lead us too far astray to consider them here. Instead we use less sophisticated tools.

---

[18] For a better comparison we use not just near and far distributions for the Northern and Southern Hemispheres, but additionally split each distribution into left and right, being left or right of the peak value in the histogram distribution. See **9-5galaxy2DF** for details.

[19] Evertsz and Mandelbrot, 1992; Vicsek, 1992; and Feder, 1988.

Thieberger, Spiegel, and Smith (1990) define three fractal measures. One is based on the integral of the number density in various dimensional spaces, and for power law densities it is easy to find without knowing how to do the integral explicitly. Another is based on self-similarity, and a third is based directly on the correlation function. To first order, these definitions all boil down to one relationship we can simply call the "fractal" dimension,

$$D_f = D - \gamma, \qquad (9.38)$$

where D is the dimension where the measurements are made and γ is the power law value obtained from the fitting of the data. Because our data are 2D, then $D = 2$.

```
gammalist = {gamnear, gamfarP, gamfarL, gamrandP, gamrandL, gamnearrnd}

dimfN = 2 - gammalist

errlist = {gamnearerr, gamfarPerr, gamfarLerr, gamrandPerr,
    gamrandLerr, gamnearerrnd}
```

The average fractal dimension between north and south seems to be unity as one might expect. This agrees with the near unity values obtained for the random model dimension. The north mean is two standard deviations of the mean higher than unity, while the south mean is two standard deviations lower. The differences between these two are mainly due to the values of near and far dimensions being systematically different between hemispheres. Overall the γ values in this case are not useful measures of fractal dimensions.

9.5 Extragalactic bending of light

In Chapter 5 we discussed the general relativistic (GR) bending of light around a compact massive object, with the sun as a classic test of GR. This theory, when applied to objects other than the sun, is called gravitational lensing. The ring effect of this theory was first predicted by Khvolsen in 1924, followed by Einstein in 1936. For this reason the visible manifestation of the ring effect is often given the name Einstein or Khvolson ring, or Einstein cross.[20] Burke and Graham-Smith (2002) cite that the first observational confirmation was obtained by Walsh, Carswell, and Weymann (1979) who were searching for optical counterparts to radio sources. Schneider, Ehlers, and Falco (1993) provide a comprehensive review up to that year, including an extensive review of the theory. The lensing literature continues to grow, and is not referenced in detail here.

There are 47 double quasars and 88 lensed quasars listed in Veron-Cetty and Veron (2010). This is out of more than 133,000 cataloged quasars, 34,000 AGNs, and 760 blazars. A spectacular image of one such object, a quasar imaged by a foreground object, is called Huchra's Lens. This object has variability in its components ascribed to microlensing, a related phenomena involving individual stars.[21]

---

[20] *Wikipedia*, 2012.

[21] See http://www.astr.ua.edu/keel/agn/qso2237.html, http://apod.nasa.gov/apod/ap100207.html, http://hyperphysics.phy-astr.gsu.edu/hbase/astro/eincros.html.

9.5.1 Deflection, magnification, and time delay

Burke and Graham-Smith (p. 327) derive the deflection equation by extending the infinitesimal deflection around a point source in the small angle approximation to obtain

$$\theta = \beta + \left(\frac{D_{SL}}{D_L D_S}\right)\left(\frac{4GM}{c^2}\right)\frac{1}{\theta}, \tag{9.39}$$

where β is the angle seen between the line to the lens and the actual object position, D_{SL} is the source-lens distance, D_S is the source-observer distance, D_L is the observer-lens distance, and M is the mass of the lens object. The angle θ is the deflection relative to the line to the position of the point lens. It is related to the familiar impact parameter,

$$b = \frac{2R_{Sch}}{\theta} = \frac{2}{\theta}\frac{2GM}{c^2}, \tag{9.40}$$

where R_{Sch} is the Schwarzschild radius of the lens mass. It should be noted that the cosmological distances used here are based on angles, not luminosity. In GR models there is a difference between the two, but given the low-order approximations we will explore this is not significant so we will not elaborate on either concept here.

The equation is quadratic in θ, so we can obtain two roots,

$$\theta_{in} = \frac{1}{2}\left(\beta - \sqrt{\frac{8D_{SL}R_{Sch}}{D_L D_S} + \beta^2}\right), \tag{9.41}$$

$$\theta_{out} = \frac{1}{2}\left(\beta + \sqrt{\frac{8D_{SL}R_{Sch}}{D_L D_S} + \beta^2}\right). \tag{9.42}$$

If the source and lens are aligned ($\beta = 0$) then there is a single "ring image",

$$\theta_o = \sqrt{\frac{2D_{SL}R_{Sch}}{D_L D_S}}. \tag{9.43}$$

Gravitational lensing can also produce a magnification of the background object. Our definition of magnification is based on the change in θ, $\delta\theta$, based on a small change in β, $\delta\beta$ as measured in the radial direction; thus

$$m = \frac{\delta\theta}{\delta\beta}. \tag{9.44}$$

For $\beta = 0$ the angular magnification is $1/2$. If, however, the object has a circular radius β, then

$$m_{in} = \frac{1}{2}\left(1 - \frac{\beta}{\sqrt{\frac{8D_{SL}R_{Sch}}{D_L D_S} + \beta^2}}\right), \tag{9.45}$$

$$m_{out} = \frac{1}{2}\left(1 + \frac{\beta}{\sqrt{\frac{8D_{SL}R_{Sch}}{D_L D_S} + \beta^2}}\right). \tag{9.46}$$

As figure (9.14) shows, an on-axis source starts with the same magnification in both rings, while as the source gets larger in angle the inner ring gets thinner until it has zero thickness, while the outer

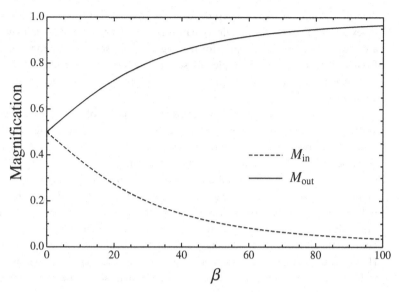

Fig. 9.14 Object magnification by a gravitational lens.

one gets thicker until its thickness is equal to the outer angular radius of the object. This is not what is expected if the lens behaved like a regular optical lens. Of course this simple case assumes the object and lens both have circular symmetry, though this is unlikely to correspond to reality. Burke and Graham-Smith briefly consider a simple model of the effect of a 3D gravitational potential in the lens on magnification, but they refer to Schneider, Ehlers and Falco (1993) for further details.

We have mentioned that sometimes components of a lensed image show time variability that is not synchronized together. This can be interpreted as a classical GR time delay,[22]

$$\tau = (1 + z) \left[\frac{D_L D_S}{2 D_{LS}} (\theta_1 - \theta_2)^2 - \frac{\psi(\theta)}{c^3} \right]. \tag{9.47}$$

Notice that because distances are defined to be those for angles not luminosity the redshift factor $(1 + z)$ must be included. Here $\theta_1 - \theta_2$ is the angular distance between the two components and $\psi(\theta)$ is the projected potential in the lens object. The same "deflection" potential appears in general expressions for the magnification.

9.5.2 Light deflection around a spinning black hole

So far we have considered the geometric approximation as applied to space–time. Although not explicit in the equations, we have dealt with a refractive index. In Chapter 7 we dealt in detail with electromagnetic propagation through a plasma, which in essence is geometrical optics in special relativity. Here we have done essentially the same thing in general relativity. In classical geometric optics, the refractive indices are related directly through Snell's law to the angles of the rays. In the plasma case, the medium is birefringent, so there are two solutions to the refractive index. If a magnetic field is present there can be three. If the refractive index becomes complex, as often happens in plasma problems,

[22] Burke and Graham-Smith, p. 340.

then the imaginary part is interpreted as absorption. But after substitution of a complex refractive index into Snell's law, the imaginary part forces an interpretation of an imaginary sine value which has a corresponding imaginary part for the angle. In the small-angle approximation of geometric optics from which we derive the simple GR solutions cited above the usual deflection angle is always real. The Schwarzschild metric is symmetric, so its small-angle deflection approximations will not have abnormalities.

In the Schwarzschild case it is assumed that the nature of the underlying object doesn't matter much as long as it has enough mass and is not relativistically degenerate, so that some sort of classical gravitational potential can be formulated. This is, of course, a weak field approximation that can be described by small angle approximations. Lensing in GR as defined above is the radial bending of space–time so that if the central object is a black hole and spinning (or orbiting another object), that motion will alter the space–time geometry, remove the symmetry (degeneracy), and the bending of light will become much more complicated. So much so that most researchers simply retreat to a computer model for the solutions. But there are some GR investigators who feel, correctly, that when doing computation it is always good to have an analytic model that can be used to check the computer results. However, unlike many such theoretical derivations, the analytic solutions we discuss in the text that follows were originally obtained using *Mathematica*, so we feel it is appropriate to try to reproduce their computations here.

Iyer and Hansen (2009) developed analytical approximations for the main relativistic deflection parameter, the total angular deflection (only in the equatorial plane), for a Kerr (spinning) black hole. In the formulae shown by Burke and Graham-Smith, the total deflection angle did not really appear explicitly as its definition was substituted by $\alpha = R_{Sch}/b$. That left an expression entirely in θ and the distances.

Compared with the Schwarzschild zero-spin case, computation of the Kerr situation is a great deal more complicated, both analytically and numerically. Once a suitable expression for $\hat{\alpha}$ (the Iyer and Hansen notation for the total angular deflection α) is obtained, then

$$\theta = \beta + \left(\frac{D_{LS}}{D_S}\right)\hat{\alpha}. \tag{9.48}$$

Before exploring the solutions for the deflection parameter, we must indicate how the changes to the theory come in. In the previous theory, α was simply related to the inverse power of θ, but that will no longer be true. Iyer and Hansen have chosen to make the impact parameter b be the independent variable, as this makes more sense in the strong field case, but there are other changes as well:

1. Because spin has a direction, the mathematics involved needs a sign parameter $s \pm 1$. The plus sign is for direct orbits (with the spin) and the minus sign for retrograde orbits (against the spin).
2. The magnitude of the spin is specified by the parameter $a = J/M$, where J/M is the angular momentum per unit mass.
3. The radial parameter to which certain distances will be normalized is $1/2$ the Schwarzschild radius, $m_\star = GM/c^2$.
4. In the case of a black hole, as the radius of photon closest approach r_o gets nearer to a critical distance $r_c = 3m_\star$, the possibility of the photon orbiting the black hole arises.[23] There is a profound effect on the deflection angle as r_o approaches the critical radius. That means the impact parameter

---

[23] See chapter 5, as well as **5-3Orbsin2Times**.

must be defined to include the spin parameter s, which is 1 for "orbits" in the same sense of the rotation, and -1 for orbits opposite to the sense of the spin. The basic equation for this limit is

$$b' = 1 - \frac{sb_c}{b}, \tag{9.49}$$

where b_c is the real root of

$$(b_c + a)^3 = 27m_\star^2 (b_c - a), \tag{9.50}$$

with the condition

$$3\sqrt{3}m_\star > b_c > 2m_\star. \tag{9.51}$$

This condition equality indicates that in the spin case space–time becomes tri-refringent. Additionally if $a > b_c$ then complex roots will be obtained.

5. Other parameters used in the formulation are

$$h = \frac{m_\star}{r_o}, \quad \omega_s = \frac{a}{sb}, \quad \omega_{sc} = \frac{a}{sb_c}, \quad \omega_o = \frac{a^2}{m_\star^2}, \tag{9.52}$$

and

$$\hat{a} = \frac{a}{m_\star} = \frac{Jc}{GM^2}. \tag{9.53}$$

6. In the limiting cases we have
   1. Flat space-time, $h \to 0$,
   2. Zero spin Schwarzschild, $\omega_s \to 0$, $\hat{a} \to 0$,
   3. Extreme Kerr, $\hat{a} \to 1$ as the maximum spin a body can attain without "breakup".

When incorporating spin into the situation, further new parameters are even more convenient. If we define the critical values of quantities with subscripts c and sc, then we have

$$r_{sc} = 3m_\star \frac{1 - \frac{a}{sb_c}}{1 + \frac{a}{sb_c}}, \quad h_{sc} = \frac{m_\star}{r_{sc}}, \quad r_o = 3m_\star \frac{1 - \frac{a}{sb}}{1 + \frac{a}{sb}}, \tag{9.54}$$

$$\frac{r_o}{Q} = \frac{1}{h_{sc}} \sqrt{\left(1 - \frac{2h}{h_{sc}}\right)\left(1 + \frac{6h}{h_{sc}}\right)}, \tag{9.55}$$

$$k^2 = \frac{\sqrt{\left(1 - \frac{2h}{h_{sc}}\right)\left(1 + \frac{6h}{h_{sc}}\right)} + \frac{6h}{h_{sc}} - 1}{2\sqrt{\left(1 - \frac{2h}{h_{sc}}\right)\left(1 + \frac{6h}{h_{sc}}\right)}}, \tag{9.56}$$

$$\psi = \arcsin \sqrt{\frac{1 - \frac{2h}{h_{sc}} - \sqrt{\left(1 - \frac{2h}{h_{sc}}\right)\left(1 + \frac{6h}{h_{sc}}\right)}}{1 - \frac{6h}{h_{sc}} - \sqrt{\left(1 - \frac{2h}{h_{sc}}\right)\left(1 + \frac{6h}{h_{sc}}\right)}}}, \tag{9.57}$$

$$\Omega_\pm = \frac{\pm\left(1 \pm \sqrt{1 - \omega_o}\right)(1 - \omega_s) \mp \omega_o/2}{\sqrt{1 - \omega_o}\left(1 \pm \sqrt{1 - \omega_o} - \frac{\omega_o h_{sc}}{4}\left(1 - \frac{2h}{h_{sc}} - \sqrt{\left(1 - \frac{2h}{h_{sc}}\right)\left(1 + \frac{6h}{h_{sc}}\right)}\right)\right)}, \tag{9.58}$$

$$n_\pm = \frac{1 - \frac{6h}{h_{sc}} - \sqrt{\left(1 - \frac{2h}{h_{sc}}\right)\left(1 + \frac{6h}{h_{sc}}\right)}}{1 - \frac{2h}{h_{sc}} - \sqrt{\left(1 - \frac{2h}{h_{sc}}\right)\left(1 + \frac{6h}{h_{sc}}\right)} - \frac{r}{\omega_o h_{sc}}\left(1 \pm \sqrt{1 - \omega_o}\right)}. \tag{9.59}$$

Finally we have the expression for the Kerr black hole as given by Iyer and Hansen,

$$\hat{\alpha} = -\pi + \frac{4}{1 - \omega_s}\sqrt{\frac{r_o}{Q}}\left(\Omega_+\left[\Pi(n_+, k^2) - \Pi(n_+, \psi, k^2)\right]\right. \tag{9.60}$$
$$\left. + \Omega_-\left[\Pi(n_-, k^2) - \Pi(n_-, \psi, k^2)\right]\right).$$

For a no-spin, Schwarzschild black hole this reduces to

$$\hat{\alpha} = -\pi + 4\sqrt{\frac{r_o}{Q}}\left[\Pi(0, k^2) - \Pi(\psi, 0, k^2)\right] = -\pi + 4\sqrt{\frac{r_o}{Q}}\left[K(k^2) - F(\psi, k^2)\right]. \tag{9.61}$$

The notation of the elliptical integrals may seem strange because often one sees only k where we have written k^2. Thus one would have to take the square root to get the numerical value to put into the function call. But the integrals are defined using only k^2, so *Mathematica* denotes this parameter as m with the understanding that the value to be entered is the square of k.

9.5.3 Computing the effect of spin on deflection

The mathematical functions built into *Mathematica* include the complete elliptical integral of the third kind, $\Pi(n, m)$, the incomplete elliptical integral of the third kind, $\Pi(n, \psi, m)$, the complete integral of the first kind, $K(m)$, and the incomplete integral of the first kind, $F(\psi, m)$. The remaining quantities are complicated functions of the parameters above that must be expressed in *Mathematica*.[24]

Iyer and Hansen present two cases, a weak deflection limit and a strong deflection limit, with asymptotic series approximations for these two. In *Mathematica* we express $m_\star$ in units of solar mass and megaparsecs. For the spin parameter, because $a = J/Mc$ its maximum is $a_{max} = m_\star$. For ease of computation we therefore express spin as a fraction of $m_\star$, $a = a/m_\star$, giving $0 < a < 1$. The impact parameter b has a minimum of b_c and an unbounded maximum, so we use b', the inverse of b, and normalize it to b_c such that $0 < b' < 1$. After determining the aforementioned parameters, the solution becomes

```
alph[bf6_, ss6_, af6_] := -π + ──────────────── √r0Q[bf6, ss6, af6]
                                1 - ωs[bf6, ss6, af6]
                                         4

(Ωplus[bf6, ss6, af6 ] (EllipticPi[ ...] -
    EllipticPi[nnminus[bf6, ss6, af6], ψψ[bf6, ss6, af6], k2[bf6, ss6, af6]])
```

Evaluation of the expression gives $\hat{\alpha}(a, s, b')$, where $s = \pm 1$ yields solutions either with or against the rotation respectively.

Figure (9.15) shows an example of the effects of spin and impact parameter on $\hat{\alpha}$. A more detailed examination of the solution including 3D plots can be found in **9-6Ecross**. When evaluating the results one should keep in mind that these are normalized to $m_\star = R_{Sch}/2$, and that we are still in the smal

---

[24] See **9-6Ecross** for details.

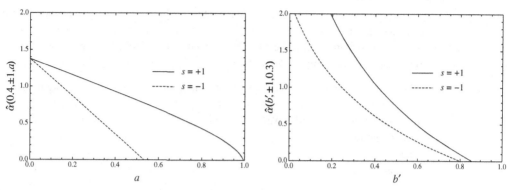

Fig. 9.15   The effect of spin (left) and impact parameter (right) on deflection angle.

angle approximation domain even when the answers are not small angles. It is also important to note that the $\hat{\alpha}$ values are not angles, but rather tangents of angles.

If we remove the normalizations applied in the calculations, then the angle of deflection can be expressed as

$$\theta = \beta + \arctan\left[\frac{D_{LS}}{D_S}\frac{m_\star}{D_L}\hat{\alpha}\right]. \tag{9.62}$$

Because $m_\star/D_L$ is always very small, we can us the small angle approximation; thus

$$\theta = \beta + \frac{D_{LS}}{D_S}\frac{m_\star}{D_L}\hat{\alpha}. \tag{9.63}$$

By plotting `ArcTan[alph[]]` one can gain a better understanding of the effect of spin on deflection.[25]

9.6 The cosmic microwave background

Ever since the expansion of the galaxies was discovered, cosmological models have predicted that in the remote past the present universe was compressed into a very small volume with properties of a GR singularity. It was proposed that perhaps the "beginning" was the result of an explosion commonly known as the "big bang."[26] Today we are much more certain that there was a big bang, and the cosmic microwave background (CMB) is the best evidence for it. The spectacular success of the WMAP satellite and its predecessor COBE has revolutionized cosmology and our view of the universe. WMAP has collected a huge archive of uniform quality, and the spectrum of the microwave sky has been established to a precision higher than probably any other continuous spectrum.

The WMAP map we will be working with shows the small temperature fluctuations for the whole sky from the average temperature of 2.73 K. As first shown unequivocally by the COBE spacecraft,[27] the CMB is amazingly uniform. In the original file obtained from the WMAP site, the fluctuations are

[25] See the homework exercises for this chapter.
[26] There has also been speculation that the end of the universe might also be a singularity caused by collapse, an idea famously presented in *The Hitchhiker's Guide to the Galaxy* by Douglas Adams.
[27] See Burke and Graham-Smith, 2002.

given as a color contour map in the familiar Mollweide projection. In this map the observed temperature fluctuations are ±200 microkelvin. The effects of our galaxy have been removed to a great degree, except for some features near the galactic center and along the galactic equator.[28]

Burke and Graham-Smith (2002) represent the post COBE and pre-WMAP views as:

1. The CMB temperature is 2.728 ± 0.002 K, and is strong evidence for a "big bang" cosmology.
2. There is a dipole term representing $\sim$360 km/s of Doppler effect relative to the CMB, presumably due to motion of the Milky Way. This motion is in the direction of 265° galactic longitude and 48° galactic latitude.
3. The CMB has spatial temperature fluctuations where $\Delta T = 30 \pm 60$ μK on scales smaller than 10 angular degrees. These fluctuations are reported to have maxima below 2° in size.

It is those particular oscillations that will be the focus of this section.

It should be noted that with the processing of the public distribution data from WMAP we will reach the limits of contemporary (2012) single desktop computer processing abilities. Although *Mathematica* has the ability to parallelize tasks across multiple machines with multiple processes, it is unlikely that many undergraduate students will have cluster computers available to them. We therefore focus on carefully selected subsets of the data to make computations manageable. Though the results will have a huge amount of noise, high-frequency spatial information is not completely lost.

9.6.1 Fourier transform analysis

Our data come from the largest color Mollweide projection map produced from seven years of WMAP data. This was then converted to a 16-bit grayscale image, and saved as a TIFF file. Because the Mollweide ellipse is not natural within the rectangular image format, we also import a "white screen" version so we can generate a deconvolution of the elliptical shape out of the Fourier transform. From these two images we sample the brightness values to create two data arrays, `wmp.csv` for the grayscale image and `wmpW.cvs` for the white screen image.

Although the images are sampled at equal intervals, these are not the coordinates in which the data were taken, but rather an Mollweide projection of the data.[29] One of the reasons for using a Mollweide system is that it is an equal area system, but this is still a distortion of the original data. The most obvious places where there will be a problem is near the galactic poles. There are also some anomalies in the pixels at the far left of the projection, but these will not be included in our analysis. As shown in **9-8MapProperties**, the horizontal axis variable x and vertical axis variable y are related to the galactic longitude λ and latitude ϕ by

$$x(\phi = 0) = 180 \frac{\sin(\lambda/2)}{\sqrt{1 + \cos(\lambda/2)}}, \tag{9.64}$$

$$y(\lambda = 0) = 90 \frac{\sin(\phi)}{\sqrt{1 + \cos(\phi)}}. \tag{9.65}$$

It should be noted that if the direction of x is taken to be positive in the same direction as increasing pixel numbering, then this is *backwards* to the convention of galactic longitude increasing to the left.

---

[28] The CMB plot in the Mollweide projection can be downloaded from http://map.gsfc.nasa.gov/media/101080/index.html. For other plots related to WMAP, see also http://lambda.gsfc.nasa.gov/product/map/current/m_images.cfm.

[29] See **9-8MapProperties** for a discussion of the Mollweide projection and how to convert data back to the original galactic longitude/latitude system.

(with north at the top). Because the image pixels number from the top of the image, if we take y as positive downward this is also backward from the above definitions. This must be kept in mind when converting pixel positions to angles:

```
arrayA = Import["wmp.csv"]; arrayWA = Import["wmpW.csv"];
...;
Do[arrayW1[[ii, jj]] = arrayWA[[ii, 1024 + jj]], {ii, 1, 2048}, {jj, 1, 2048}];
```

The Fourier transform of our two data sets is simple:

```
ftimage = Fourier[array1]; ftWimage = Fourier[arrayW1];
```

For deconvolution Fourier theory says to divide the observed data transform by the instrument response transform. Here is where Fourier techniques can be problematic. If there are any zeros in the Fourier transform of the instrument function there will be trouble:

```
newft = ftimage / ftWimage; newps = (Abs[newft])^2;
```

An image of the deconvolution can be generated via `ArrayPlot[]`.

The resulting power spectrum is redundant in four quadrants, so we only need to work with positive frequencies. We must, however, be careful about which quadrant we choose to extract the power spectrum components. We must be sure the order of image pixel indexing is being followed. `ArrayPlot[]` follows the conventions of image display routines, so the index in the upper left corner is $(1, 1)$ with subsequent numbering down for y and to the right for x. This means if we want the power spectrum to be selected with frequency increasing in the direction of the index order we must choose the lower right quadrant.

To account for the Mollweide projection, we create the nonlinear frequency scales for galactic longitude and latitude.

```
exL = 2048. / 180. λ1; fλ1 = exL[[1]] / 2; fλ = fλ1 / 1024 ii;
```

```
whL[φ2_] :=
Module[{x}, x = FindRoot[π Sin[φ2] == 2 θ1 + Sin[2 θ1], {θ1, 1}]; 1024 Sin[x[[1, 2]]]]
fφ0 = fλ[[1]]; fφ[ii1_] := fφ0 whL[ π / 2048. ii1];
```

We then generate two slices of the power spectra for comparison, one north–south, and one east–west:

```
psQ = Table[0., {ii, 1, 1024}, {jj, 1, 1024}];

Do[psQ[[ii, jj]] = newps[[ii, jj + 1024]], {ii, 1, 1024}, {jj, 1, 1024}];

slicelatA = Table[{Log[10, fφ], Log[10, psQ[[ii, 1]]]}, {ii, 1, 1024}];

slicelongA = Table[{Log[10, fλ], Log[10, psQ[[1, ii]]]}, {ii, 1, 1024}];
```

The results can be seen in figure (9.16).

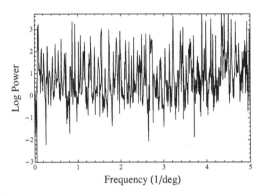 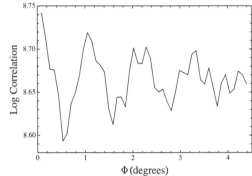

Fig. 9.16   Power spectra of CMB along longitude (left) and latitude (right).

The longitudinal (east–west) trace likely includes residual galactic features, so we assume the latitude (north–south) trace is the better representation of the power spectrum, then our results cover frequencies from 0.5 to about 100. This corresponds to a range of angles from about $2°$ to $1/100$ of a degree. The range of $0.2°$–$2.0°$ is where WMAP and COBE investigators have reported the appearance of CMB ripples with good signal to noise. If we have detected such features among the noise, they correspond to the correlated peaks that are best defined in the latitude spectrum at $\log_{10}$ frequencies of 0.32, 0.42, and 0.49 with relative powers of 100, 31, and 25 respectively. This spectral region has fairly good signal to noise, but given the appearance of other peaks in the spectrum and the extreme nonlinearity of our frequency calibration, either the positions or the relative powers could be off.

Our two power spectra are very noisy, and the agreement between them is not very good, even on a log scale, particularly at high frequencies. Part of the difficulty is introduced by the sampling and deconvolution process, as division by small instrument function entries with large errors that are near zero will produce hugely erroneous features in the power spectrum. Given the existence of the "equatorial" galactic features in the longitudinal power spectrum it is not surprising that there is a considerable difference between the spectra. Besides, we have not calibrated the fluctuation power in terms of actual temperature fluctuation.

The lack of detailed agreement between the two angular directions highlights some of the difficulties of power spectrum analysis: aliasing, leakage, noise-induced false peaks, etc. Details of these can be found in many books and papers about Fourier analysis. But Fourier transforms are not the only (and perhaps not the best) way to analyze this map of fluctuations to get critical information about their properties. Instead, mathematical constructs called wavelets provide a different and more powerful way of looking at signals and their properties.

9.6.2 Wavelet analysis

Once again we introduce a subject that has an immense amount of modern literature available for which we have no intention of covering rigorously. We include them because *Mathematica* provides a comprehensive set of tools for wavelet analysis. It is hoped that our primitive use will spur students and instructors to make further application of these important concepts.

At first encounter wavelets appear more difficult to understand than Fourier transforms. That is why so many books in (and out of) print claim to be a "friendly" guide to wavelets. To add to the confusion new wavelets and the algorithms for constructing them are discovered all the time. In keeping with those trends, *Mathematica* even lets you design your own. The point is that one often has to choose the wavelet to match the problem, and that leads to vast confusion as most novices do not know where to start, though there is a thriving wavelet community online to help in that regard.

In searching for references, it is important to recognize that wavelets have been known for many years under many different names. It has taken years for investigators from many diverse fields to realize that they were all talking about the same thing. In recent times, the term *multiresolution analysis* has been used to distinguish the process from the constructs. The astute reader will notice that we are considering only Discrete Wavelet Transforms, although *Mathematica* can calculate and display the continuous version for functions that have not been digitized.

For the person starting out, a fairly condensed summary appears at http://en.wikipedia.org/wiki/Wavelet#History, and this is a good place to start. Among texts "The World According to Wavelets" by Hubbard (1998) is outstanding. This text will show why all the fuss about wavelets and the relationship to Fourier analysis that makes wavelets work properly. Hubbard has a good set of references to the early history of the subject. For a view of the present and future, there is "A Really Friendly Wavelet Guide" by Sweldens (2012).[30] The level of mathematics is about the same as Hubbard's, but the description leads to one of the most sophisticated constructions of modern times, the Fast Lifting Wavelet Transform. Following that is Kaiser (1994), where applications of traditional wavelets to physical situations such as electromagnetic waves are considered. Finally, Bratelli and Jorgensen (2002) includes many topics not contained in Kaiser, including lifting wavelets under their proper name Daubechies-Sweldens lifting.[31]

There are several features of wavelets that make it worthwhile to consider them:

1. Wavelet analysis is not a replacement for Fourier analysis, as the methods disclose different aspects of your data. Fourier analysis produces spectral information, but without any time localization. Wavelets straddle the time-frequency domain by "localizing" the energy.
2. Instead of frequency, wavelet analysis uses scale as the length of time containing the same number of oscillations.
3. Wavelets operate in the time-domain without explicit reference to frequency. Spectral information is indicated through the width of the scale. The center of scale is the "time" localization of the energy on that scale.
4. In discrete versions of wavelets, the Nyquist frequency (as used in DFTs) becomes a Nyquist scale as the smallest range in time over which an oscillation can be detected.
5. Because wavelets get stretched or compressed, the same number of oscillations occur at small scales as large ones.
6. Wavelets extract (filter) the energy at a particular scale from the previous stage so that one has two time domains. One (called high pass) contains the energy distribution in time at that scale. The other (called low pass) contains an average residual energy distribution in time that gets passed onto the next stage.
7. The sum of the wavelet coefficients at a single scale is analogous to a power spectrum integral in frequency over the equivalent bandwidth.
8. Each scale is twice as long as the previous one, and so represents an octave in frequency.
9. The higher the level setting, the larger the final scale achieved in the process.

---

[30] See http://www.polyvalens.com/blog/

[31] For additional references, see also www.wavelet.org, www.waveletstutorial.com, and http://www.conceptualwavelets.com.

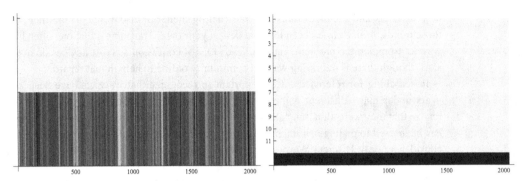

Fig. 9.17   Wavelet scalograms for the cosmic microwave background.

We illustrate some of these concepts by applying wavelets to the CMB data. Like the previous data we start with equally spaced samples. Without justifying the choice we apply what is known as a Daubechies filter to the raw WMAP data. The Daubechies filter is used because it produces the best "time" domain fidelity of the fluctuation locations. In the first order, the Daubechies filter becomes the original Haar filter, the oldest wavelet known and the default wavelet used by all *Mathematica* routines. We will begin with the north–south data along the prime meridian ($\lambda = 0$):

```
array2 = Table[0., {ii, 1, 2048}];

Do[array2[[ii]] = arrayA[[ii, 2048]], {ii, 1, 2048}];

wavelet1 = DiscreteWaveletTransform[array2, DaubechiesWavelet[1], 1]
```

Visualization of the results of a wavelet process is an important part of their application. In one dimension, one of the most useful plots is called a scalogram. *Mathematica* has other modes of display available, but the scalogram makes the most sense, particularly to novices. The left image in figure (9.17) shows the scalogram of the first two signal representations, with the top one being the first level subdivision high-pass filter. The bottom one is the original signal minus the high-pass component. Here black indicates the maximum amplitude, while white indicates zero amplitude.

In regular wavelet analysis in *Mathematica* there can be up to 11 scale levels that can be calculated. This means the final high pass strip will have 2^{10} larger scale than the original high pass strip. At the highest limit, we can make the wavelet $N = 16$, which is the highest order with the best locating properties, and the extraction levels at 11.

```
wavelet3 = DiscreteWaveletTransform[array2, DaubechiesWavelet[16], 11]
```

The right image of figure (9.17) shows the result at this maximum level. Remember that each successive high-pass strip has a factor of two larger scale than the previous one, which can be seen in the increase of line width as you move down the diagram. The result is really faint showing that most of the signal is in the average value. Notice there is no structure at all left in the final signal at 12 in the scalogram. That indicates all of the fluctuation data has been removed by the process. Notice also that except for the last strip only the high band strips are displayed.

The strongest fluctuation is very near the galactic center (centered on pixel 1024) in strip 8. This corresponds to a scale:

$$\frac{180}{2048}2^8 = 22.5°. \tag{9.66}$$

The appearance of the region around the galactic center is especially interesting because galactic features were supposed to have been removed in the processing, but here is evidence that the removal was not complete within about 22 degrees of the galactic equator. In fact if you look at the original CMB image you can perhaps see some "linear" features along the galactic equator that may be residuals from the cleaning process. That also would explain why the power spectra in the two orthogonal directions are so drastically different.

9.6.3 Wavelet packets

The characteristic restriction to scales that are separated by octaves was seen as a disadvantage in the diagnostic use of wavelets, so a method known as wavelet packets was devised. As explained by Kaiser (1994) and Brateli and Jorgensen (2002), in regular wavelet analysis, at each level (stage) of the process it is the low pass result that gets decomposed by removal of the high pass information. In wavelet packets both the low-pass and high-pass get decomposed. The result is a series of strips with scales intermediate to octaves. These results are much more complicated than previous ones.[32]

For our purposes we restrict our range of study to the region from 0 to 2 degrees to isolate the region of interest where CMB oscillations were found by WMAP and COBE investigators:

```
r = 11;

waveletP3 = DiscreteWaveletPacketTransform[array2, DaubechiesWavelet[16], r]

energyP3 = waveletP3["EnergyFraction"];

listEP3 = Table[{90/2^r 2^r 2^(ii/2^r), Log[10, energyP3[[ii + 1, 2]]]}, {ii, 1, 2^r - 1}];
```

In the wavelet analysis, as seen in figure (9.18), we find peaks in the energy spectra at scales of $0.8°$, $0.55°$, and $0.2°$.[33] These may not be exact, because even if the Daubechies is very stable in the time domain, the uncertainty principle prevents determination from being exact. The relative amplitudes are even less reliable because of the internal normalization of power at each scale, particularly for the $0.2°$ scale. But given the mean for the original data is $2.7\,K$, we can estimate the oscillations to be something like 40, 10, and 10 μK respectively. These are in reasonable agreement ratio wise with the power spectrum results. But here the scale positions are superior to the power spectrum results, which depended on a very nonlinear frequency calibration. On the other hand, the power ratios are probably better estimated from the power spectrum since the wavelet normalization is a bit unusual.

It has taken a considerable amount of effort to tease out the cosmological oscillations, but in spite of the difficulties we seem to have detected them even if marginally. Although additional processing such as averaging the power or energy spectra can be applied to the power spectrum or the power scale results to make them smoother, we leave that to the reader.

---

[32] It is often useful to look at a tree structure of how each level is decomposed. Examples of these can be found in **9-7CMBR**.

[33] Analysis of an equatorial strip yields similar results. For details see **9-7CMBR**.

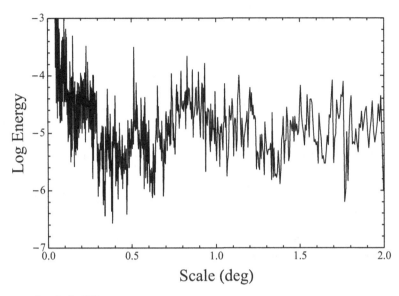

Fig. 9.18 Wavelet package transform for the CMB.

9.7 Cosmological models

Before Einstein published his theories of relativity, most cosmological models were Newtonian, and although some good questions were posed by Olbers and others, prior to Einstein most discussions were quite philosophical rather than scientific. Not only did the relativistic view change cosmological thought, but it also nurtured an observational interest in using large telescopes to study the elliptical and spiral nebulae that seemed to be in a different class from those nebulae associated with clusters of stars. This close association between theory and observation has remained a hallmark of cosmology ever since. In the present day, along with the data flood we mentioned earlier, the theory continues to thrive as well, with no lack of ideas to try to explain the surprising observations that continue to pop up. Cosmology probably has more written upon it than any other astrophysical topic. Here we take only a broad overview of these models.

9.7.1 Cosmological parameters

Prior to general relativity and the discovery of the expansion of the galaxies, most Newtonian models had to struggle with the notion of a static universe. Even after GR, Einstein struggled with the same issue. He solved it by postulating that atomic positive and negative charges did not balance. This led to a repulsive term that just balanced gravity. As a result this added a term to his equations he called the cosmological constant Λ.

Once the expansion velocity was observed, then Newtonian models could be constructed without concern about catastrophic collapse, at least for the present. However even models ignoring Λ need to indicate something about the expansion. To first order[34] the expansion is dominated by a linear term in redshift $z = \Delta\lambda/\lambda$. The constant is of course called Hubble's constant, H_o such that

$$H_o = \frac{cz}{d}. \tag{9.67}$$

As seen in Section 9.1, there is some evidence for an additional quadratic term in the least squares fit of the data. Following Kolb and Turner (1990), we started with the usual formula that makes the Hubble parameter proportional to the slope,

$$z = \frac{H_o d_L}{c} + \frac{1}{2c}(q_o - 1)(H_o d_L)^2 + \cdots \tag{9.68}$$

with the new parameter q_o called the deceleration parameter. As in regular Newtonian physics, this would indicate some slight deceleration or acceleration. In this case the coefficient was negative, so it is interpreted as deceleration.

Because looking out in distance implies looking back into time, if the expansion increases with distance in reality it has been slowing down with time at a constant rate. On the other hand, if the Hubble parameter, rather than being a constant, is defined by a power series in z we have[35]

$$H_o d_L = cz + \frac{1}{2}(q_o - 1)(cz)^2 + \cdots \tag{9.69}$$

Because z is better determined than d_L, this equation is better statistically.

There are two problems with these kinematic equations. First, the distance determination has been, until very recent times, uncertain. The uncertainty has been large enough that one finds an h fudge factor substituted into the equations in the literature. Kolb and Turner use $H_o = 100h$ (km/s)/Mpsc. Between the most recent Hubble telescope and COBE-WMAP work, we find[36]:

1. 73.8 ± 2.4 (km/s)/Mpsc. (2011 HST, various objects)
2. 71.0 ± 2.5 (km/s)/Mpsc. (2010 WMAP, 7 year data)
3. 72.6 ± 3.1 (km/s)/Mpsc. (2010 HST, gravitational lensing)
4. 72 ± 8 (km/s)/Mpsc. (2001 HST, Cepheid variables)

To these we add our own value,

$$H_o = 73.23 \pm 0.16 \text{ (km/s)/Mpsc}, \tag{9.70}$$

obtained from the NED (2009) galaxy catalog. The second coefficient is $(-2.6 \pm 0.3) \times 10^{-6}$, with the implication that q_o is slightly larger than 1 for that data. But because of the fluctuations in the residuals, presumably due to clumping by dark matter, this value is suspect as a global value.

The second issue is because we have assumed that the special relativity corrections were all that were needed to obtain the observational z values. We also assumed that the sun's orbital velocity had to be subtracted as well. As discussed by Burke and Graham-Smith (2002), there are first-order GR corrections that can also be applied to the distances, and these will affect the more distant object results. Since the distances are usually obtained based on the luminosity of objects, we have to convert these luminosity distances to GR "coordinate" distances D. That is, we set

$$dL = dC(1 + z). \tag{9.71}$$

---

[34] See **9-1HubbleLaw**.
[35] Kolb and Turner, 1990.
[36] *Wikipedia*, 2012.

But the actual dependence of D is on the Hubble parameter and z, therefore it is model dependent, which we do not want. A more proper approach is to redo the regression and see if we get a better value of q_o than 1:

```
newdata = Import["NEDRdata.csv"]; numline = Length[newdata];
```

. . .

```
h1 = c2 / coeffB[[1]]; b0 = coeffB[[2]] / c2 h1 ;
```

This yields

$$H_o = 161.0 \pm 1.4 \text{ (km/s)/Mpcs}, \qquad (1 - q_o) = (-7.15 \pm 0.93) \times 10^{-6}. \qquad (9.72)$$

Thus while the Hubble constant is rescaled higher as one expects, the second-order coefficient is proportionally smaller by a factor of 3. Taken at face value this means on the GR distance scale the universe is flatter than on even the luminosity distance scale. To explain this extraordinary flatness of the universe today, an abrupt period of expansion is postulated by the "inflationary scenario" of cosmology. This period happened in the time between the "beginning" and the formation of the CMB, long before the formation of the elements and the galaxies.

If indeed $q_o \sim 1$, then it means that the present expansion is almost purely Newtonian, and self gravity is slowing down the expansion of the galaxies. This also means that the nearby universe is now cold-matter–dominated, which includes dark matter.

9.7.2 Newtonian models

The "constant" part of the expansion itself is characterized by the Hubble parameter regardless of which model is chosen. Thus

$$H_o = \left(\frac{\dot{R}}{R} \right)_{t_o}. \qquad (9.73)$$

The function $R(t)$ is a scaling function that follows the expansion. The Hubble parameter indicates that the velocity of expansion divided by the present radius is invariant.

In the Newtonian case with an inverse square law, there is an acceleration (deceleration) given by

$$\frac{d^2 r}{dt^2} = -\frac{GM}{r^2}, \qquad (9.74)$$

which can be integrated to find the gravitational potential,

$$PE = -\frac{GM}{r}. \qquad (9.75)$$

In the same way, the kinetic energy is

$$KE = \int \dot{R} d\dot{R} = \frac{1}{2} \dot{R}^2. \qquad (9.76)$$

The total energy is then

$$E = KE + PE = \frac{1}{2} \dot{R}^2 - \frac{GM}{r}. \qquad (9.77)$$

For a flat universe, $E = 0$. With this and the boundary condition $R(0) = 0$ we can solve the differential equation to find

$$R(t) = \left(\sqrt{\frac{9GM}{2}} t \right)^{2/3}. \tag{9.78}$$

Applying conservation of mass, we assume the density changes as the volume expands. Thus

$$\rho(t) = \rho_o \left(\frac{R_o}{R(t)} \right)^3. \tag{9.79}$$

If we let

$$\rho(t) = \frac{3M}{4\pi R^3}, \tag{9.80}$$

and let $\rho_c = \rho_o$ as the density when $R_o = 1$, then the density can be expressed in terms of the Hubble constant as

$$\rho_c = \left(\frac{3}{8\pi G} \right) H_o^2. \tag{9.81}$$

This is used to define another parameter Ω, with

$$\Omega = \frac{\rho_m}{\rho_c}, \tag{9.82}$$

where ρ_m is the characteristic model density, analogous to ρ_c.

There are four parameters that arise in cosmological theory: Λ, H_o, Ω, and q_o. In Newtonian theory q_o has a special place because it is defined as a deceleration parameter to go with the minus sign in the $-GM/R$ term. To make q_o positive when $\ddot{R}$ is negative, we have

$$q_o = \frac{R(t)\ddot{R}(t)}{\dot{R}^2(t)}. \tag{9.83}$$

It turns out that for $\Lambda = 0$, $\Omega = 2q_o$, while our results on galaxies indicate that q_o is very close to unity. Is there something wrong here? The observations thus indicate that our part of the universe is "bound," since our weighted results are dominated by accurate redshifts and are therefore "local." We can see in our results the effects of clustering on the redshifts, so it is likely that the presence of dark matter binds the galaxies and galaxy clusters together.

Another property of the flat Newtonian model is that we can estimate the age of the universe. Although the Hubble parameter is in the standard form of (km/s)/mpcs, if we convert the megaparsecs to kilometers then the Hubble parameter is a unit of time. For the flat case this gives a cosmic age of

$$T_u = \frac{2}{3H_o}. \tag{9.84}$$

Using our value of $H_o = 73.23$, this gives an age of 8.8 billion years. Although this value is not highly accurate, it is of the correct order of magnitude.

9.7.3 Relativistic models

We now consider the rudiments of cosmological theory. The fundamental basis of modern theory is the Friedmann–Robertson–Walker, or hot big bang model.[37] Kolb and Turner (1990) present a nice

---

[37] Burke and Graham-Smith (2002); Kolb and Turner (1990).

discussion of the geometric and kinematic interpretation of the Robertson–Walker metric to which the reader is referred for details. Recalling the previously used Schwarzschild metric, you will see some similarities, as both are for spherical "co-moving" coordinates,

$$ds^2 = dt^2 - R^2(t)\left(\frac{dr^2}{1 - kr^2} + r^2 d\theta^2 + r^2 \sin^2\theta\, d\phi^2\right). \tag{9.85}$$

The main difference here is the inclusion of $R(t)$ as a scale factor, making the metric time dependent. Kolb and Turner use this metric to develop the essentials of standard cosmology. This starts with deriving the Friedmann equation that governs the evolution of $R(t)$, as if the metric came first. Historically, the Friedmann equation came first, then the metric was found by Robertson and Walker.[38] What Friedmann did was to incorporate the density law given above into the Newtonian equation for $R(t)$ and replace the energy expression with a curvature term that expressed the fact that the energy, in a classical mechanics sense, is what distinguished flat (parabolic), positive curvature (bound) and negative curvature (unbound). space. This was replaced by the k parameter, with $k = 0$ for flat, $k = -1$ for unbound and $k = 1$ for bound. Thus, the original Friedmann equation with $\Lambda = 0$ was

$$\dot{R}^2 - \frac{8\pi}{3}\rho R^2 = -kc^2. \tag{9.86}$$

If we include the cosmological constant, this becomes[39]

$$\dot{R}^2 - \frac{8\pi}{3}\left(\rho + \rho\Lambda\right)R^2 = -kc^2. \tag{9.87}$$

The selection of the density relationship depends on a bit of thermodynamics to relate the density to pressure. Pressure through the stress-tensor is a natural concept in GR and density and pressure are related through an equation of state. The simplest equation of state is a direct proportionality between pressure and density such that $p = w\rho$, where w is some constant.

Because the Hubble parameter is defined as $H_o = \dot{R}/R$, one does not need to solve a differential equation to glean useful information. If we take the derivative of this equation, then[40]

$$\frac{\ddot{R}}{R} = -\frac{4\pi G}{3}\left(\rho - 2\rho\Lambda + 3p\right), \tag{9.88}$$

where ρ is the material density and $3p$ is the radiation density, $w = -1$ for the dark energy equation of state,[41] and

$$\rho\Lambda = \frac{c^2}{8\pi G}\Lambda. \tag{9.89}$$

Defining R as the present scale radius and k as the curvature, then

$$\frac{k}{H^2 R^2} = \Omega - 1. \tag{9.90}$$

[38] Burke and Graham-Smith, p. 301.
[39] Obergaulinger (2006). See http://www.mpa-garching.mpg.de/lectures/ADSEM/SS06_Obergaulinger.pdf.
[40] Obergaulinger (2006).
[41] For regular matter, one would expect $w > 0$. Here, however, cosmic expansion requires an equation of state for dark energy to have $w < 0$, such that dark energy works to expand the universe rather than (gravitationally) contract it. The simplest model (that of a cosmological constant) has $w = -1$. Other models such as quintessence give $w \neq -1$, or even varying in time. Current observations cannot distinguish between these models, so we here use the cosmological constant.

If a zero subscript indicates the present, then

$$q_o = \frac{\Omega_o}{2}\left(1 - \frac{3p}{\rho}\right) = \frac{\Omega_o}{2}(1 + 3w).$$
                                                                          (9.91)

To solve these differential equations one needs only to specify the R dependence on the density or pressure, and the k value. There are several simple density relations which can be assumed:

1. $\rho \propto R^{-3(1+w)}$, general model
2. $\rho \propto R^{-4}$, radiation dominated
3. $\rho \propto R^{-3}$, matter dominated
4. $\rho \propto R^0$, dark energy dominated

For dark matter, the equation of state is typically taken as $w = 0$, but recently Serra and Romero (2011) showed that values of $w \approx -0.3$. On the other hand, a theory paper by Müller (2005) gives

$$-1.50 \times 10^{-6} < w < 1.13 \times 10^{-6}$$
                                                                          (9.92)

for no entropy, or

$$-8.78 \times 10^{-3} < w < 1.86 \times 10^{-3}$$
                                                                          (9.93)

if the adiabatic sound speed vanishes. The jury is still out on this one, so we will take $w = 0$.

Substituting the various density relations into equation (9.87), we can compare their evolution. Most of these can be solved in *Mathematica* simply through DSolve or NDSolve[]. The parametrized form of the Friedmann equation we can use is

```
friedmann = rR'[t]² - 8 π/3 ( krad/rR[t]⁴ + kbarDM/rR[t]³ + kde ) rR[t]² + κ c²
```

where kde, krad, and kbarDM represent the presence of dark energy, radiation, and matter respectively. For example, for a negative curvature radiation dominated model,

```
friedmann10 = friedmann /. {krad → 1, kbarDM → 0, kde → 0, κ → -1}

sol10 = DSolve[{friedmann10 == 0, rR[1] == r0}, rR[t], t]

negrad[t_] = sol10[[2, 1, 2]] /. {c → 1, r0 → 1}
```

Two solutions, the matter dominated universes with curvature, cannot be handled directly with DSolve, so we must restructure the equation a bit. In Chapter 5 we explored a number of ways to deal with the relativistic two-body problem, including those involving squares of the first derivatives and powers of inverse radii. We face a similar issue here. Starting with equation (9.87), we substitute $u = 1/R$; thus

$$\dot{u}^2 - \frac{8\pi G}{3}(\rho(u) + \Lambda\rho(u))\,u^2 = -kc^2u^4.$$
                                                                          (9.94)

In *Mathematica* this becomes:

```
friedmannU = u'[t]² - 8 π/3 ( krad u[t]⁴ + kbarDM u[t]³ + kde ) u[t]² + κ c² u[t]⁴
```

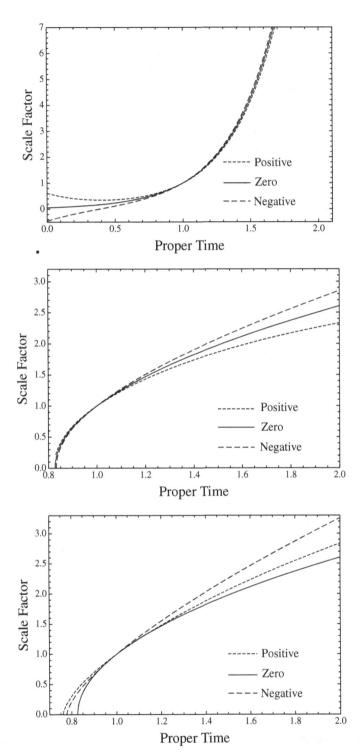

Fig. 9.19 Evolution of energy (top), radiation (center), and matter (bottom) dominated universes.

The solutions can then be obtained via NDSolve. For example, for the positive curvature case,

```
friedmannU1 = friedmannU /. {krad → 0, kbarDM → 1, kde → 0, κ → 1}

friedmann8 = friedmannU1 /. c → 1

solu3 = NDSolve[{friedmann8 == 0, u[1] == 1}, u[t], {t, 0, 2}]
```

In the negative curvature case,

```
friedmannU2 = friedmannU /. {krad → 0, kbarDM → 1, kde → 0, κ → -1}

friedmann11 = friedmannU2 /. c → 1

solu4 = NDSolve[{friedmann11 == 0, u[1] == 1}, u[t], {t, 0, 2}]
```

Something unusual happens here. *Mathematica* responds with two different interpolating functions, one for each sign of the derivative. Looking carefully at the arguments of the functions you can see the first root corresponds to an expanding model, while the second root corresponds to a collapsing model. While the latter could be used to represent a "big crunch" model, we are here interested in the first one:

```
posbar[t1_] = 1 / solu3[[1, 1, 2]] /. t → t1
```

The results of these different models can be seen in figure (9.19). As expected, the dark energy dominated models differ radically from the matter and radiation driven ones. However, even in the matter and radiation models different curvatures have differing effects. In general, radiation, matter, and dark energy all have measurable effects on both the history and future of cosmic evolution.

9.7.4 A final model

We now put all the elements of our simple models together to produce three final models, one for each curvature. The proportion of each type of density will follow modern values (Obergaulinger, 2006). Specifically, $krad = 10^{-2}$, $kbarDM = 0.26$, and $kde = 0.74$. As with the matter cases there are two solutions, one for expansion and one for collapse. We will consider the expansion solution, as this agrees with the Hubble flow parameters. The result is seen in figure (9.20). It is clear that although these models are similar, the curvature effects both the past age and future expansion of the universe.

Interpreting the time axis as proper time and not coordinate time, we can test the flatness of our final model using a least square fit. We will assume the maximum look-back time (on the plot scale) is at most 0.005. For the zero curvature case,

```
friedmannUzero = friedmannU /. {krad → 10⁻², kbarDM → 0.26, kde → 0.74, κ → 0, c → 1}

. . .

lmzero = LinearModelFit[ptimezero, {tp, tp²}, tp, IncludeConstantBasis → False]
```

which yields

$$R(t) = 6.20t - 2.67t^2. \tag{9.95}$$

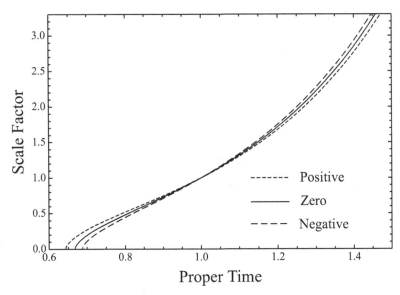

Evolution of the universe using modern parameters.

From the relative size of the second-order coefficient to the first, we must conclude that our models are not flat enough to match our present universe. These were, of course, simple models where we assumed constant proportions of radiation, matter, and dark energy. To make a better model we would need to consider the time-dependent proportions of the various constituents, and hence consider the thermodynamics of the universe as a whole. This is a subject far beyond the scope of our present efforts.[42]

Exercises

9.1 In **9-1HubbleLaw** the weighting of the fit z versus distance was the square of the weights in the weight list. Play "what if" with the exponents of the weights to find what noninteger values produces the best correlation of the relationship.

9.2 A superluminal jet is observed to have an apparent transverse velocity of $6.2c$. Determine the possible range of the actual speed of the jet and the possible range of the angle θ from our line of sight.

9.3 Assume that the galactic center is an old quasar. According to Burke and Graham-Smith (p. 200), the luminosity of Sgr A* is 10^{-7} that of a moderately active radio galaxy such as M87. Use the plot in **9-2quasars** that gives the range of absolute magnitude for a moderate radio source ($\sim$1500).

1. Select a value for an M87-like galaxy from the mean of the Mv range, then find what Mv corresponds to 10^{-7} of that average galaxy.

---

[42] We have only scratched the surface of Kolb and Turner, which is recommended for those readers interested in pursuing our model further.

2. Use the regression for radio quiet quasars to plot the evolutionary curve versus z for the galactic center and compare with a "normal" quasar curve.

9.4 Create a plot of `ArcTan[alph[]]` for the deflection angle of a spinning black hole.

1. Show that the against-spin case initially increases in angle, but ends up minus.

2. Show that the difference of angle between with and against spin cases become a maximum at minimum impact parameter and maximum spin.

9.5 In **1-5ImageRestore** we presented and analyzed an image of Comet Hyakutake (comet8). In a fashion parallel to the CMB analysis in **9-7CMB**, compute, display, and interpret the 2D power spectrum of the comet image in terms of the fine structure of the image.

9.6 The observable universe has a radius of about 14 gigaparsecs, and an average density of about 10^{-30} g/cc. The cosmic microwave background originates from a time (known as decoupling) when the universe had a temperature of about 3000 K.

1. Model the universe as a uniform ideal gas with a temperature equal to that of the cosmic background, and estimate the radius of the universe at the time of decoupling.

2. A better approach is to assume a FRW model, where the scale factor of the universe at the time of origin is related to the observed redshift,

$$a(t) = \frac{1}{1 + z}.$$

Analysis of the WMAP seven year data gives the redshift of decoupling as $z = 1090.88 \pm 0.72$. Calculate the radius of the universe at decoupling.

9.7 Consider a universe consisting of regular matter and purely Newtonian gravity. Demonstrate that a "static" universe is unstable; therefore the matter of such a universe will either continually expand or contract. Hint: Suppose you tried to construct a balanced universe of perfectly placed equal mass galaxies.

9.8 A de Sitter universe is one in which contains a cosmological constant Λ, but no regular matter. For a de Sitter universe the density and pressure are given by

$$\rho = \frac{\Lambda}{8\pi G}, \quad p = -\frac{\Lambda}{8\pi G}.$$

It is also assumed to be flat; thus $k = 0$.

1. From the Friedmann equation (9.87), derive the value of Λ in terms of the Hubble parameter H_o.

2. Show that a pure de Sitter universe has no big bang. That is, there is no solution $R = 0$ for finite time.

9.9 In **9-9Models** we quoted a possible w value of -0.3 for the dark matter equation of state. This makes the R exponent -1.8. Make this substitution in the Friedmann models and compare with the $w = 0$ results.

9.10 Find "flicker" data for a quasar other than 3C273 and analyze it using the methods in **9-3Q3C273**. Compare with 3C273.

9.11 Using the methods and data of **9-7CMB**, Fourier analyze a region of the CMB that is off the galactic equator or off the galactic meridian. Compare with the spectra in **9-7CMB**.

9.12 Using the methods and data of **9-7CMB**, wavelet analyze a region of the CMB that is off the galactic equator or off the galactic meridian. Compare with the scalograms in **9-7CMB**.

9.13 In **9-9Models**, change the proportions of radiation, matter, and dark energy in the "final model" by trial and error to see what numbers give the "observed" acceleration to Hubble parameter ratio.

References

Abraham, R., and Marsden, J.E. 1980. *Foundations of Mechanics*. Reading, MA: The Benjamin/ Cummings. Advanced Book Program.

Adams, Douglas. 1986. *The Ultimate Hitchhiker's Guide*. New York: Wings Books.

Aitchison, J., and Brown, J.A.C. 1957. *The Lognormal Distribution*. New York: Cambridge University Press.

Aller, Lawrence H. 1963. *Astrophysics: The Atmospheres of the Sun and Stars*, 2nd ed. New York: Ronald Press.

Ambartsumian, V. 1958. *Theoretical Astrophysics*. London: Pergamon Press.

Andrews, H.C., and Hunt, B.R. 1977. *Digital Image Restoration*. Prentice-Hall Signal Processing Series. Englewood Cliffs, NJ: Prentice-Hall.

Angione, R.J., and Smith, H.J. 1983. 3C 273 Historical Data Base. *Astronomical Journal*, **90**, 2474–2486.

Angélil, R., and Saha, P. 2010. Relativistic Redshift Effects and the Galactic Center Stars. *Astrophysical Journal*, **711**.

Angélil, R., Saha, P., and Merritt, D. 2010. Toward Relativistic Orbit Fitting of Galactic Center Stars and Pulsars. *Astrophysical Journal*, **720**, 1303–1310.

Arfkin, George B., and Weber, Hans J. 2001. *Mathematical Methods for Physicists*, 5th ed. San Diego: Harcourt/Academic Press.

Austen, Jane. 2006. *Pride and Prejudice*, Edited by Pat Rogers. The Cambridge Edition of the Works of Jane Austen. Cambridge: Cambridge University Press.

Bahcall, N.A. 1988. *Large-Scale Structure and Motion Traced by Galaxy Clusters*. Princeton, NJ: Princeton University Press.

Baker, G.L., and Gollub, J.P. 1996. *Chaotic Dynamics*, 2nd ed. Cambridge, U.K.: Cambridge University Press.

Baldwin, Ralph B. 1944. Certain Effects of Stellar Rotation. *Popular Astronomy*, **52**, 134–138.

Baugh, C.M., and Efstathiou, G. 1993. The Three-Dimensional Power Spectrum Measured from the APM Galaxy Survey-I. Use of the Angular Correlation Function. *MNRAS*, **265**, 145–156.

Baumann, Gerd. 2000. *Symmetry Analysis of Differential Equations with Mathematica*. New York: Springer-Verlag.

Beutler, F.J. 1961. Sampling Theorems and Bases in a Hilbert Space. *Information Control*, **4**.

Bolstad, William M. 2004. *Introduction to Bayesian Statistics*. Hoboken, NJ: John Wiley & Sons.

Borne, K.D., and 91 others. 2009. *Astroinformatics: A 21st Century Approach to Astronomy*. Tech. rept. Submitted to 2010 Decadal Survey.

Borwein, J.M, and Bailey, David H. 2004. *Mathematics by Experiment: Plausible Reasoning in the 21st Century*. Natick, MA: A.K. Peters.

Borwein, J.M, Bailey, David H., and Girgensohn, Roland. 2004. *Experimentation in Mathematics: Computational Paths to Discovery*. Natick, MA: A.K. Peters.

Bowers, Richard L., and Wilson, James R. 1991. *Numerical Modeling in Applied Physics and Astrophysics*. Boston: Jones and Bartlett.

Bracewell, R. 1965. *The Fourier Transform and Its Applications*. New York: McGraw-Hill.

Brandt, J.C. 1970. *Introduction to the Solar Wind*. San Francisco: W.H. Freeman.

Bratteli, O., and Jorgensen, P. 2002. *Wavelets through a Looking Glass: The World of the Spectrum*. Applied and Numerical Harmonic Analysis. Boston: Birkhäuser.

Brigham, E. Oran. 1974. *The Fast Fourier Transform*. Englewood Cliffs, NJ: Prentice-Hall.

Buchler, J. Robert, and Eichhorn, Heinrich. 1987. Chaotic Phenomena in Astrophysics. *Annals of the New York Academy of Sciences*, **497**, 83.

Buchler, J.R., and Eichhorn, H. 1987. *Chaotic Phenomena in Astrophysics*, Vol. 155. New York: New York Academy of Sciences.

Budden, K.G. 1961. *Radio Waves in the Ionosphere*. Cambridge, U.K.: Cambridge Press.

Burke, B.F., and Graham-Smith, F. 2002. *An Introduction to Radio Astronomy*. 2nd ed. Cambridge: Cambridge University Press.

Böhm-Vitense, Erika. 1989. *Introduction to Stellar Astrophysics, Vol. 2: Stellar Atmospheres*. Cambridge: Cambridge University Press.

Böhm-Vitense, Erika. 1992. *Introduction to Stellar Astrophysics, Vol. 3: Stellar Structure and Evolution*. Cambridge: Cambridge University Press.

Callen, Herbert B. 1985. *Thermodynamics and an Introduction to Thermostatistics*. 2nd ed. New York: John Wiley & Sons.

Cartan, E. 1966. *The Theory of Spinors*. Cambridge, MA: MIT Press.

Chandra, Suresh. 2005. *Relation between Spectroscopic Constants with Limited Dunham Coefficients Pramana*, **65**(6), 1133–1137.

Chandrasekhar, Subrahmanyan. 1934. The Radiative Equilibrium of Extended Stellar Atmospheres. *MNRAS*, **94**, 443.

Chandrasekhar, Subrahmanyan. 1942, 1943. *Principles of Stellar Dynamics, Expanded Edition*. New York: Dover.

Chandrasekhar, Subrahmanyan. 1943a. Dynamical Friction, Part 1. *Astrophysical Journal*, **97**(No. 2), 255–262.

Chandrasekhar, Subrahmanyan. 1943b. Dynamical Friction, Part 2. *Astrophysical Journal*, **97**(No. 2).

Chandrasekhar, Subrahmanyan. 1943c. Dynamical Friction, Part 3. *Astrophysical Journal*, **98**(No. 1).

Chandrasekhar, Subrahmanyan. 1960. *Radiative Transfer*. New York: Dover.

Chandrasekhar, Subrahmanyan. 1967. *An Introduction to the Study of Stellar Structure*. New York: Dover Publications.

Chapman, G., et al. 1989. *The Complete Monty Python's Flying Circus: All the Words*, Vol. 1. New York: Pantheon Books.

Cincotta, P.M., Mendez, M., and Nunez, J.A. 1995. Astronomical Time Series Analysis. I. A Search for Periodicity Using Information Entropy. *Astrophysical Journal*, **449**, 231–235.

Coles, W.D., Fakan, J.C., and Laurence, J.C. 1966. *Superconducting Magnetic Bottle for Plasma Physics Experiments. NASA* (NASA Technical TN D-3595).

Collins, George W. 1978. *The Virial Theorem in Stellar Astrophysics*. Astronomy and Astrophysics Series, Vol. 7. Tucson, AZ: Pachart.

Contopoulos, George. 1966. *Problems of Stellar Dynamics*. Providence, RI: American Mathematical Society, pp. 169–258.

Contopoulos, G. 1987. *Stochasticity in Galactic Models*. New York: New York Academy of Sciences.

Cox, Arthur N. 2000. *Allen's Astrophysical Quantities*, 4th ed. New York: Springer-Verlag.

Crandall, R.E. 1994. *Projects in Scientific Computation*. TELOS, The Electronic Library of Science. Santa Clara, CA: Springer-Verlag.

Crandall, R.E. 1996. *Topics in Advanced Scientific Computation*. TELOS, The Electronic Library of Science. Santa Clara, CA: Springer-Verlag.

Crandall, R.E. 1999. New Representation of the Madelung Constant. *Experimental Mathematics*, **8**, 367–379.

Crocce, M., Cabre, A., and Gaztanaga, E. 2011. Modeling the Angular Correlation Function and Its Full Covariance in Photometric Galaxy Surveys. *MNRAS*, **414**, 329–349.

Danby, John M.A. 1988. *Fundamentals of Celestial Mechanics*. 2nd rev and enlarged edn. Richmond, VA: Willmann-Bell.

Däppen, Werner. 2000. *Ch.3 -Atoms and Molecules*. 2nd ed., New York: Springer-Verlag, pp. 27–51.

Daubechies, I., and Sweldens, W. 1998. Factoring Wavelet Transforms into Lifting Steps. *Journal Fourier Analysis and Application*, **4**, 247–269.

Davies, Kenneth. 1966. *Ionospheric Radio Propagation*. New York: Dover Publications.

Davis, Harold T. 1962. *Introduction to Non-linear Differential and Integral Equations*. New York: Dover Publications.

Deeming, T.J. 1975. Fourier Analysis with Unequally-Spaced Data. *Astrophysics and Space Science*, **36**(1), 137–158.

Dekel, A. 1988. *Theoretical Implications of Superclustering*. Princeton, NJ: Princeton University Press.

Diacu, F., and Holmes P., 1996. *Celestial Encounters*. Princeton, NJ: Princeton University Press.

Dingle, R.B, Arndt, D., and Roy, S.K. 1957. The Integrals Cp and Dp and their tabulation. *Applied Science Research*, **6-B**, 144–155.

Dirac, P.A.M. 1975. *General Theory of Relativity*. New York: Wiley Interscience.

Drazin, P.G. 1992. *Non-linear Systems*. Cambridge Texts in Applied Mathematics. Cambridge, U.K.: Cambridge University Press.

Dubin, Daniel. 2003. *Numerical and Analytical Methods for Scientists and Engineers using* Mathematica. Hoboken, NJ: John Wiley Sons.

Eddington, Arthur S. 1923 (reprinted 1960). *The Mathematical Theory of Relativity*. Cambridge, U.K.: Cambridge University Press.

Eddy, J.A. 1978. *The New Solar Physics*. The AAAS Selected Symposia Series, Vol. 17. Boulder, CO: Westview Press for the American Association for the Advancement of Science.

Faber, S.M., and Burstein, D. 1988. *Motions of Galaxies in the Neighborhood of the Local Group*. Princeton, NJ: Princeton University Press.

Feller, William. 1950. *An Introduction to Probability Theory and Its Applications*. Vol. I, 3rd ed. New York: John Wiley & Sons.

Feller, William. 1966. *An Introduction to Probability Theory and Its Applications*, Vol. II. New York: John Wiley and Son.

Feynman, R.P., Morinigo, F.B., and Wagner, W.G. 1995. *Feynman Lectures on Gravitation*. The Advanced Book Program. Reading, MA: Addison-Wesley.

Fischel, D. 1976. Oversampling of Digitized Images. *Astronomical Journal*, **81**.

Friedman, Harold, L. 1985. *A Course in Statistical Mechanics*. Englewood Cliffs, NJ: Prentice-Hall.

Friedrich, M., Finsterbusch, R., Tokar, K.M., and Spœcker, H. 1991. A Further Generalization of the Sen and Wyller Magnetoionic Theory. *Advances in Space Research*, **11**(10), 105–108.

Fry, J.N. 1987. *On the Origin of Large Scale Cosmological Structure*. New York: New York Academy of Sciences.

Funsten, H.O., and 13 others. 2009. Structures and Spectral Variations of the Outer Heliosphere in IBEX Energetic Neutral Atom Maps. *Science*, **326**(13 November 2009), 964–966.

Fuselier, S.A., and 14 others. 2009. Width and Variation of the ENA Flux Ribbon Observed by the Interstellar Boundary Explorer. *Science*, **326**(13 November 2009), 962–964.

Gibbs, J.W. 1966. "On the Equilibrium of Heterogeneous Substances" in The Scientific Papers of J. Willard Gibbs: Vol. 1. *Thermodynamics*, H.A. Bumstead, Van Name, and Ralph Gibbs Dover, New York, 55–371.

Galison, P. 2003. *Einstein's Time, Poincaré's Maps Empires of Time*. New York: W.W. Norton.

Gillessen, S., Eisenhauer, F., Trippe, S., Alexander, T., Genzel, R., Martins, F., and Ott, T. 2009. Monitoring Stellar Orbits Around the Massive Black Hole in the Galactic Center. *Astrophysical Journal*, **692**(2), 1075–1109.

Goldstein, Herbert, Poole, C.P. Jr., and Safko, J.L. 2002. *Classical Mechanics*, 3rd edn. New York: Addison-Wesley.

Goldstein, S.J., and Meisel, D.D. 1969. Observations of the Crab Pulsar During an Occultation by the Solar Corona. *Nature*, **224**, 349–350.

Gray, David F. 1992. *The Observation and Analysis of Stellar Photospheres*, 2nd ed. Cambridge Astrophysics Series. Cambridge: Cambridge University Press.

Green, Robin M. 1985. *Spherical Astronomy*. Cambridge: Cambridge University Press.

Griffin, R.F. 2010. Spectroscopic Binary Orbits from Photoelectric Radial Velocities Paper 215. *The Observatory: A Review of Astronomy*, **130**(1219), 349–363.

Grossman, Nathaniel. 1996. *The Sheer Joy of Celestial Mechanics*. Boston: Birkäuser.

Groth, H.G., and Wellman, P. 1970. *Spectrum Formation in Stars with Steady-State Extended Atmospheres*. IAU Collequia: International Astronomical Union.

Gurnett, Don A., Bhattacharjee, Amitava. 2005. *Introduction to Plasma Physics With Space and Laboratory Applications*. Cambridge: Cambridge University Press.

Haardt, F., and 15 others. 1998. The Hidden X-ray Seyfert Nucleus in 3C273:BeppoSAX Results. *Astronomy and Astrophysics*, **340**, 35–46.

Hamming, R.W. 1977. *Digital Filters*. Englewood Cliffs, NJ: Prentice-Hall.

Harris, John W., and Stöcker, Horst. 1998. *Handbook of Mathematics and Computational Science*. New York: Springer Verlag.

Hartmann, L., and Burkert, A. 2007. On the Structure of the Orion A Cloud and the Formation of the Orion Nebula Cluster. *Astrophysical Journal*, **654**, 988–997.

Harwit, Martin. 1988. *Astrophysical Concepts*, 2nd ed. Astronomy and Astrophysics Library. New York: Springer-Verlag.

Hasani, Sadri. 1999. *Mathematical Physics: A Modern Introduction*. New York: Springer-Verlag.

Hassani, Sadri. 2003. *Mathematical Methods Using Mathematica: For Students of Physics and Related Fields*. Undergraduate Texts in Contemporary Physics. New York: Springer.

Hawking, S. W. 1976. Black Holes and Thermodynamics. *Physical Review D*, **13**(2), 191–197. PRD.

Haynes, M.P., and Giovanelli, Ri. 1988. *Large Scale Structure in the Local Universe: The Pisces-Perseus Super Cluster*. Princeton, NJ: Princeton University Press.

Heggie, D., and Hut, P. 2003. *The Gravitational Million-Body Problem*. Cambridge: Cambridge University Press.

Herbst, Eric. 1995. Chemistry in the Interstellar Medium. *Annual Review of Physical Chemistry*, 27–53.

Herbst, Eric, and van Dishoeck, Ewine F. 2009. *Complex Organic Interstellar Molecules*, Vol. 47, Annual Review of Astronomy and Astrophysics: Annual Reviews, 427–480.

Herman, J.R., and Goldberg, R.A., 1978. *Sun, Weather, Climate*. NASA Special Publications, Vol. SP-426. Washington, DC: National Aeronautics and Space Administration.

Herzberg, Gerard. 1950. *Molecular Spectra and Molecular Structure I. Spectra of Diatomic Molecules* 2nd ed. New York: Van Nostrand Reinhold.

Hilborn, Robert C. 1994. *Chaos and Non-linear Dynamics: An Introduction for Scientists and Engineers*. New York: Oxford University Press.

Hillenbrand, L.A., and Hartmann, L.W. 1998. A Preliminary Study of the Orion Nebula Cluster Structure and Dynamics. *Astrophysical Journal*, **492**, 540–553.

Hook, I.M., McMahon, R.G., Boyle, B.J., and Irwin, M.J. 1994. The Variability of Optically Selected Quasars. *Monthly Notices of the Royal Astronomical Society*, **268**, 305–320.

Hubbard, B.B. 1998. *The World According to Wavelets;The Story of a Mathematical Technique in the Making*, 2nd ed. Wellesley, MA: A.K. Peters.

Hubbard, J.H., and West, B.H. 1991. *Differential Equations:A Dynamical Approach Ordinary Differential Equations*. Texts in Applied Mathematics, Vol. 5. New York: Springer-Verlag.

Hubbard, J.H., and West, B.H. 1995. *Differential Equations: A Dynamical Approach Higher-Dimension Systems*. Texts in Applied Mathematics, Vol. 18. New York: Springer-Verlag.

Hulse, R.A., and Taylor, J.H. 1975. Discovery of a Pulsar in a Binary System. *Astrophysical Journal (Letters)*, **195**, L51.

Hundhausen, A.J. 1972. *Coronal Expansion and Solar Wind*. Berlin: Springer-Verlag.

Hundhausen, A.J. 1978. *Streams, Sectors and Solar Magnetism*. Boulder, CO.: AAAS, Chap. 4.

Hundhausen, A.J. 1995. *The Solar Wind*. Cambridge: Cambridge University Press.

Hunt, B.R. 1978. *Digital Image Processing*. Prentice-Hall Signal Processing Series. Englewood Cliffs, NJ: Prentice-Hall. Chap. 4, pp. 169–237.

Iyer, S.V., and Hansen, E.C. 2009. Light's Bending Angle in the Equatorial Plane of a Kerr Black Hole. *Physical Review D*, **80**(12).

Jefimenko, Oleg D. 1989. *Electricity and Magnetism*, 2nd ed. Star City, WV, 26505: Electret Scientific Co.

Kaiser, Gerald. 1994. *A Friendly Guide to Wavelets*. Boston: Birkhäuser.

Kataoka, J., and 6 others. 2002. RXTE Observations of 3C 273 between 1996 and 2000: Variability Time-scale and Jet Power. *MNRAS*, **336**, 932–944.

Kelso, J.M. 1964. *Radio Ray Propagation in the Ionosphere*. New York: McGraw-Hill.

King, I.R. 1966. The Structure of Star Clusters. III. Some Simple Dynamical Models. *Astronomical Journal*, **71**(1), 64–75.

Kittel, Charles. 1986. *Introduction to Solid State Physics*, 6th ed. New York: John Wiley & Sons.

Kittel, Charles, and Kroemer, Herbert. 1980. *Thermal Physics*. New York: W.H. Freeman.

Kivelson, Margaret G., and Russell, Christopher T. Editors. 1995. *Introduction to Space Physics*. Cambridge: Cambridge University.

Kolb, E.W., and Turner, M.S. 1990. *The Early Universe*. Frontiers in Physics. Redwood City, CA: Addison-Wesley. Advanced Book Program.

Kosirev, N. 1934. Radiative Equilibrium of the Extended Photosphere. *MNRAS*, **94**, 430.

Krasner, S. 1990a. *The Ubiquity of Chaos*, Vol. 89-15S. Washington, DC: American Association for Advancement of Science.

Krasner, Saul. 1990b. *The Ubiquity of Chaos*. Washington, DC: American Association for Advancement of Science.

Kraus, Stefan. 2009. Tracing the Binary Orbit of the Young, Massive High-eccentricity Binary System Theta 1 Orionis C. *The ESO Messenger*, **136**(June 2009), 44–47.

Krimigis, S.M., and 4 others. 2009. Imaging the Interaction of the Heliosphere with the Interstellar Medium from Saturn with Cassini. *Science*, **326**(13 November 2009), 971–973.

Lamb, D.Q., and Van Horn, H.M. 1975. Evolution of Crystallizing Pure 12 C White Dwarfs. *Astrophysical Journal*, **200**, 306–323.

Lang, Kenneth R. 1992. *Astrophysical Data: Planets and Stars*. New York: Springer-Verlag.

Lanm, H. Y.-F. 1979. *Analog and Digital Filters: Design and Realization*. Englewood Cliffs, NJ: Prentice-Hall.

Laurendeau, Normand M. 2010. *Statistical Thermodynamics*. Cambridge: Cambridge University Press.

Limber, D.N. 1952. Anaysis of Counts of the ExtraGalactic Nebulae in Terms of a Fluctuating Density Field. *Astrophysical Journal*, **117**, 134–144.

Limber, D.N. 1954. Analysis of Counts of the Extragalactic Nebulae in Terms of a Fluctuating Density Field. II. *Astrophysical Journal*, **119**, 655–681.

Linnik, Yu. V. 1961. *Method of Least Squares and Principles of the Theory of Observations*. New York: Pergamon Press.

Lomb, N.R. 1976. Least-Squares Frequency Analysis of Unequally Spaced Data. *Astrophysics and Space Science*, **39**, 447–463.

Lucy, L.B. 1974. An Iterative Technique for the Rectification of Observed Distributions. *Astronomical Journal*, **79**(4), 745–754.

Madore, B.F. 2012. *Level 5: A Knowledge Base for Extragalactic Astronomy and Cosmology*. ned.ipac.caltech.edu/level5/.

Madore, B.F, and Steer, I.P. 2008. *NASA/IPAC Extragalactic Database Master List of Galaxy Distances*. ned.ipac.caltech.edu/Library/Distances/.

Maller, A.H., McIntosh, D.H., Katz, N., and Weinberg, M.D. 2005. The Galaxy Angular Correlation Functions and Power Spectrum from the Two Micron All Sky Survey. *Astrophysical Journal*, **619**, 147–160.

Manchester, R.N., and Taylor, J.H. 1977. *Pulsars*. A Series of Books in Astronomy and Astrophysics. San Francisco: W.H. Freeman.

Margenau, Henry, and Murphy, George M. 1956. *The Mathematics of Physics and Chemistry*, Vol. 1, 2nd ed. Princeton, NJ: D. Van Nostrand Co.

Margenau, Henry, and Murphy, George M. 1964. *The Mathematics of Physics and Chemistry*, Vol. 2. Princeton, NJ: D. Van Nostrand Co.

Mason, Brian D., McAlister Harold A., and Hartkopf, William I. 1995. Binary Star Orbits from Speckle Interferometry. VII The Multiple System Xi Ursae Majoris. *Astrophysical Journal*, **109**(January 1995), 332–340.

Mason, B.D., Wycoff, G.L., Hartkopf, W.I., Douglas, G.G., and Worley, C.E. 2001. *Washington Double Star Catalog 2001.0*.

McComas, D.J., and 35 others. 2009. Global Observations of the Interstellar Insteration from the Interstellar Boundary Explorer (IBEX). *Science*, **326**(13 November 2009), 959–962.

McLennan, James A. 1989. *Introduction to Nonequilibrium Statistical Mechanics*. Prentice-Hall Advanced Reference Series. Englewood Cliffs, NJ: Prentice-Hall.

Meeus, Jean. 1998. *Astronomical Algorithms*, 2nd ed. Richmond, VA: Willmann-Bell.

Meisel, D.D. 1978. Fourier Transforms of Data Sampled at Unequal Observational Intervals. *Astrophysical Journal*, **83**(5), 538–545.

Meisel, D.D. 1979. Fourier Transforms of Data Sampled in Equally Spaced Intervals. *Astrophysical Journal*, **84**(1), 116–126.

Meisel, D.D., Saunders, B.A., Frank, Z.A., and Packard, M.L. 1982. The Helium 10830 Angstrom Line in Early Type Stars: An Atlas of Fabry-Perot Scans. *Astrophysical Journal*, **263**, 759–776.

Meisel, David D., Janches, D., and Mathews, J.D. 2002. The Size Distribution of Arecibo Interstellar Particles and Its implications. *Astrophysical Journal*, **579**(2002 Nov. 10), 895–904.

Mertins, A. 1999. *Signal Analysis*. Chichester, U.K.: John Wiley & Sons.

Mestel, L., and Ruderman, M.A. 1967. The Energy Content of a White Dwarf and Its Rate of Cooling. *MNRAS*, **136**, 27.

Meyer, Stuart L. 1975. *Data Analysis for Scientists and Engineers*. Wiley Series in Probability and MatheMatical Statistics. New York: John Wiley & Sons.

Mihalas, Dimitri. 1978. *Stellar Atmospheres*. 2nd ed. A Series of Books in Astronomy and Astrophysics. San Francisco: W.H. Freeman.

Mills, Allan P. Jr. 2011. *The Problem of Sticking of Slow Positronium Atoms Incident upon a Cold Surface*. ebookbrowse.com/icpa-15-isps-lecture-mills-pdf-d51817945 (2011).

Milton, K.A. 2011. Resource Letter VWCPF-1:Van der Waals and Casimir-Polder forces. arxiv at http://arkiv.org/abs/cond-mat/0410424.

Misner, C.W., Thorne, K.S., and Wheeler, J.A. 1973. *Gravitation*. San Francisco: W.H. Freeman.

Möbius, E., and 19 others. 2009. Direct Observations of Interstellar H,He and O by the Interstellar Boundary Explorer. *Science*, **326**(13 November), 969–971.

Mohling, Franz. 1982. *Statistical Mechanics: Methods and Applications*. New York: Halsted Press John Wiley.

Montgomery, D.C., and Peck, Elizabeth A. 1990. *Multicollinearity in Regression*. New York: McGraw-Hill, Chap. 15.

Moore, F.K. 1966. *Models of Gas Flows with Chemical and Radiative Effects*. Providence, RI: American Mathematical Society.

Müller, C.M. 2005. Cosmological Bounds on the Equation of State of Dark Matter. *Physical Review D*, **71**(4).

Nash, David. 2011. *The Astronomy Nexus Web Site*. www.astronaxus.com.

Nelson, R.D., Lide, D.R., and Maryott, A.R. 1967. *Selected Values of Electric Dipole Moments for Molecules in the Gas Phase*. United States Department of Commerce.

Obergaulinger, Martin. 2006 (8/09/ 2012). *Inflation, Dark Energy, Dark Matter*. www.mpa-garching .mpg.de/lectures/ADSEM/SS06-Obergaulinger.pdf.

Observatory, Applications Department U.S. Naval. 1998–2005. *Multiyear Interactive Computer Almanac 1800-2050*. Richmond, VA.: Willmann-Bell.

O'Keefe, John A. 1966. *Stability of a Rotating Liquid Mass*. Providence, RI: American Mathematical Society, pp. 155–169.

Opher, M., Stone, E.C., and Gombosi, T.I. 2007. The Orientation of the Local Magnetic Field. *Science*, **316**, 875.

Oppenheim, Alan V. 1975. *Digital Signal Processing*. Englewood Cliffs, NJ: Prentice-Hall.

Orfanides, S.J. 1988. *Optimum Signal Processing*, 2nd ed. New York: Macmillan.

Ott, Edward. 1993. *Chaos in Dynamical Systems*. Cambridge: Cambridge University Press.

Pack, J.L., and Phelps, A.V. 1959. Electron Collision Frequencies in Nitrogen and the Lower Atmosphere. *Physical Review Letters*, **3**.

Paltani, S., Courvoisier, T.J.L., and Walter, R. 1998. The Blue-Bump of 3C273. *Astronomy and Astrophysics*, **340**, 47–61.

Parker, E.N. 1963. *Interplanetary Dynamical Processes*. New York: Wiley Interscience.

Parker, E.N. 1978. *Solar Physics in Broad Perspective*. Boulder, Co.: AAAS, Chap. 1.

Peebles, P.J.E. 1980. *The Large Scale Structure of the Universe*. Princeton NJ: Princeton University Press.

Penrose, R., and Rindler, W. 1984, 1986. *Spinors and Spacetime*, Vols. I–II. Cambridge, UK: Cambridge University Press.

Percival, I.C. 1987. *Chaos in Hamiltonian Systems*. Princeton, NJ: Princeton University Press.

Peterson, I. 1993. *Newton's Clock*. New York: W.H. Freeman.

Phelps, A.V. 1960. Propagation Constants for Electromagnetic Waves in a Refracting Mdeioum in a Magnetic Field. *Proceedings of the Cambridge Philosophical Society*, **27**.

Plischke, Michael, and Bergersen Birger. 1989. *Equilibrium Statistical Physics*. Prentice-Hall Advanced Reference Series. Englewood Cliffs, NJ: Prentice-Hall.

Poincaré, H. 1993. *New Methods of Celestial Mechanics*, Parts 1–3. History of Modern Physics and Astronomy, Vol. 13. New York: American Institute of Physics.

Pollard, Harry. 1966a. *Mathematical Introduction to Celestial Mechanics*. Prentice-Hall Mathematics Series. Englewood Cliffs, NJ: Prentice-Hall.

Pollard, Harry. 1966b. *Qualitative Methods in the N-Body Problem*. Providence, RI: American Mathematical Society, pp. 259–291.

Prasanna, A.R., Narlikar, J.V., and Vishveshwrar, C.V. 1980. *Gravitation, Quanta, and the Universe*. In Einstein Centenary Symposium, Ahmedabad, 1979. New York: Halsted Press.

Press, William H., Teukolsky, Saul A., Vetterling, William T., and Flannery, Brian P. 2002. *Numerical Recipes: The Art of Scientific Computing CDROM v.2.10*. Cambridge: Cambridge University Press.

Prialnik, Dina. 2000. *An Introduction to the Theory of Stellar Structure and Evolution*. Cambridge: Cambridge University Press.

Qaisar, S.M., Fesquest, L., and Renaudin, M. 2006. Spectral Analysis of a Signal Driven Sampling Scheme. *14th European Signal Processing Conference*, Florence, Italy.

Rabiner, L.R., and Gold, B. 1975. *Theory and Application of Digital Signal Processing*. Englewood Cliffs NJ: Prentice-Hall.

Rasband, S. Neil. 1990. *Chaotic Dynamics of Nonlinear Systems*. New York: Wiley-Interscience.

Ratcliffe, J.A. 1959. *The Magnetoionic Theory*. Cambridge, U.K.: Cambridge University Press.

Reif, Fredrick. 1965. *Fundamentals of Statistical and Thermal Physics*. Reissued 2009 by Waveland Press. Long Grove, IL: Waveland Press.

Rensch, R., and Koel, B.E. 2003. *Surfaces and Films*. New York: AIP Press Springer, Chap. 25.

Riley, K.F., Hobson, M.P., and Bence, S.J. 1997. *Mathematical Methods for Physics and Engineering*. Cambridge: Cambridge University Press.

Robertson, Howard P. 1938. The Two-Body Problem in General Relativity. *Annals of Mathematics*, **39**, 101–104.

Robertson, Howard P., and Noonan, Thomas W. 1968. *Relativity and Cosmology*. Saunders Physics Books. Philadelphia: W.B. Saunders Co.

Rodriquez, L.F., Gomez, L., Loinard, L., Lizano, S., Allen, C., Poveda, A., and Menten, K.M. 2008. New Observations of the Large Proper Motions of Radio Sources in the Orion BN/Kl Region. *RevMexAA (Serie de Conferencias)*, **34**, 75–78.

Rovithis-Livaniou, H. 2004. The Gravity Darkening Effect: From von Zeipel Up to Date. *Aerospace Research in Bulgaria* **20**, 90–98 (2005).

Rubin, V.C., and Coyne, G.V. 1988. *Large-Scale Motions in the Universe*. Princeton Series in Physics. Princeton, NJ: Princeton University Press.

Rybicki, George B. Theoretical Methods of Treating Line Formation Problems in Steady-State Extended Atmospheres. In: Groth, H.G., and Wellman, P. (eds), *Spectrum Formation in Stars with Steady-State Extended Atmospheres IAU No2*, vol. NBS Special Publication 332. National Bureau of Standards.

Rybicki, George B., and Press, W.H. 1992. Interpolation, Realization, and Reconstruction of Noisy, Irregularly Sampled Data. *Astrophysical Journal*, **398**, 169–176.

Savedoff, M.P., Van Horn, H.M., and Vila, S.C. 1969. Late Phases of Stellar Evolution I: Pure Iron Stars. *Astrophysical Journal*, **155**(Jan 1969), 221–245.

Scargle, J.D. 1981. Studies in Astronomical Time Series Analysis I. Modeling Random Processes in the Time Domain. *Astrophysical Journal Supplement Series*, **45**, 1–71.

Scargle, J.D. 1982. Studies in Astronomical Time Series Analysis. II. Statistical Aspects of Spectral Analysis of Unevenly Spaced Data. *Astrophysical Journal*, **263**, 835–853.

Scargle, J.D. 1989. Studies in Astronomical Time Series Analysis III Fourier Transforms, Autcorrelation Functions and Cross-Correlation Functions of Unevenly Spaced Data. *Astrophysical Journal*, **343**, 874–887.

Scargle, J.D. 1990. Studies in Astronomical Time Series IV. Modeling Chaotic and Random Processes with Linear Filters. *Astrophysical Journal*, **359**, 469–482.

Scargle, J.D. 1997. *Astronomical Time Series Analysis: New Methods for Studying Periodic and Aperiodic Systems*. Dordrect: Kluwer, p. 1.

Scargle, J.D. 1998. Studies in Astronomical Time Series Analysis. *Astrophysical Journal*, **504**, 405–418.

Scargle, J.D. 2003. *Advanced Tools for Astronomical Times Series and Image Analysis*. New York: Springer, pp. 293–308.

Scheuer, P.A.G. 1968. Amplitude Variations in Pulsed Radio Sources. *Nature*, **218**, 920.

Schneider, P., Ehlers, J., and Falco, E.E. 1993. *Garvitational Lenses*. New York: Springer-Verlag.

Schwandron, N.A., and 23 others. 2009. Comparison of Interstellar Boundary Explorer Observations with 3D Global Heliospheric Models. *Science*, **326**(13 November 2009), 966–968.

Schwarzschild, M. 1954. Mass Distribution and Mass-luminosity Ratio in Galaxies. *Astronomical Journal*, **59**.

Schwarzschild, M. 1987. *Galactic Models with Moderate Stochasticity*. New York: New York Academy of Sciences.

Schwarzschild, Martin. 1958. *Structure and Evolution of Stars*. New York: Dover Publications (Originally published by Princeton University Press).

Sebok, W.L. 1986. The Angular Correlation Function of Galaxies as a Function of Magnitude. *Astrophysical Journal Supplement Series*, **62**, 301–330.

Seidelmann, P. Kenneth. 1992. *Explanatory Supplement to the Astronomical Almanac*. Mill Valley, CA: University Science Books.

Serra, A.L., and Romero, M.J.L.D. 2011. Measuring the Dark Matter Equation of State. *MNRAS*, **415**, L74–L77.

Shai, D.E., Cole, M.W., and Lammert, P.E. 2007. Adsorption of Quantum Gases on Curved Surfaces. *Journal of Low Temperature Physics*, **147**(1–2), 59–79.

Shen, S.F. 1966. *Shock Waves in Rarefied Gases*. Providence, RI: American Mathematical Society.

Shu, Frank H. 1991. *The Physics of Astrophysics*, Vol. 1: *Radiation*. A Series of Books in Astronomy. Mill Valley, CA: University Science Books.

Shu, Frank H. 1992. *The Physics of Astrophysics*, Vol. II: *Gas Dynamics*. A Series of Books in Astronomy. Mill Valley, CA: University Science Books.

Simonetti, J.H. Cordes, J.M., and Heeschen, D.S. 1985. Flicker of Extragalactic Radio Sources at Two Frequencies. *Astrophysical Journal*, **296**, 46–59.

Slettebak, Arne. 1970. *Stellar Rotation: Proceedings of the IAU Colloquium held at the Ohio State University*. Dordrecht The Netherland: Kluwer Academic Publishers, D. Reidel Publishing.

Smart, W. M. 1960. *Text-Book on Spherical Astronomy*, 4th ed. Cambridge: Cambridge University Press.

Sobolev, V. 1960. *Moving Envelopes of Stars*. Cambridge, MA: Harvard University Press.

Söderhjelm, Staffan. 1999. Visual Binary Orbits and Masses Post Hipparchos. *Astronomy and Astrophysics*, **341**, 121–140.

Soldi, S., and 14 others. 2008. The multiwavelength variability of 3C273. *Astronomy and Astrophysics*, **486**, 411–425.

Soleng, Harald H. 2011. *Tensors in Physics*. Harald H. Soleng.

Souokoulis, Costas M., and Economou, Eleftherios N. 2003. *Ch. 24 Solid State Physics*. New York: Springer-Verlag AIP Press.

Spitzer, L. 1987. *Dynamical Evolution of Globular Clusters*. Princeton, NJ: Princeton University Press.

Steer, I. 2011. *NED-D*.

Stowe, Keith. 1984. *Introduction to Statistical Mechanics and Thermodynamics*. New York: John Wiley & Sons.

Strogatz, S.H. 1994. *Nonlinear Dynamics and Chaos*. Reading, MA: Addison-Wesley.

Sturrock, Peter A. 1996. *Plasma Physics: An introduction to the theory of Astrophysical, Geophysical and Laboratory Plasmas*. Cambridge: Cambridge University Press.

Sturrock, Peter A., and Scargle, J.D. 2009. A Bayesian Assessment of P-values for Significance Estimation of Power Spectra and an Alternative Procedure with Application to Solar Neutrino Data. *Astrophysical Journal*, **706**(2009 Nov. 20), 393–398.

Sweldens, Wim. 2012. *Polyvalens: A Really Friendley Wavelet Guide*. www.polyvalens.com/blog/?page_id=15.

Szatmáry, K., Vinkó, J., and Gál, J. 1994. Application of wavelet analysis in variable star research. *Astronomy and Astrophysics Supplement Series*, **108**, 377–394.

Tanenbaum, B. Samuel. 1967. *Plasma Physics*. McGraw-Hill Physical and Quantum Electronics Series. New York: McGraw-Hill.

Terrell, J. 1977. Size Limits on Fluctuating Astronomical Sources. *Astrophysical Journal (Letters)*, **213**, L93–L97.

Terrell, J., and Olsen, K.H. 1970. Power Spectrum of 3C 273 Light Fluctuations. *Astrophysical Journal*, **161**, 399–413.

Thieberger, R., Spiegel, E.A., and Smith, L.A. 1990. *The Dimensions of Cosmic Fractals*. Washington, DC: The Ubiquity of Chaos, ed. Saul Krasner. American Association for the Advancement of Science Washington DC, pp. 197–217.

Thompson, A.R., Moran, J.M., and Swenson, G.W. Jr. 2001. *Interferometry and Synthesis in Radio Astronomy*. New York: John Wiley & Sons.

Trusler, J.P.M., and Wakeham, W.A. 2003. *Thermodynamics and Thermophysics, Ch. 26*. New York: AIP Press /Springer.

Tyagi, S. 2004. *"New Series Representation for Madelung Constant."* Progress of Theoretical Physics, **114**(3), 517–521 (2005).

Van de Hulst, H.C., Müller, C.A., and Oort, J.H. 1954. The Spiral Structure of the Outer Part of the Galactic System Derived from the Hydrogen Emission at 21 cm Wavelength. *Bulletin of the Astronomical Institute of the the Netherlands (BAN)*, **12**(May 1954), 117–149.

van de Weijgaert, R. 2007. *Measuring the Two Point Correlation Function*. www.astro.rug.nl/∼welygaert/tim1publication/iss2007/computerII.pdf.

van der Ziel, A. 1970. *Noise; Sources, Characterization,Measurement*. Englewood Cliffs, NJ: Prentice-Hall.

Van Horn, Hugh. 1968a. *Implications of White Dwarf Crystallization for the Chemical Composition of the Planetary Nuclei*. Dordrecht, Netherlands: International Astronomical Union, pp. 425–427.

Van Horn, Hugh. 1968b. Crystallization of White Dwarfs. *Astrophysical Journal*, **151**(Jan 1968), 227–238.

Veron-Cetty, M.P., and Veron, P. 2010. A Catalog of Quasars and Active Nuclei, 13th ed. *Astronomy and Astrophysics*, **518**.

Vio, R., Cristiani, S., Lessi, O., and Provenzale, A. 1992. Time Series Analysis in Astronomy: An Application to Quasar Variability Studies. *Astrophysical Journal*, **391**, 518–530.

Vityazev, V.V. 1996a. Time-Series of Unequally Spaced Data: Intercomparison between the Shuster Periodogram and the LS Spectra. *Astronomical and Astrophysical Transactions*, **12**, 139–158.

Vityazev, V.V. 1996b. Time-Series of Unequally Spaced Data: The Statistical Properties of the Schuster Periodogram. *Astronomical and Astrophysical Transactions*, **12**.

Vollmer, M. 2009. Newton's Law of Cooling Revisited. *European Journal of Physics*, **30**, 1063–1084.

Wadsworth Harrison M., Jr. 1990. *Handbook of Statistical Methods for Scientists and Engineers*. New York: McGraw-Hill.

Walsh, D. Carswell, R.F., and Weymann, R.J. 1979. *Nature*, **279**.

Wax, N. 1957. *Selected Papers on Noise and Stochastic Processes*. New York: Dover Publications.

Weinstein, Eric. 2007a. *MathWorld-Madelung Constants*. Mathworld.wolfram.com/Madelung Constants.html.

Weinstein, Eric. 2007b. *MadelungConstants.nb (downloadable working Mathematica notebook)*. mathworld.wolfram.com/notebooks/Constants/MadelungConstants.nb.

Weinstein, Eric W. 2003. *The CRC Concise Encyclopedia of Mathematica*, 2nd ed. Boca Raton, Fl: Chapman and Hall/CRC.

Wellin, Paul, Gaylord, Richard, and Kamin, Samuel. 2005. *An Introduction to programming with Mathematica*. Cambridge: Cambridge University Press.

Wiener, N. 1949. *Extrapolation, Interpolation, and Smoothing of Stationary Time Series*. Cambridge, MA: MIT Press.

Wikipedia. 2012 (9 May 2012). *WKB Approximation*. en.wikipedia.org/wiki/WKB_approximation.

Wilcox, J.Z., and Wilcox, T.J. 1995. Algorithm for Extraction of Periodic Signals for Sparse Irregularly Sampled Data. *Astronomy and Astrophysics Supplement Series*, **112**, 395–405.

Will, Clifford M. 1993. *Theory and Experiment in Gravitational Physics*. Revised 2nd ed. Cambridge: Cambridge University Press.

Wilson, B.A., Dame, T.M., Mashender, M.R.W., and Thaddeus, P. 2005. A Uniform CO Survey of the Molecular Clouds in Orion and Monoceros. *Astronomy and Astrophysics*, **430**, 523–539.

Wisdom, J. 1987. *Chaotic Behavior in the Solar System*. Princeton, NJ: Princeton University Press.

Wyller, A.A., and Sen, H.K. 1960. On the Generalization of the Appleton-Hartree Magnetoionic Formulas. *Journal of Geophysical Research*, **65**(No. 12), 3931–3950.

Zimmerman, Robert L., and Olness, Fredrick I. 2002. *Mathematica for Physics*, 2nd ed. Reading, MA: Addison-Wesley.

Index